Advanced Materials *and* Design *for* Electromagnetic Interference Shielding

Advanced Materials *and* Design *for* Electromagnetic Interference Shielding

Xingcun Colin Tong

CRC Press
Taylor & Francis Group
Boca Raton London New York

CRC Press is an imprint of the
Taylor & Francis Group, an **informa** business

CRC Press
Taylor & Francis Group
6000 Broken Sound Parkway NW, Suite 300
Boca Raton, FL 33487-2742

© 2009 by Taylor & Francis Group, LLC
CRC Press is an imprint of Taylor & Francis Group, an Informa business

No claim to original U.S. Government works
Printed in the United States of America on acid-free paper
10 9 8 7 6 5 4 3 2 1

International Standard Book Number-13: 978-1-4200-7358-4 (Hardcover)

This book contains information obtained from authentic and highly regarded sources. Reasonable efforts have been made to publish reliable data and information, but the author and publisher cannot assume responsibility for the validity of all materials or the consequences of their use. The authors and publishers have attempted to trace the copyright holders of all material reproduced in this publication and apologize to copyright holders if permission to publish in this form has not been obtained. If any copyright material has not been acknowledged please write and let us know so we may rectify in any future reprint.

Except as permitted under U.S. Copyright Law, no part of this book may be reprinted, reproduced, transmitted, or utilized in any form by any electronic, mechanical, or other means, now known or hereafter invented, including photocopying, microfilming, and recording, or in any information storage or retrieval system, without written permission from the publishers.

For permission to photocopy or use material electronically from this work, please access www.copyright.com (http://www.copyright.com/) or contact the Copyright Clearance Center, Inc. (CCC), 222 Rosewood Drive, Danvers, MA 01923, 978-750-8400. CCC is a not-for-profit organization that provides licenses and registration for a variety of users. For organizations that have been granted a photocopy license by the CCC, a separate system of payment has been arranged.

Trademark Notice: Product or corporate names may be trademarks or registered trademarks, and are used only for identification and explanation without intent to infringe.

Library of Congress Cataloging-in-Publication Data

Tong, Colin.
   Advanced materials and design for electromagnetic interference shielding / Colin Tong.
     p. cm.
   "A CRC title."
   Includes bibliographical references and index.
   ISBN 978-1-4200-7358-4 (alk. paper)
     1. Shielding (Electricity) 2. Electronic apparatus and appliances--Materials. 3. Electromagnetic compatibility. 4. Electromagnetic interference. I. Title.

TK7867.8.T66 2009
621.382'24--dc22   2008025450

Visit the Taylor & Francis Web site at
http://www.taylorandfrancis.com

and the CRC Press Web site at
http://www.crcpress.com

Contents

Preface xiii

Author xv

Chapter 1   Electromagnetic Interference Shielding Fundamentals and Design Guide 1

1.1   Concepts of Electromagnetic Interference and Electromagnetic Compatibility 1
     1.1.1   EMI Problems and Common Concerns 2
     1.1.2   EMI Emissions and Controlling Methods 3
          1.1.2.1   Conducted EMI Emission 4
          1.1.2.2   Radiated EMI Emission 5
     1.1.3   EMC Regulations and Testing Standards 8
          1.1.3.1   The International Standard 9
          1.1.3.2   FCC Standard 9
          1.1.3.3   European Standard 10
          1.1.3.4   Military Standard 10
1.2   EMI Shielding Principles 11
     1.2.1   Shielding Effects in the E-Field and H-Field 11
     1.2.2   Shielding Effectiveness 13
          1.2.2.1   Absorption Loss 14
          1.2.2.2   Reflection Loss 15
          1.2.2.3   Multiple Reflection Correction Factor 16
     1.2.3   Shielding Modeling in Practice 18
          1.2.3.1   Metallic Enclosure 18
          1.2.3.2   EMI Gasket 19
          1.2.3.3   Shield with Apertures 20
          1.2.3.4   Cavity Resonance 23
1.3   EMC Design Guidelines and EMI Shielding Methodology 24
     1.3.1   Design for EMC 24
     1.3.2   Shielding Methods and Materials Selection 27
          1.3.2.1   Shielding Housing and Enclosures 27
          1.3.2.2   Gasketing 29
          1.3.2.3   Integral Assembly 30
          1.3.2.4   Corrosion and Material Galvanic Compatibility 31
     1.3.3   Environmental Compliance 32
          1.3.3.1   RoHS Directive 33
          1.3.3.2   WEEE Directive 34
References 35

Chapter 2   Characterization Methodology of EMI Shielding Materials 37

2.1 Shielding Effectiveness Measurement 37
      2.1.1 Testing Methods Based on MIL-STD-285 38
      2.1.2 Modified Radiation Method for Shielding Effectiveness Testing Based on MIL-G-83528 39
      2.1.3 Dual Mode Stirred Chamber 39
      2.1.4 Transverse Electromagnetic (TEM) Cell 40
      2.1.5 Circular Coaxial Holder 42
      2.1.6 Dual Chamber Test Fixture Based on ASTM ES7-83 42
      2.1.7 Transfer Impedance Methods 42
            2.1.7.1 Transfer Impedance of Coaxial Cables 43
            2.1.7.2 Transfer Impedance Method for Shielding Effectiveness Measurement of Conductive Gaskets 43
2.2 Electrical and Thermal Conductivities 45
      2.2.1 Electrical Conductivity and Contact Resistance 45
      2.2.2 Thermal Conductivity 46
2.3 Permeability and Permittivity 47
      2.3.1 Permeability 47
      2.3.2 Permittivity 48
      2.3.3 Characterization of Permeability and Permittivity 48
2.4 Mechanical Properties 49
      2.4.1 Uniaxial Tensile Testing 49
      2.4.2 Hardness 50
      2.4.3 Poisson's Ratio 50
      2.4.4 Stress Relaxation 50
      2.4.5 Contact Force 51
      2.4.6 Friction Force 52
      2.4.7 Permanent Set 52
2.5 Surface Finish and Contact Interface Compatibility 52
      2.5.1 Corrosion and Oxidation Protection 53
            2.5.1.1 Noble Finish Selection 53
            2.5.1.2 Nonnoble Finishes 54
      2.5.2 Solderability of Surface Finishes 55
      2.5.3 Effects of Mating Cycles and Operating Environments on Contact Finishes 57
      2.5.4 Galvanic Corrosion and Contact Interface Compatibility 58
2.6 Formability and Manufacturability 58
      2.6.1 Bending Formability 59
      2.6.2 Effect of Strain Hardening on Formability 60
      2.6.3 Anisotropy Coefficient 61
      2.6.4 Springback during Metal Strip Forming 62
2.7 Environmental Performance Evaluation 62
      2.7.1 Thermal Cycling 63
      2.7.2 Thermal Aging 63
      2.7.3 Gaseous Testing 63

Contents   vii

    2.7.4   Pore Corrosion and Fretting Corrosion 63
    2.7.5   Temperature and Humidity Testing 64
    2.7.6   Dust Sensitivity Test 64
    2.7.7   Vibration and Shock Test 65
References 65

Chapter 3   EMI Shielding Enclosure and Access 67

3.1   Enclosure Design and Materials Selection 67
3.2   Magnetic Field Shielding Enclosure 73
    3.2.1   Basics of Magnetic Shielding 73
    3.2.2   Magnetic Shielding Materials Selection 74
    3.2.3   Magnetic Shielding Enclosure Design Consideration 76
3.3   Shielding Enclosure Integrity 78
3.4   Specialized Materials Used for Shielding Enclosure 82
    3.4.1   EMI Gaskets 82
        3.4.1.1   Metal Strip Gaskets 82
        3.4.1.2   Knitted Metal Wire Gaskets 83
        3.4.1.3   Metal-Plated Fabric-over-Foam (FOF) 84
        3.4.1.4   Electrically Conductive Elastomer (ECE) Gaskets 84
    3.4.2   Magnetic Screening Materials 84
    3.4.3   Shielding Tapes 85
    3.4.4   Thermoformable Alloys 85
    3.4.5   Honeycomb Materials 85
    3.4.6   Painted or Plated Plastic Enclosures 86
    3.4.7   Conductive Composite Shielding Materials 86
3.5   Summary 86
References 87

Chapter 4   Metal-Formed EMI Gaskets and Connectors 89

4.1   Introduction 89
4.2   Metal Strip Selection and Performance Requirement 90
4.3   Copper Beryllium Alloys 91
    4.3.1   Phase Constitution and Primary Processing of CuBe Strips 91
    4.3.2   Performance and Availability of CuBe Alloys 95
        4.3.2.1   High-Strength Alloys 95
        4.3.2.2   High-Conductivity CuBe Alloys 97
        4.3.2.3   Nickel Beryllium 97
4.4   Copper–Nickel–Tin Spinodal Alloys 97
    4.4.1   Composition and Physical Properties 97
    4.4.2   Strip Gauges and Temper Designations 99
    4.4.3   Mechanical Properties and Bending Formability 100
    4.4.4   Stress-Relaxation Resistance 101
    4.4.5   Heat Treatment and Spinodal Decomposition 101
    4.4.6   Surface Cleaning, Plating, and Soldering 103

		4.4.7	Elastic Performance 103
		4.4.8	Shielding Effectiveness 104
		4.4.9	Fatigue Strength 104
		4.4.10	Some Considerations to Improve Performance of CuNiSn Gasketing 105
4.5	Copper–Titanium Alloy 107		
4.6	Stainless Steel 108		
4.7	Other Materials Options 110		
		4.7.1	Phosphor Bronze 110
		4.7.2	Brass 111
		4.7.3	Nickel Silver 111
		4.7.4	Other High Strength and High Conductive Copper Alloys 112
4.8	Design Guideline of Metal Gaskets and Connectors 113		
		4.8.1	Primary Approaches 114
		4.8.2	Some Special Design Options 115
		4.8.3	Gasketing Design Guideline 116
4.9	Fabrication Process and Types of Metal Gaskets and Connectors 118		
		4.9.1	Progressive Die Forming 118
		4.9.2	Roll Forming 119
		4.9.3	Multiple-Slide Stamping 120
		4.9.4	Photoetching 120
		4.9.5	Typical Part Profiles 120
4.10	Mounting Methods and Surface Mating Assurance 122		
4.11	Summary 124		
References 124			

Chapter 5   Conductive Elastomer and Flexible Graphite Gaskets 127

5.1	Introduction 127		
5.2	Raw Material Selection and Conductive Elastomer Fabrication 128		
		5.2.1	Base Binder Materials 128
			5.2.1.1 Silicone 128
			5.2.1.2 Fluorosilicone 129
			5.2.1.3 Ethylene Propylene Diene Monomer (EPDM) 129
			5.2.1.4 Fluorocarbon 129
			5.2.1.5 Natural Rubber and Butadiene-Acrylonitrile 129
		5.2.2	Conductive Fillers 129
		5.2.3	Fabrication of Conductive Elastomer Materials 130
5.3	Conduction Mechanism and Processing Optimization of Conductive Elastomer Materials 131		
		5.3.1	Conduction Mechanism and Process Parameters 131
		5.3.2	Appearance Properties and Performance Evaluation 133
			5.3.2.1 Physical and Mechanical Properties 133
			5.3.2.2 Electrical Properties 135
			5.3.2.3 Thermal Properties 137
5.4	Conductive Elastomer Gasket Design Guideline 137		
		5.4.1	Flange Joint Geometry and Unevenness 138

Contents

		5.4.2	Applied Clamping/Compressive Force and Deformation Range 139
		5.4.3	Gasket Profile and Materials Selection 141
		5.4.4	Flange Materials and Joint Surface Treatment 145
5.5	Conductive Elastomer Gasket Fabrication 146
		5.5.1	Extrusion 146
		5.5.2	Molding 147
		5.5.3	Form-in-Place and Screen Printing 148
		5.5.4	Reinforced Shielding Gaskets with Oriented Wires and Metal Meshes 149
5.6	Conductive Elastomer Gasket Installation and Application 149
5.7	Comparable Flexible Graphite Gaskets 152
		5.7.1	Flexible Graphite and Its Properties 153
		5.7.2	Fabrication and Application of Flexible Graphite Gaskets 155
5.8	Summary 157
References 157

Chapter 6 Conductive Foam and Ventilation Structure 159

6.1	Introduction 159
6.2	Waveguide Aperture and Ventilation Panel 159
	6.2.1	Waveguide Aperture and Its Design 160
	6.2.2	Materials Selection and Fabrication Options 163
6.3	Conductive Foam and Integral Window Structure 165
	6.3.1	Conductive Plastic Foam and Vent Panels 165
	6.3.2	Integral EMI Window Structure 168
6.4	Metallized Fabrics and Fabric-over-Foam 169
	6.4.1	Metallized Fabrics 170
	6.4.2	Fabric-over-Foam 171
6.5	Summary 175
References 175

Chapter 7 Board-Level Shielding Materials and Components 177

7.1	Introduction 177
7.2	Board-Level Shielding Design and Materials Selection 178
	7.2.1	Basic Principles of Board Circuit EMC Design 178
		7.2.1.1	PCB Layout 178
		7.2.1.2	Power Decoupling 179
		7.2.1.3	Trace Separation 179
		7.2.1.4	Grounding Techniques 180
		7.2.1.5	Chassis Construction and Cabling 180
	7.2.2	Board-Level Shielding Design with Proper Materials Selection 181
7.3	Board-Level Shielding Components and Their Manufacturing Technology 185

	7.3.1	Metal Cans	186
		7.3.1.1 Single-Piece Shielding Cans	186
		7.3.1.2 Two-Piece Cans	189
		7.3.1.3 Multilevel Shielding and Multicompartment Cans	190
	7.3.2	Metal-Coated Thermoform Shields	190
		7.3.2.1 snapSHOT Shield	191
		7.3.2.2 Form/Met Shield	192
		7.3.2.3 Formable Shielding Film	196
7.4	Conductive Coating Methods Used for EMI Shielding Components		196
	7.4.1	Electroplating	197
		7.4.1.1 Electroless Plating	197
		7.4.1.2 Electrolytic Plating	198
		7.4.1.3 Tin Plating and Tin Whisker Growth in EMI Shielding Systems	199
		7.4.1.4 Immersion Surface Finishes	202
	7.4.2	Conductive Paints	204
	7.4.3	Vapor Deposition	205
	7.4.4	Conductive Coating Application and Design Consideration for EMI Shielding	206
7.5	Board-Level Shielding with Enhanced Heat Dissipating		207
	7.5.1	Minimizing EMI from Heat Sinks	207
	7.5.2	Combination of Board-Level Shielding and Heat Dissipation	207
	7.5.3	Thermal Interface Materials (TIMs) for Thermal-Enhanced Board-Level Shielding	208
7.6	Summary		211
References			212

Chapter 8 Composite Materials and Hybrid Structures for EMI Shielding ... 215

8.1	Introduction		215
8.2	Knitted Wire/Elastomer Gaskets		215
	8.2.1	Fabrication Process and Materials Selection	216
	8.2.2	Knitted Wire Gasket Performance	218
	8.2.3	Typical Gasket Types and Mounting Methods	219
8.3	EMI Shielding Tapes		221
8.4	Conductive Fiber/Whisker Reinforced Composites		224
	8.4.1	Materials Selection and Process of Conductive Composites	225
	8.4.2	Carbon Fiber/Whisker Reinforced Materials	226
	8.4.3	Metal Fiber/Whisker Reinforced Composites	227
	8.4.4	Nanofiber Reinforced Polymer Composites	229
8.5	Hybrid Flexible Structures for EMI Shielding		230
8.6	Electroless Metal Deposition for Reinforcements of Composite Shielding Materials		231

Contents   xi

8.7   Summary 234
References 234

Chapter 9   Absorber Materials 237

9.1   Introduction 237
9.2   Microwave Absorber Materials 238
      9.2.1   Resonant Absorbers 240
            9.2.1.1   Dallenbach Tuned Layer Absorbers 240
            9.2.1.2   Salisbury Screens 241
            9.2.1.3   Jaumann Layers 242
      9.2.2   Graded Dielectric Absorbers: Impedance Matching 242
            9.2.2.1   Pyramidal Absorbers 242
            9.2.2.2   Tapered Loading Absorbers 243
            9.2.2.3   Matching Layer Absorbers 244
      9.2.3   Cavity Damping Absorbers 244
9.3   Anechoic Chambers 245
      9.3.1   Antennas 245
      9.3.2   Absorber Materials Used in Anechoic Chambers 246
9.4   Dielectric Materials for Absorber Applications 246
9.5   Electromagnetic Wave Absorbers 249
9.6   Absorbing Materials Selection and Absorber Applications 251
      9.6.1   Absorbing Materials and Absorber Types 251
      9.6.2   Applications 252
            9.6.2.1   Military 252
            9.6.2.2   Commercial 253
9.7   Summary 253
References 254

Chapter 10   Grounding and Cable-Level Shielding Materials 257

10.1   Introduction 257
10.2   Cable Assembly and Its EMI Shielding Design 257
10.3   Cable-Level Shielding Materials 261
      10.3.1   Metallic Shielding 261
      10.3.2   Conductive Heat–Shrinkable Shielding 263
      10.3.3   Ferrite Shielding Materials 264
10.4   Bonding and Grounding 267
      10.4.1   Bonding 268
      10.4.2   Grounding 269
            10.4.2.1   Shields Grounded at One Point 270
            10.4.2.2   Multigrounded Shields 271
            10.4.2.3   Hybrid Grounding 271
10.5   Summary 271
References 272

Chapter 11 Special Shielding Materials in Aerospace and Nuclear Industries 275

11.1 Introduction 275
11.2 Lightweight Shielding Materials for Aerospace Applications 275
    11.2.1 Lightweight Radiation Shielding Material 276
    11.2.2 Fiber Reinforced Composite Materials 277
    11.2.3 Nanomaterials 279
    11.2.4 Foam Structures 280
11.3 Radiation Shielding for Space Power Systems 281
    11.3.1 Neutron Shielding Design and Materials Selection 282
    11.3.2 Shielding Materials for Earth Neighborhood Infrastructure 282
    11.3.3 Active Electromagnetic Shielding for Space Radiation 283
    11.3.4 Optimization of Space Radiation Shielding 285
11.4 Nuclear Shielding Materials 286
11.5 Summary 289
References 291

Chapter 12 Perspectives and Future Trends 293

12.1 Introduction 293
12.2 Early Design-in for EMC and Optimal Design of EMI Shielding 293
12.3 Advanced Materials Selection for EMI Shielding 297
    12.3.1 Materials Selection for EMI Shielding Enclosure/Cover/Barrier 298
    12.3.2 EMI Gasketing Material Selection 299
    12.3.3 Other Shielding Materials Selection 301
12.4 Future Trends and Applications 302
    12.4.1 EMI Shielding Design Techniques 302
    12.4.2 Integrated Circuit-Level Shielding 303
    12.4.3 Printed Circuit Board (PCB)-Level Shielding 305
        12.4.3.1 Lightweight Shields for PCB-Level Shielding 305
        12.4.3.2 Mold-in-Place Combination Gaskets for Multicompartment Shields 306
        12.4.3.3 Rotary Form-in-Place Gaskets 307
    12.4.4 EMI Shielding Modules or Enclosures 308
        12.4.4.1 Conductive Foam with Z-Axis Conductivity 308
        12.4.4.2 Dent-Resistant Vent Panels 309
        12.4.4.3 Nanocomposite Shielding Materials 309
        12.4.4.4 Ultrasoft Sculpted Fabric-over-Foam 310
    12.4.5 Interconnecting Level Shielding 310
    12.4.6 EMI Control or Immunity by Software 311
12.5 Summary 311
References 312

Index 315

Preface

Electromagnetic compatibility (EMC) has been widely recognized as a part of electronic design, especially since implementation of the EMC directive (89/336/EEC) in 1992. National agencies in many countries have developed EMC standards and regulations to minimize the problems caused by failure to deal with electromagnetic interference (EMI). Historically, a concern about EMI can be dated back to military products in the 1930s. Since then great efforts have been made to develop EMI shielding approaches. As digital devices appeared, EMI became a great concern. In the 1970s, problems associated with EMI and EMC design became an issue for products that were beginning to be used within the commercial marketplace. EMC is now a major factor in the design of all electrical products, and EMI shielding products have been widely used in the military, aerospace industry, and commercial marketplaces.

With the development of EMC design and EMI shielding approaches, numerous textbooks and articles have been published that cover all aspects of theoretical physics related to EMC. However, comprehensive work devoted to the practicalities of EMI shielding materials and design is rare. To meet the need in this area, this book aims to comprehensively introduce the design guideline, materials selection, characterization methodology, manufacturing technology, and future prospective of EMI shielding. Chapter 1 gives the necessary overview of EMI shielding theory and product design guideline. Chapter 2 is an extensive review of the characterization methodology of EMI shielding and materials used. Chapter 3 through Chapter 11 introduce particular EMI shielding materials and component designs, including enclosures, metal-formed gaskets, conductive elastomer and flexible graphite components, conductive foam and ventilation structures, board-level shielding materials, composite materials and hybrid structures, absorber materials, grounding and cable-level shielding materials, and aerospace and nuclear shielding materials. Finally, Chapter 12 presents a perspective of future trends in EMI shielding materials and design.

It is a great pleasure to gratefully acknowledge the help and support I have received from my colleagues who have contributed to my understanding of EMC and EMI shielding approaches.

Author

Colin Tong is a senior staff scientist at Laird Technologies. He has over twenty years of experience in research and development, characterization, design and manufacturing of advanced materials, metallurgical products, electronic packaging, and electromagnetic interference shielding, having worked at Honeywell, Materials and Electrochemical Research Corporation (MER), and the University of Michigan.

He holds a PhD and an MEng in materials science and engineering, and a BEng in mechanical engineering. His research and development activities have resulted in more than thirty papers and several patents.

Dr. Tong is a member of The Minerals, Metals & Materials Society (TMS), IEEE, and ASM International, and received the Henry Marion Howe Medal from ASM International for his research work on advanced composite materials.

1 Electromagnetic Interference Shielding Fundamentals and Design Guide

Electromagnetic compatibility (EMC) is a field of science and engineering concerned with the design and operation of electrical and electronic systems/devices in a manner that makes them immune to certain amounts of electromagnetic interference (EMI), while at the same time keeping their EMI emissions within specific limits. With the rapid increase in the use of telecommunications, digital systems, fast processors, and the introduction of new design practices, EMC has been brought to the forefront of advanced design. This includes minimizing EMI generation at its source, reducing or eliminating coupling paths by proper layout; shielding, filtering, and grounding practices; designing hardware with an inherent immunity to EMI; and adopting defensive programming practices to develop software that has a high level of immunity to EMI (Christopoulos, 1995). In these approaches, EMI shielding is an important and effective method in EMC design. Certainly, advanced materials and process technology is the key to achieving successful EMI shielding.

This chapter will present an overview of the concepts of EMI and EMC, EMI shielding principles, EMC design guidelines, and EMI shielding materials and methodology.

1.1 CONCEPTS OF ELECTROMAGNETIC INTERFERENCE AND ELECTROMAGNETIC COMPATIBILITY

EMC and EMI concepts have been developed by EMC regulations and IEEE journals, and defined as follows:

> *Electromagnetic compatibility* (EMC) is the capability of electrical and electronic systems, equipment, and devices to operate in their intended electromagnetic environment within a defined margin of safety, and at design levels or performance, without suffering or causing unacceptable degradation as a result of EMI. EMC can generally be achieved by suppressing EMI and immunizing susceptibility of the systems and devices.
> *Electromagnetic interference* (EMI) is the process by which disruptive electromagnetic energy is transmitted from one electronic device to another via radiated or conducted paths or both. *Suppression* is the process of reducing or eliminating EMI energy. It may include shielding and filtering.

Susceptibility is a relative measure of a device's or a system's propensity to be disrupted or damaged by EMI exposure to an incident field or signal. It indicates the lack of immunity. *Immunity* is a relative measure of a device's or system's ability to withstand EMI exposure while maintaining a predefined performance level. Radiated immunity is a product's relative ability to withstand electromagnetic energy that arrives via free space propagation. Conducted immunity is a product's relative ability to withstand electromagnetic energy that penetrates it through external cables, power cords, and input/output (I/O) interconnects. *Electromagnetic discharge* (ESD) is a transfer of electric charge between bodies of different electrostatic potential in proximity or through direct contact (Montrose, 1999).

These terms are helpful to comprehensively understand the EMC requirements of various electrical and electronic systems. The crucial point is that many of these requirements are enforced by legislation, and such legal requirements can be expected to increase in the future, with severe penalties for infringement. On the other hand, even in situations where a legal requirement is not involved, the advanced techniques of EMC and EMI control will help to solve EMI problems in any new electrical product development and production.

1.1.1 EMI Problems and Common Concerns

EMI problems have impacted almost all electrical and electronic systems from daily life, to military activity, to space exploration. In the case of a laptop computer, its power system generates broadband EMI energy. This radiated energy is propagated and picked up by the TV antenna, wireless remote control units, and any other powers that might cause abnormal performance. In addition, EMI energy can be conducted by the incoming power line through common impedance coupling, which can cause temporary malfunctioning of the receiver. In general, EMI problems can be caused by electromagnetic energy at any frequency in the spectrum, as shown in Figure 1.1. Most problems are caused by energy in the radio frequency range, which extends

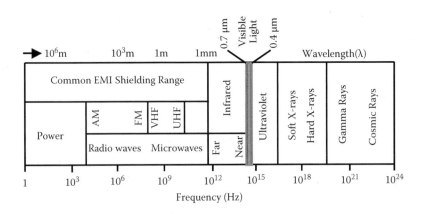

FIGURE 1.1 Electromagnetic spectrum.

from about 100 kHz to 1 GHz. Radio frequency interference (RFI) is a specific form of EMI. Microwave, light, heat, x-rays, and cosmic rays are other specific types of electromagnetic energy. Other common interference includes electromagnetic pulse (EMP), which is a kind of broadband; high-intensity short duration burst of electromagnetic energy, such as lightning or nuclear explosion; and electrostatic discharge (ESD), which is a transient phenomenon involving static electricity friction.

There have been many facets that increase concerns about EMI, such as (Chatterton, 1992; Molyneux-Child, 1997):

1. The proliferation of electronic equipment and various devices in both industrial and domestic environments has given rise to a great number of sources and receptors of EMI, which increases the potential for interference.
2. Miniaturization of equipment enables sources and receptors to be placed together. This shortens the propagation path and increases the opportunity for interference.
3. Miniaturization of components in electronic systems increases their susceptibility to interference. Because the units are smaller and more portable, equipment such as mobile phones and laptop computers can now be used anywhere instead of only in a controlled environment like an office. This leads to EMI problems.
4. Developments in interconnection technology have lowered the threshold for EMI. High-speed digital circuitry generally generates more interference than traditional analog circuitry.
5. Reliability and EMC compliance have become important marketing features of electronic equipment in an increasingly competitive industry.
6. Electronic discharge has been a major concern due to the destruction of microchips during handling. ESD also poses problems for aircraft and automobiles.
7. The need to ensure security of data has been one of the main driving forces in the growth of the shielding market.
8. People are becoming aware of the potential health risks associated with all types of EMI radiation.
9. Military systems involve protection against EMP and electronic countermeasures including microwave weapons and stealth technology. Considerations of an array of communications and radars in battle conditions are also important.

1.1.2 EMI Emissions and Controlling Methods

Three interactive constituents are needed for EMI to occur, as shown in Figure 1.2: (1) a source or emitter (culprit) of electromagnetic energy; (2) a susceptor or victim, which is susceptible to that particular amplitude and frequency of energy; and (3) a propagation or coupling path that is present between the source and the susceptor along which the energy can transfer. EMC or EMI control is normally achieved by reducing EMI emissions from the source, modifying or diverting the propagation path of the electromagnetic energy, and improving the immunity of the susceptor.

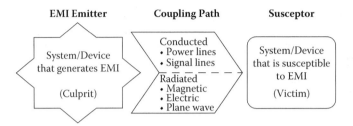

FIGURE 1.2 Coupling modes between the EMI source and susceptor.

EMI can be dominated by radiation or conduction, depending upon the type of coupling or propagation path involved. However, conduction always accompanies some radiation, and vice versa. Radiated interference happens when a component predominately emits energy, which is transferred to the receptor through the space. The source of the radiated interference may be part of the same system as the receptor or a completely electrically isolated unit. Conducted interference occurs when the source is connected to the receptor by power or signal cables, and the interference is transferred from one unit to the other along the cables. Conducted interference usually affects the main supply to and from the system/device and can be controlled using filters. If the conducted emissions are reduced, the relative radiated emissions are also often reduced. However, the dominant radiated interference can affect any signal path within and outside the system/device and is much more difficult to shield.

1.1.2.1 Conducted EMI Emission

Conducted EMI emission is usually defined as undesirable electromagnetic energy coupled out of an emitter or into a susceptor via any of its respective connecting wires or cables. There are two principal modes of propagation—differential and common mode—used to characterize the conducted EMI emission. The emission of the differential mode EMI takes place between conductor pairs, which form a conventional return circuit, such as negative/positive conductors and line phase/neutral conductors. The differential EMI is the direct result of the fundamental operation of the switching converter. The emission of the common mode EMI occurs between a group of conductors and either ground or another group of conductors. The path for the common mode EMI usually includes parasitic capacitive or inductive coupling. The origin of the common mode EMI is either magnetic or electric. Common mode EMI is electrically generated when a circuit with large dV/dt (voltage slew rate) has a significant parasitic capacitance to ground. Magnetically generated common mode EMI appears when a circuit loop with large dI/dt (current slew rate) in it has significant mutual coupling to a group of nearby conductors. In addition, there is an important energy exchange between two modes. This effect is known as differential–common mode conversion. A great many theoretical and analytical models of the conducted EMI emission, including EMI propagation models and EMI generation models, have been developed to control the conducted EMI emission and optimize the design of EMI suppression systems and components, such as EMI filters. Details can be found in any related literature (Holmes, 2003; Nave, 1991).

1.1.2.2 Radiated EMI Emission

Radiated EMI emission starts from an emitting source, propagates via a radiating path, and reaches a susceptible receiver. The strength of the radiated EMI is determined by the source, the media surrounding the source, and the distance between the source and the susceptor.

The basic component of EMI is an electromagnetic wave. Any source of the radiated EMI energy generates expanding spherical wave fronts, which travel in all directions from the source. At any point the wave consists of an electric field (E) and a magnetic field (H), which are perpendicular to each other and to the direction of propagation. The field characteristics and the relative magnitude between the H-field and the E-field depend upon how far away the wave is from its source and on the nature of the generating source itself. In fact, the space surrounding the source can be divided in two regions. The region close to the source is called the near, or induction field, in which the field properties are determined primarily by the source characteristics. At a distance greater than $\lambda/2\pi$ (λ is the wavelength), the region is called the far, or radiation field. In this region, the properties of the field depend mainly upon the medium through which the field is propagating. The region around $\lambda/2\pi$ between the near and far fields is called the transition region (Ott, 1988).

The ratio of E to H (E/H) is the wave impedance, Z. In the far field, E/H equals the characteristic impedance of the medium. For air or free space $E/H = Z_0 = 377\ \Omega$, the wave is indicated to be a plane wave. In the far field, all radiated EMI waves essentially lose their curvature, and the surface containing E and H becomes a plane instead of a section of a sphere. In the near field, the E/H is determined by the characteristics of the source and the distance from the source to where the field is observed. If the source has high current and low voltage ($E/H < 377\ \Omega$), and may be generated by a loop, a transformer, or power lines for instance, the near field is predominantly magnetic. Conversely, if the source has low current and high voltage ($E/H > 377\ \Omega$), the near field is predominantly electric. With the distance increasing away from the source, both E- and H-fields begin to attenuate, eventually becoming the plane wave.

Depending on the characteristics of the source, the radiated EMI emission can be classified as differential mode radiation or common mode radiation. Differential mode radiation results from the flow of electromagnetic current loops within a system's structure. Common mode radiation is caused by unintentional voltage drops in a circuit, which causes some grounded parts of the circuit to rise above the referenced real ground potential. These two modes can be simply represented by the small loop radiation and dipole radiation, separately, as shown in Figure 1.3.

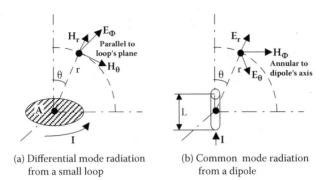

(a) Differential mode radiation from a small loop

(b) Common mode radiation from a dipole

FIGURE 1.3 EMI radiation from two basic circuit configurations: (a) differential mode radiation from a small loop, (b) common-mode radiation from a dipole.

For differential mode radiation by a small loop shown in Figure 1.3a, the E and H can be described approximately as (Mardiguian, 2001):

$$E_\phi (V/m) = \pi Z_0 \frac{IA}{\lambda^2 r} \sqrt{1 + \left(\frac{\lambda}{2\pi r}\right)^2} \sin\theta \tag{1.1}$$

$$H_\theta (A/m) = \pi \frac{IA}{\lambda^2 r} \sqrt{1 - \left(\frac{\lambda}{2\pi r}\right)^2 + \left(\frac{\lambda}{2\pi r}\right)^4} \sin\theta \tag{1.2}$$

where I is loop current (A, amper); A is loop area (m^2); λ is wavelength (m), $\lambda = 300/f$, f is frequency (MHz); r is distance to observation point (m); and Z_0 is free space impedance, 377 Ω.

In the near field, $r < \lambda/2\pi$, Equations 1.1 and 1.2 can be simplified as:

$$E(V/m) = \frac{Z_0 IA}{2\lambda r^2} = 0.63 \frac{IAf}{r^2} \tag{1.3}$$

$$H(A/m) = \frac{IA}{4\pi r^3} = 0.08 \frac{IA}{r^3} \tag{1.4}$$

Therefore the wave impedance can be obtained:

$$Z = E/H = 7.88 fr \tag{1.5}$$

In the far field, $r > \lambda/2\pi$, Equations 1.1 and 1.2 can be simplified as:

$$E(V/m) = \frac{Z_0 \pi IA}{\lambda^2 r} = 0.013 \frac{IAf^2}{r} \tag{1.6}$$

$$H(A/m) = \frac{\pi IA}{\lambda^2 r} = \frac{E}{120\pi} \tag{1.7}$$

For common mode radiation by a dipole, such as a straight wire or cable acting as an antenna, as shown in Figure 1.3a, the E and H can be predicted approximately as (Mardiguian, 2001):

$$E_\theta(V/m) = Z_0 \frac{IL}{2\lambda r} \sqrt{1 - \left(\frac{\lambda}{2\pi r}\right)^2 + \left(\frac{\lambda}{2\pi r}\right)^4} \sin\theta \qquad (1.8)$$

$$H_\phi(A/m) = \frac{IL}{2\lambda r} \sqrt{1 + \left(\frac{\lambda}{2\pi r}\right)^2} \sin\theta \qquad (1.9)$$

where I is loop current (A, amper); L is dipole length (m); λ is wavelength (m), $\lambda = 300/f$, f is frequency (MHz); r is distance to observation point (m); and Z_0 is free space impedance, 377 Ω.

In the near field, $r < \lambda/2\pi$, Equations 1.8 and 1.9 can be simplified as:

$$E(V/m) = \frac{Z_0 IL\lambda}{8\pi^2 r^2} = 1434 \frac{IL}{r^3 f} \qquad (1.10)$$

$$H(A/m) = \frac{IL}{4\pi r^2} = 0.08 \frac{IL}{r^2} \qquad (1.11)$$

Therefore, the wave impedance can be obtained:

$$Z = E/H = 17925 \frac{1}{fr} \qquad (1.12)$$

In the far field, $r > \lambda/2\pi$, Equations 1.8 and 1.9 can be simplified as:

$$E(V/m) = \frac{Z_0 IL}{2\lambda r} = 0.628 \frac{ILf}{r} \qquad (1.13)$$

$$H(A/m) = \frac{IL}{2\lambda r} = \frac{E}{120\pi} \qquad (1.14)$$

Equations 1.1 through 1.14 have been used to predict the radiated EMI emissions in an electronic system. As with other EMI modeling methodologies, these predictions may still raise some concerns about the accuracy, feasibility, and generality. However, they will help to fundamentally understand the EMI emission process and its factors, and therefore, find the right routine to optimize the EMC design and effectively control EMI problems.

Within a complex system, EMI energy may leave a source or enter susceptible equipment by conduction, coupling, or radiation. EMI can occur between one part of the equipment and another part, such as between a power supply and nearby circuitry, or between separate equipment.

Reducing either the *E*-field or *H*-field can lower EMI energy emissions and is the basis of the following suppression techniques for EMI control:

- Reduce the intensity, including the voltage and current drive levels, of the EMI source.
- Provide differential and common mode filtering for high speed signals, or use balanced differential pairs with impedance-matched signals to control the conducted EMI emission, which might come from signal lines, antenna leads, power cables, and even by ground connections between EMI susceptible equipment.
- Reduce the EMI energy being coupled between components, circuits, or equipment having some mutual impedance, through which currents or voltages in one circuit can cause currents or voltages in another circuit. The mutual impedance may be conductive capacitive, inductive, or any combination of these.
- Reduce or shield radiated EMI emission from openings of any kind in equipment enclosures; ventilation, access, cable, or meter holes; around the edges of doors and hatches; drawers and panels; imperfect joints in the enclosures; and leads and cables entering a susceptible device.
- EMI shielding is always the final defense for EMC and is an effective method for controlling EMI problems.

In general, multiple techniques of suppression or shielding can be required to effectively control EMI and achieve an EMC-compliant system, such as enclosure shielding, gasketing, grounding, filtering, decoupling, isolation and separation, circuit impedance control, I/O interconnect design, as well as proper PCB layout. Depending on the complexity of the system, speed of operation, and EMC requirements, the use of EMI shielding can be minimized by proper electronic design.

1.1.3 EMC REGULATIONS AND TESTING STANDARDS

All electronic systems have the potential for EMI problems. Moreover, susceptibility to radiated and conducted EMI emissions can impair system performance in the presence of lightning, electrostatic discharge, and normal transmissions from radio stations. In the case of the military, these systems also must be able to withstand electromagnetic pulses (EMP) resulting from the detonation of nuclear devices. As a result of these performance requirements, EMC regulations and testing standards have been established and covered in detail, and are continuously evolving (Christopoulos 1995; CISPR11 2004). The setting of EMC regulations and testing standards is a complex and lengthy process because it cuts across technical, trade, and national prestige issues. There are standards set by international organizations, national bodies, trade associations, insurance companies, and so forth. These standards are, in most cases, similar, but there are also many differences in the specific limits set by various organizations, the methods of measurement and testing, and the route to compliance. However, complying with an EMC standard is no guarantee that the product will not experience upsets due to EMI. It is thus important to examine how far beyond meeting EMC regulations one has to go to produce a marketable and EMC-compliant product at a reasonable cost (Christopoulos, 1995).

There are several major standards frameworks: the international standard, the Federal Communications Commission (FCC) standard in the United States, the European standard, and the military standard.

1.1.3.1 The International Standard

The International Electrotechnical Commission (IEC) has the most impact on all EMC regulations and testing standards. Within the IEC, three technical committees are related to EMC: CISPR (International Special Committee on Radio Interference), TC77 (concerned mainly with EMC in electrical equipment and networks), and TC65 (concerned mainly with immunity standards). The output of the CISPR committee is in the form of standards with the designation CISPR10 to CISPR23 covering a wide range of EMC-related topics. TC77 is responsible for a series of publications with the designation 555 covering harmonics and fliker, and designation 1000 covering low-frequency phenomena in power networks. TC65 has produced a series of standards with the designation 801 covering issues related to immunity. The recommendations contained in CISPR 22 for conducted and radiated emissions of information technology equipment have formed the foundation for most of the major national standards. Unfortunately, there has been incompatibility between the requirements of different countries because (a) the recommendations proposed by CISPR cannot become law unless the individual member countries take the appropriate action themselves and (b) individual countries have made adjustments to the recommendations before adopting them as national standards.

1.1.3.2 FCC Standard

In the United States, Congress has delegated authority to the Federal Communications Commission (FCC) to regulate civilian radio, communications, and interference. An individual or organization must comply with all applicable FCC standards and equipment authorization requirements to legally import or market a product in the United States. FCC regulations specify EMC standards, test methods (ANSI C63.4), equipment authorization procedures, and marking requirements. Part 15 is the most often applied FCC EMI standard; it covers digital computing devices, plus garage door openers, radio-controlled toys, cordless telephones, and other intentional low-power transmitters. The regulations concern conducted and radiated interference and refers to two classes of equipment. Class A equipment is used in commercial, industrial, and business premises, while class B equipment is intended for residential use. Regulations for class B equipment (approximately 10 dB) tend to be more stringent. Any device with timing (clock) pulses in excess of 10 kHz is covered by FCC regulations. The upper frequency tends to be of the order of ten times the highest clock frequency. Hence, with clock frequencies now exceeding 100 MHz, EMC regulations will cover emissions well into the gigahertz range. Conducted interference covers frequencies ranging between 450 kHz and 30 MHz; these regulations are aimed at controlling interference current in power leads. Since specifications are given in terms of voltages measured on such leads, it is necessary to more closely define the impedance across which such measurements are made. This is done by interposing a line impedance stabilization network (LISN) between the equipment to be tested and the power supply network. This consists of a combination of capacitors, resistors, and inductors, so that for frequencies up to 30 MHz the impedance to ground is essentially constant and equal to approximately 50 Ω (Christopoulos, 1995).

Compliance with FCC regulations is mandatory for all equipment in use in the United States. The coordinating organization for standardization in the United States is the American National Standards Institute (ANSI) to which other interested bodies, such as the IEEE EMC society, contribute EMC-related standards.

1.1.3.3 European Standard

The European Standards Committee (ESC), through the European Committee for Electrotechnical Standards (CENELEC), is responsible for the adoption of suitable standards. These standards cover broad terms and broad classes of equipment, that is, class 1 (residential, commercial, and light industry) and class 2 (heavy industry); basic standards (setting out test procedures); and product standards (covering specific types of apparatus, e.g., information technology equipment). The directive of May 3, 1989, concerning the requirements for electromagnetic compatibility, was amended for a transitional period with the full force of the directive coming into effect on January 1, 1996. All member countries are required to enact legislation to implement the directive. The directive is all embracing, covering emission and susceptibility for a wide range of equipment and imposes duties on manufacturers whether or not appropriate standards exist. In order to eliminate trade barriers, the FCC and European CENELEC have signed an agreement in 1997 by which EMC qualification tests conducted in the United States by an FCC-approved lab will be accepted by European countries, and vice versa. Therefore, U.S. manufacturers and EMC laboratories can test and approve in the United States equipment bound for the European market (Mardiguian, 2001).

1.1.3.4 Military Standard

Military systems present special EMI problems to the EMC system. Increased use of electronic communications for integrated command, control, communications, and intelligence between the various services (land, air, and sea), the hostile nature of the battle environment, the need for portability and secrecy, and the deliberate use of electromagnetic energy for jamming has created extremely complex EMC systems and EMI problems. Electronic systems designed for the military market generally have to comply with MIL-STD-461, which imposes limits on the susceptibility of a system to conduct and radiate EMI, as well as specifying its maximum conducted and radiated emissions. The companion document, MIL-STD-462, specifies the testing methods. Mil-STD-461 has been recognized and applied by many defense organizations outside the United States, as well as some nonmilitary agencies.

Commercially conducted emission limits of the FCC, ESC, and CISRR 22 are more severe than those for military equipment because the ambient noise in which commercial systems operate must be kept to a minimum since there are generally no susceptibility requirements for this type of equipment. By contrast, military equipment must be able to withstand conducted and radiated radio frequency (RF) noise without malfunction. Unlike the regulations for conducted emissions, however, commercially radiated limits are less stringent than the military standards. The more stringent military standard for radiated emissions takes into account the closer proximity of electronic equipment in aircraft, ships, tanks, and so forth, compared with commercial installations, which are typically more widely dispersed. It should be noted that measurement distances for testing equipment to the various radiated emission regulations and testing standards are based on models of actual equipment locations.

1.2 EMI SHIELDING PRINCIPLES

EMI shielding means to use a shield (a shaped conducting material) to partially or completely envelop an electronic circuit, that is, an EMI emitter or susceptor. Therefore, it limits the amount of EMI radiation from the external environment that can penetrate the circuit and, conversely, it influences how much EMI energy generated by the circuit can escape into the external environment. A variety of materials have been used for shielding with a wide range of electrical conductivity, magnetic permeability, and geometries. Shields invariably contain apertures or openings used for access and ventilation, as well as a number of joints and seams for practical manufacturing. Shields for practical equipment also allow attachment to wires or pipes used for signaling and services. All these components constitute breaches of the shield integrity and play a decisive part in the overall performance of the shield (Christopoulos, 1995). Beyond the intelligent circuit design, shielding is the only means of suppressing EMI or hardening for susceptibility, and does not slow down the operation of high-speed systems. Shielding not only reduces EMI radiations, but it also provides an isolated ground reference that effectively reduces internal crosstalk and circuit path coupling and overall common mode coupling. Shields can be used at different space levels, such as integrated circuit (IC), printed circuit board (PCB), systems level, shielded rooms, and even shielding buildings. Regardless of the size of the enclosure, the shielding design principles are essentially the same. In addition, shielding is also used to limit ESD and ESP, and absorb microwave energy (Instrument Specialties, 1998).

1.2.1 SHIELDING EFFECTS IN THE E-FIELD AND H-FIELD

Shielding effect is provided by a conductive barrier or enclosure that harmlessly reflects or transmits EMI into the ground. In 1821, Michael Faraday first introduced the concept that an enclosed conductive housing has a zero electrical field. This principle is known as *Faraday cage*, which forms the basis of today's shielding technology. If sensitive equipment is enclosed within a thin, conductive, spherical shell that is placed in an *E*-field, as shown in Figure 1.4, it will be shielded because the current setup by the electromagnetic wave is not conducted into the inside of the shell. This is not because the shell has completely absorbed the field, but because the *E*-field has caused electronic charges of different polarity along the shell. These charges will, in turn, generate an electrical field that will tend to cancel the original field inside the shell (Bjorklof, 1999; Bridge, 1988). The thickness of the shell can be very small when the wave frequency is high enough. The electromagnetic current takes the path

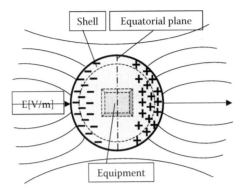

FIGURE 1.4 Charge distribution on a spherical conducting shell contained on equipment in an electrical field. (Modified from Bridges, 1988.)

of least resistance and will run right around the exterior of the conductive shell. However, if there is any joint gap in the shell, the current will prefer to run around the one surface of the gap onto the inside of the shell rather than to jump even the smallest gap to the other surface of the gap. In another words, any apertures or breaks in a shield will limit its shielding effectiveness.

When it comes to H-fields, Faraday's effect will disappear. However, the H-field shielding can be achieved by means of shields made of a soft magnetic material with high permeability, $\mu \gg 1$, and sufficient thickness to attenuate the magnetic field in the shielding shell by providing a low reluctance, as illustrated in Figure 1.5. That is, the spherical shell of magnetic material with good permeability will reduce the H-field intensity inside because the H-field tends to remain in the magnetic material layer as the magnetic material offers a low-reluctance path (Bjorklof, 1999;

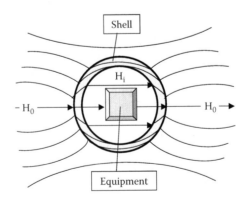

FIGURE 1.5 A spherical magnet shell with good permeability will reduce the field strength ($H_i \ll H_0$) inside because the field tends to remain in the magnetic layer as it offers a low-reluctance path. (Modified from Duffin, 1968.)

Electromagnetic Interference Shielding Fundamentals and Design Guide

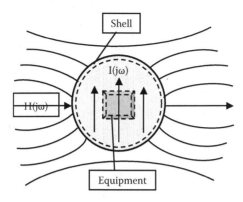

FIGURE 1.6 A thin metallic shell with good conductivity will reduce an alternating magnetic field strength because the currents I(jω) induced by the field H(jω) tend to generate a field of opposite direction from the original. (Modified from Bjorklof, 1999.)

Duffin, 1968). Alternatively, a thin shield made of a conductive material with low permeability can also provide effective shielding for *H*-fields at high frequencies. This is because an alternating *H*-field will induce *eddy currents* in the shielding screen, assuming that the shield has adequate conductivity, as shown in Figure 1.6. These eddy currents will themselves create an alternating *H*-field of the opposite orientation inside the shell. The effect will increase as the frequency increases, resulting in high shielding effectiveness at high frequencies (Bjorklof, 1999). Therefore, it is relatively difficult to shield against low-frequency *H*-fields. Whereas magnetic absorption shielding typically needs the installation of thick shields constructed of fairly expensive magnetic materials, conductive shields based on the induced current principle may be reasonably effective at powerline frequencies. For example, aluminum screens are commonly used to protect against 50 and 60 Hz magnetic fields generated by transformers and other sources (Bjorklof, 1999). In this case, any apertures, openings, or breaks in the shield will limit its effectiveness, as the principle behind the *H*-field shielding via induced currents presumes that such currents can only flow when there are no obstacles in their path. Therefore, it is essential that any apertures should be arranged in such a way as to minimize their effect on the currents.

1.2.2 Shielding Effectiveness

Shielding can be specified in the terms of reduction in magnetic (and electric) field or plane-wave strength caused by shielding. The effectiveness of a shield and its resulting EMI attenuation are based on the frequency, the distance of the shield from the source, the thickness of the shield, and the shield material. Shielding effectiveness (SE) is normally expressed in decibels (dB) as a function of the logarithm of the ratio of the incident and exit electric (*E*), magnetic (*H*), or plane-wave field intensities (*F*): $SE\ (dB) = 20 \log (E_0/E_1)$, $SE\ (dB) = 20 \log (H_0/H_1)$, or $SE\ (dB) = 20 \log (F_0/F_1)$, respectively. With any kind of electromagnetic interference, there are three mechanisms contributing to the effectiveness of a shield. Part of the incident radiation is

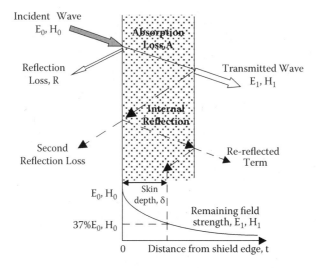

FIGURE 1.7 Graphical representation of EMI shielding.

reflected from the front surface of the shield, part is absorbed within the shield material, and part is reflected from the shield rear surface to the front where it can aid or hinder the effectiveness of the shield depending on its phase relationship with the incident wave, as shown in Figure 1.7. Therefore, the total shielding effectiveness of a shielding material (SE) equals the sum of the absorption factor (A), the reflection factor (R), and the correction factor to account for multiple reflections in thin shields:

$$SE = R + A + B \tag{1.15}$$

All the terms in Equation 1.15 are expressed in dB. The multiple reflection factor B can be neglected if the absorption loss A is greater than 10 dB. In practical calculation, B can also be neglected for electric fields and plane waves.

1.2.2.1 Absorption Loss

Absorption losses A are a function of the physical characteristics of the shield and are independent of the type of source field. Therefore, the absorption term A is the same for all three waves. When an electronic wave passes through a medium such as a shield, its amplitude decreases exponentially, as shown in Figure 1.7. This decay or absorption loss occurs because currents induced in the medium produce ohmic losses and heating of the material, and E_1 and H_1 can be expressed as (Ott 1988): $E_1 = E_0\,e^{-t/\delta}$ and $H_1 = H_0\,e^{-t/\delta}$. The distance required for the wave to be attenuated to $1/e$ or 37% is defined as the skin depth, δ. Therefore, the absorption term A is given by the expression:

$$A = 20\,(t/\delta)\log(e) = 8.69\,(t/\delta) = 131\,t\sqrt{f\mu\sigma} \tag{1.16}$$

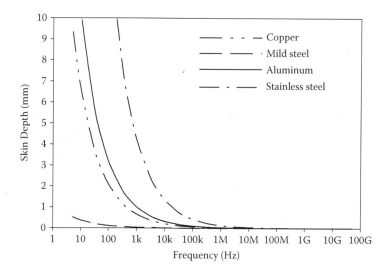

FIGURE 1.8 Skin-depth variation with frequency for copper, aluminum, steel, and stainless steel.

where A is the absorption or penetration loss expressed in decibels; t is the thickness of the shield in mm; f is frequency in MHz; μ is relative permeability (1 to copper); σ is conductivity relative to copper in IACS. The skin depth δ can be expressed as:

$$\delta = \frac{1}{\sqrt{\pi f \mu \sigma}} \quad (1.17)$$

The absorption loss of one skin depth in a shield is approximately 9 dB. Skin effect is especially important at low frequencies, where the fields experienced are more likely to be predominantly magnetic with lower wave impedance than 377 Ω.

Figure 1.8 shows the skin depth variation with frequency for copper, aluminum, steel, and stainless steel. Copper and aluminum have over five times the conductivity of steel, so are very good at stopping electric fields; but they have a relative permeability of one (the same as air). Typical mild steel has a relative permeability of around 300 at low frequencies, falling to one as frequencies increase above 100 kHz, and its higher permeability gives it a reduced skin depth, making a reasonable thickness of mild steel better than aluminum for shielding low frequencies. Different grades of steel (especially stainless) have different conductivity and permeability, and their skin depth varies considerably as a result. From the absorption loss point of view, a good material for a shield will have high conductivity and high permeability, and sufficient thickness to achieve the required number of skin depths at the lowest frequency of concern.

1.2.2.2 Reflection Loss

The reflection loss is related to the relative mismatch between the incident wave and the surface impedance of the shield. The computation of reflection losses can be greatly simplified by considering shielding effectiveness for incident electric fields as a separate problem from that of electric, magnetic, or plane waves. The equations for the three principle fields are given by the expressions (Ott, 1988):

$$R_E = 321.8 + 10\log\frac{\sigma}{f^3 r^2 \mu} \tag{1.18}$$

$$R_H = 14.6 + 10\log\frac{fr^2\sigma}{\mu} \tag{1.19}$$

$$R_P = 168 - 10\log\frac{f\mu}{\sigma} \tag{1.20}$$

where R_E, R_H, and R_P are the reflection losses for the electric, magnetic, and plane wave fields, respectively, expressed in dB; σ is the relative conductivity referred to copper; f is the frequency in Hz; μ is the relative permeability referred to free space; and r is the distance from the source to the shielding in m.

1.2.2.3 Multiple Reflection Correction Factor

The factor B can be mathematically positive or negative (in practice it is always negative), and becomes insignificant when the absorption loss $A > 6\ dB$. It is usually only important when metals are thin and at low frequencies (i.e., below approximately 20 kHz). The formulation of factor B can be expressed as (Vasaka, 1956):

$$B(dB) = 20\log\left|1 - \frac{(K-1)^2}{(K+1)^2}(10^{-A/10})(e^{-i227A})\right| \tag{1.21}$$

where A is the absorption loss (dB); $K = Z_S/Z_H = 1.3(\mu/fr^2\sigma)^{1/2}$; Z_S is the shield impedance; and Z_H is the impedance of the incident magnetic field. When $Z_H \ll Z_S$, the multiple reflection factor for magnetic fields in a shield of t and skin depth δ can be simplified as (Ott, 1988):

$$B = 20\log(1 - e^{-2t/\delta}) \tag{1.22}$$

Consequently, the total shielding effectiveness for electric, magnetic, and plane-wave fields can be obtained by Equation 1.15 with a combination of the related equations of absorption and reflection losses, as well as correction factor B. When B is neglected, the plots of reflection loss, absorption loss, and total shielding effectiveness for copper and iron are shown in Figure 1.9 and Figure 1.10. These illustrations give a good physical representation of the behavior of the component parts of an electromagnetic wave. They also illustrate why it is so much more difficult to shield magnetic fields than electric fields or plane waves. Here, copper offers more shielding effectiveness than iron in all cases except for absorption loss due to the high permeability of iron. That is, for a given thickness, magnetic materials such as iron or steel provide higher absorption losses than nonmagnetic materials such as copper. The absorption losses are low at low frequencies and rise gradually to high levels at

Electromagnetic Interference Shielding Fundamentals and Design Guide

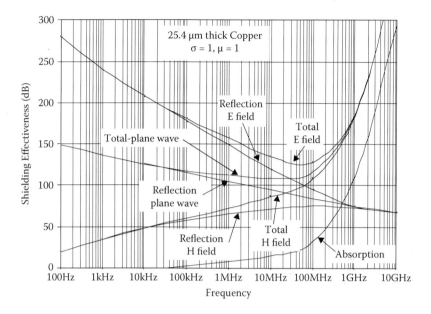

FIGURE 1.9 Shielding effectiveness of a copper shield.

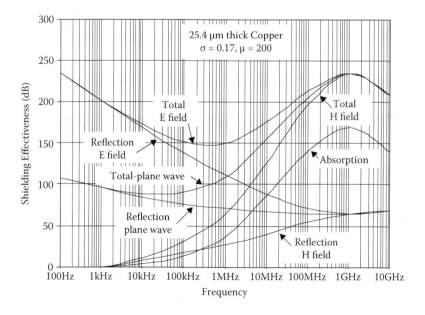

FIGURE 1.10 Shielding effectiveness of an iron shield.

high frequencies. In general, absorption performance can be improved by increasing the thickness of the shield or by using a metal with high permeability. Usually, metals with high permeability exhibit low electrical conductivity. And only magnetic materials effectively shield against magnetic fields at low frequencies. If magnetic shielding is required, particularly at frequencies below 14 kHz, it is customary to neglect all terms in Equation (1.15) except the absorption term A. Conversely, if only electric field or plane-wave protection is required, reflection is an important factor to consider in the design. From Figure 1.9 and Figure 1.10, it is worth noting that when frequency is around 1 GHz, the total shield effectiveness of the material in all three fields tends to be the same. Moreover at higher frequencies where both reflection and absorption losses increase, the higher electrically conductive materials (like copper) provide better electrical and magnetic shielding.

In a summary, the amount of reflection loss depends on the impedance of the electromagnetic wave and the shield. When an electromagnetic wave encounters a shield, if the wave's impedance differs significantly from that of the shield, the wave is partially reflected back. Conversely, if the wave's and the shield's impedance are closely matched, the EMI energy will pass through the shield with minimal reflection. An electrically dominant wave (*E*-field) in the near field has high impedance (greater than 377 Ω). Higher conductive metals have low impedance and, therefore, are successful at reflecting back electrically dominant waves because of the impedance mismatch. Reflection of EMI is the primary shielding mechanism for electrically dominant waves. Magnetically dominant waves (*H*-fields), on the other hand, have low impedance (less than 377 Ω). With these waves, absorption plays an important role in shielding. Magnetic waves are more difficult to shield; however, their energy generally diminishes as the distance from the source increases. Over greater distances (far field), the electric field component dominates the wave, and it is this electrical component that must be dealt with through EMI shielding. The relative thickness of the shielding material has an effect on the amount of the wave's energy that is absorbed. However, thickness has little effect on the amount of the wave's energy that is reflected. Both the absorption and reflection of EMI are important considerations in selection of a shielding method. Attenuation or reduction of E-fields can be handled effectively by metal sheet, foil, and metal coatings since reflection loss is very large relative to absorption loss. Primary reflection occurs at the front surface of the shield, so that very thin sections of conductive material provide good reflection properties. Moreover, multilayer shields provide increased reflection loss due to the additional reflecting surfaces. Conversely, where absorption plays a key role in attenuation, strong magnetic fields are better handled by thick magnetic materials or thick-skin shields, such as those provided by conductive paints (GE Plastics, n.d.).

1.2.3 Shielding Modeling in Practice

1.2.3.1 Metallic Enclosure

The shielding effectiveness of a metallic enclosure is related to its geometry and material properties. The amount of current flow at any depth in the shield and the rate of decay are governed by the conductivity of the metal and its permeability.

The residual current appearing on the opposite face is the one responsible for generating the field that exists on the other side. The thickness plays an important role in shielding. When skin depth is considered, however, it turns out that thickness is only critical at low frequencies. At high frequencies, even metal foils are effective shields. The current density in thick shields is the same as for thin shields. A secondary reflection occurs at the far side of the shield for all thicknesses. The only difference with thin shields is that a large part of the reflected wave may appear on the front surface. This wave can add to or subtract from the primary reflected wave depending on the phase relationship between them. For this reason, the correction factor should be involved in the shielding calculations to account for reflections from the far surface of a thin shield.

Depending upon whether the EMI source is internal or external to the enclosure, the reflected component may or may not add to the overall shielding. If the emitter is external to the shield, then the reflected electromagnetic energy propagates away from the enclosure and adds to the shielding effectiveness. If the enclosure contains the emitter, the reflected electromagnetic energy is always inside the enclosure and as a result does not add fully to the overall shielding effectiveness. Therefore, shielding effectiveness calculations for an enclosure containing the emitter can be simplified since the absorption loss (A) provides most of the shield attenuation and is approximated by Equation 1.16. Shielding effectiveness calculations for an enclosure containing a susceptor comprised of the absorption and reflection losses provide attenuation. The attenuation for the emitter external to the shield can be approximated by the combination of Equation 1.16 and Equations 1.18 to 1.20.

Here the calculated shielding effectiveness values basically indicate best-case shielding effectiveness since they do not include the effects of apertures or other discontinuities. These discontinuities account for most of the leakage in a shielding enclosure. As it is not practical to build an all-welded enclosure, most enclosures require various openings for controls, access panels, ventilation, viewing, and so forth.

1.2.3.2 EMI Gasket

The function of an EMI gasket is to preserve continuity of current flow in the gap or slot of a shield. If the gasket is made of a material identical to the walls of the shielded enclosure, the current distribution in the gasket will also be the same assuming it could perfectly fill the slot. This is not practical due to mechanical considerations. Electromagnetic leakage through the seam can occur via the material directly or at the interface between the gasket and the shield. If an air gap exists in the seam, the flow of current will be diverted to those points or areas that are in contact without the gap. A change in the direction of the flow of current alters the current distribution in the shield as well as in the gasket. A high-resistance joint does not behave much differently than open seams. It simply alters the distribution of current somewhat. It is important in gasket design to make the electrical properties of the gasket as similar to the shield as possible, maintain a high degree of electrical conductivity at the interface, and avoid air or high-resistance gaps. Other particular factors include junction geometry, contact resistance, and applied force, which will affect gasket performance (Chomerics, 2000).

1.2.3.3 Shield with Apertures

Instead of a solid shield, practical equipment cases or enclosures usually have apertures such as windows, vents, seams, and joints that degrade the shielding effectiveness of the conductive materials. In such cases, the method for calculating the shielding effectiveness of the integral structure would be (1) calculate the shielding effectiveness for a panel of the conductive material at each frequency of interest; (2) calculate the shielding effectiveness of the aperture at each of the same frequencies; and (3) use the lower shielding effectiveness value as the integral shielding effectiveness at each frequency.

Generally, there are several types of apertures, including round holes, slots, mesh screens, and waveguide or honeycomb structures as shown in Figure 1.11. The method of combining their effects is similar to the method of calculating total resistance produced by several parallel resistors (Evens, 1997):

$$\frac{1}{SE_{Total}} = \frac{1}{SE_1} + \frac{1}{SE_2} + \frac{1}{SE_3} \cdots \quad (1.23)$$

The methods for determining individual shielding effectiveness for various types of apertures are described next.

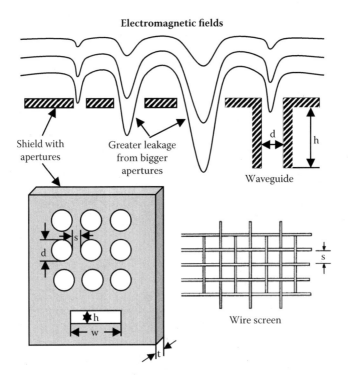

FIGURE 1.11 Effect of waveguide and apertures on shielding effectiveness.

Conductive panel with apertures. For a rectangular-shaped slot as shown in Figure 1.11 (bottom left) the shielding effectiveness can be expressed as (White, 1988):

$$SE\ (dB) \approx 100 - 20 \log (wf_{MH}) + 20 \log[1 + \ln (w/h)] + 30\ (t/w),$$
$$\text{when } w < \lambda/2 \quad (1.24)$$

$$SE\ (dB) \approx 0, \text{ when } w \geq \lambda/2 \quad (1.25)$$

where w and h are slot width and height in mm; and t is the depth of slot, usually thickness of material in mm. The shielding effectiveness of a single panel with n slots can be approximately expressed as (Instrument Specialties, 1998):

$$SE_{dB} \approx 20 \log (\lambda/2w) - 20 \log n|_{w \leq \lambda/2} \quad (1.26)$$

where w is length of slot (meters) and $w > h$ and $w \gg t$; λ is wavelength in meters; and n is number of apertures within $\lambda/2$.

For a round hole as shown in Figure 1.11 bottom left, the shielding effectiveness can be expressed as (White, 1988):

$$SE\ (dB) \approx 102 - 20 \log (df_{MH}) + 30(t/d), \text{ when } d < \lambda/2 \quad (1.27)$$

$$SE\ (dB) \approx 0, \text{ when } d \geq \lambda/2 \quad (1.28)$$

where d is diameter in mm; t is the depth of hole, usually thickness of material in mm. For multiple holes with a number of n as shown in Figure 1.11 bottom left, if $s < \lambda/2$, $d > \lambda/2$, and $s/d < 1$, the shielding effectiveness can be expressed as (Canter, 1999):

$$SE\ (dB) \approx 20 \log (\lambda/2d) - 10 \log n \quad (1.29)$$

Equations 1.24 to 1.29 can be solved in terms of width (w) or diameter (d) to determine what size aperture is required for a given attenuation. In general, however, apertures should be smaller than $\lambda/50$ and never bigger than $\lambda/20$. To achieve acceptable attenuation values at a frequency of 100 MHz, typical for high-speed digital devices, apertures should not exceed 10 mm. The type of shielding used and how it is applied are determined to a large extent by the function of the aperture. Apertures are required for displays, fans, keyboards, ventilation, connectors, parting seams, and so forth. In the case of a cathode ray tube (CRT) or other displays, the best optical qualities are provided by thin film-reflective materials that are sputtered or vacuum deposited on the screen. At frequencies of approximately 10 MHz, however, skin effect begins to degrade the shielding effectiveness of these materials (Instrument Specialties, 1998).

Wire screen. Another approach that can be used as either a window or a vent panel is knitted or woven wire screening. These discontinuous or nonuniform materials can be used up to about 150 MHz, beyond which skin effect across the material becomes excessive, and it begins to fail. A better solution is perforated metal materials commonly used in microwave ovens but now available in optical grades. As an alternative, the knitted or woven screen can be plated to improve shielding effectiveness.

To avoid the visual distraction from moiré patterns that result from superimposing one repetitive image over another, consideration should be given to shielding the printed circuit boards or the back side of the CRT itself.

The shielding effectiveness of the wire screen, as shown in Figure 1.11 (bottom right), can be expressed as (White, 1998):

- For plane waves,

$$SE\ (dB) \approx 20 \log (\lambda/2s),\ \text{when}\ s \leq \lambda/2 \tag{1.30}$$

$$SE\ (dB) \approx 0,\ \text{when}\ s \geq \lambda/2 \tag{1.31}$$

- For near fields,

$$\textit{Magnetic fields}: SE(dB) = 20 \log (\pi r/s) \tag{1.32}$$

$$\textit{Electric fields}: SE(dB) = 20 \log [\lambda^2/(4\pi rs)] \tag{1.33}$$

where r is the distance from the EMI source to the screen in meters; and λ is the wavelength in meters; s is distance between wires in meters. Equations 1.30 to 1.33 are valid when $g \geq 10^{-6}\ \lambda$. When s is a tiny fraction of a wavelength, such as $s \leq 10^{-6}\ \lambda$, the screen looks like a solid piece of thin metal (Evans 1997). Therefore, the shielding effectiveness can be calculated with the previous equations for near fields and far fields. Here, material conductivity is equal to that of the wire materials times its percentage of optical coverage.

Waveguide. Additional attenuation can be obtained from a hole if it is shaped to form a waveguide, as shown in Figure 1.11 (top). A waveguide has a cutoff frequency below which it works as an attenuator. The attenuation is a function of the length h of the waveguide. For a circular waveguide, the cutoff frequency is (MHz) (Pozar, 1988):

$$f_c = \frac{5.5c}{\pi d} = \frac{1.75 \times 10^8}{d} \tag{1.34}$$

where d is the diameter in meter, c is the speed of light in vacuum, and ε is the relative perttivity. For a rectangular waveguide the cutoff frequency (MHz) is (Pozar, 1988):

$$f_c = \frac{c}{2w} = \frac{1.50 \times 10^8}{w} \tag{1.35}$$

where w is the largest dimension of the waveguide cross-section in meters.

When the operating frequency is much less than the cutoff frequency, the magnetic field shielding effectiveness or absorbing loss of a round waveguide frequency can be expressed as (Ott, 1988):

$$SE\ (dB) = 32\ h/d \tag{1.36}$$

Electromagnetic Interference Shielding Fundamentals and Design Guide

where d is the diameter and h is the length of the waveguide. For a rectangular waveguide (Ott, 1988):

$$SE\ (dB) = 27.2\ h/w \quad (1.37)$$

where w is the largest linear dimension of the waveguide cross-section and h is the length.

For shielded ventilation panels, honeycomb materials provide the best airflow with almost the best attenuation characteristics. The honeycomb is constructed of small waveguides assembled in parallel. This produces a material that is approximately 97% open area and, for frequencies less than 1/3 the cutoff frequency, has a single-cell attenuation value of (Instrument Specialties, 1998):

$$SE\ (dB) \approx 30\ h/d \quad (1.38)$$

where d is the diameter of a cell tube and h is the thickness of the material, that is, length of cell tube.

For honeycomb panels with a cutoff frequency fc the absorption loss for a single waveguide is reduced by a factor related to the total number of openings in the panel. The total shielding effectiveness including both obserbing loss and reflection factor can be expressed as (Cater, 1999):

$$SE\ (dB) = 20\ \log\ (fc/f) + 27.3\ h/d - 10\ \log\ n \quad (1.39)$$

where n is the total number of cells, and f is frequency and usually not more than $fc/10$ for an effective shielding.

Consequently, an effective shield must topologically be as tight as possible. The presence of intentional or inevitably apertures—that is, everything from doors, windows, ventilation holes, and inlets for panel instrumentation to seam gaps and cable throughputs—will lower the shielding effectiveness of a panel or enclosure. Moreover, multiple small apertures will provide better protection than a single large one. A window with multiple waveguides, such as a tube or a honeycomb window, can be an effective tool for preventing electromagnetic fields where the cutoff frequency of those waveguides is considerably higher than the frequency span of the electromagnetic field. In addition, honeycomb materials can be constructed in drip-proof and vision-blocking configurations for special applications without appreciable change in airflow characteristics. A less costly approach for applications where only moderate shielding is required is to use perforated materials or woven or knitted screening. The screen, especially the knitted, tends to fail at the higher frequencies, because the frequency is usually as close or bigger than the waveguide cutoff frequency.

1.2.3.4 Cavity Resonance

Cavity resonance results from a shielding enclosure that forms a reflecting cubicle of interior dimensions, which are each some multiple of half-wavelengths. Under these conditions, interior fields are extremely nonuniform. The resonance frequencies in a rectangular box are given by (mardiguian, 2001):

$$f_{res} = \frac{c}{2}\sqrt{(m/a)^2 + (n/b)^2 + (l/c)^2} = 1.5 \times 10^8 \sqrt{(m/a)^2 + (n/b)^2 + (l/c)^2} \quad (1.40)$$

where a, b, and c are width, height, and length, respectively, in meters. The terms m, n, and l are integer numbers, which can take any value, but no more than one at a time equal to zero. Otherwise, an empty metallic box could exhibit resonances, which could result in a negative shielding effectiveness.

Therefore, establishing proper protection against EMI requires an awareness of the electromagnetic topology and efficient use of appropriate barriers or enclosures. Moreover, the topology must not be compromised by holes cut into the barrier to permit cable screens or separate grounding cables to pass through, because such holes will allow ground currents to slip into the protected zone. All screens should be grounded at the barrier to prevent the introduction of unwanted disturbances. The grounding, like the shielding, should be as tight as possible. A capacitor connected between the signal line and the screen, or a feedthrough capacitor mounted in the screen itself, can be employed to shunt away high frequency disturbances superimposed on the signal line. If a number of cables are to be connected to a shielded apparatus, a feedthrough plate with efficient connectors and filters should be used (Bjorklof, 1999).

1.3 EMC DESIGN GUIDELINES AND EMI SHIELDING METHODOLOGY

1.3.1 Design for EMC

EMC design is an integral part of electrical, mechanical/shielding and system design for any electronic equipment. The design involves control of all electronic circuits, power supplies, interconnections, and electromechanical components, as well as the mechanical, shielding, and constructional details of a product. Figure 1.12 illustrates a general flow map for EMC design. Only a broad system/device EMC overview and evaluation would ensure that the specification and procedures adopted are consistent with the requirements of EMC. Conflicts between electrical and mechanical construction details (which affect mechanical strength, appearance, and cost) and details regarding shielding (which affect EMC) should be resolved early in the design together with basic input/output and grounding arrangements (Christopoulos 1995).

An EMC design for electronic equipment must meet both an emission and immunity requirement that can be achieved via (a) suppression—reducing the EMI at its source; (b) isolation—isolating the radiated or susceptible circuits by filtering, grounding, and shielding; and (c) desensitization—increasing the immunity of any susceptible circuits. These three procedures should be carried on throughout the entire electrical, mechanical, and system design of the equipment and implemented as early as possible within the design program (Instrument Specialties, 1998).

In general, the engine that drives electronic equipment is the IC, located on a PCB. The IC and PCB are comprised of potential EMI sources, as well as components and circuits sensitive to EMI energy. Therefore, design of the PCB is the most critical EMC-influencing factor for any system, since all EMI active devices are located on the board. As the electronic device is faster, smaller, and digital rather than analog, the PCB design must create a circuit board that has the lowest possible EMI, combined with the highest possible operating/processing speeds.

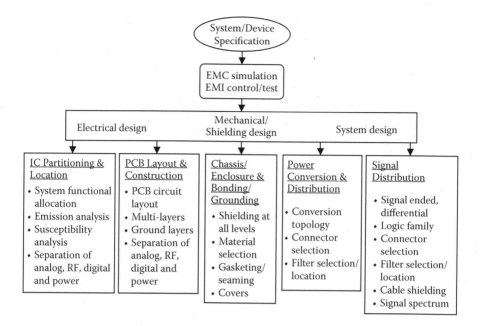

FIGURE 1.12 General EMC design flow map.

However, the faster the logic speed, the greater the required circuit bandwidth, and the more difficult it is to control radiated emissions and susceptibility. As discussed in Section 1.1.2, radiation emitted by electronic devices results from both differential and common mode currents. In semiconductor devices, differential mode currents flowing synchronously through signal and power distribution loops produce time-variant electromagnetic fields, which may be propagated along a conducting medium or by radiation through space. On simple one- or two-layer PCBs, loops are formed by the logic signals being transferred from one device to another that return by means of the power distribution rails. Loops are also created by PCB traces or rails that supply power to these devices. Common-mode radiation results from voltage drops in the system that create common-mode potential with respect to grounding. In addition, parasitic capacitive coupling, a hard-to-control phenomenon that occurs between all conductive materials, makes external cables act like antennas. The radiated EMI levels created by the active circuit loops on the board are proportional to the square of the highest created frequencies (plane wave). These frequencies are determined by the data pulse rise time and contain significant EMI energy at typically ten to fifteen times the operating speed. The rise time also determines the circuit bandwidth. For small circuits whose dimensions are less than the dimensions at resonance, the plane-wave emission levels generated by these loops may be calculated by Equation 1.6. To some extent, the higher frequencies can be reduced by a low-pass filter device, but the major reductions in emission levels come from PCB layout and shielding. To create a high-speed operational board that meets its design requirements, the design has to handle the following major, interrelated operational design problems: operating speed, EMI/EMC/ESD, timing/propagation delay, waveform distortion, and crosstalk. These problems are increased with increasing operation speeds. The increased speed also requires increased bandwidth, which in turn increases the susceptibility of the system. Improvements in the operational design are accomplished by the proper design and layout of the PCB. Timing problems are reduced by making the boards smaller and/or by locating critical high-speed components closer together. Waveform distortion is reduced in multilayer PCBs by utilizing transmission line trace configurations for high-speed data transfer. Crosstalk is reduced by careful layout and separation of the traces and components on the PCB (Instrument Specialties, 1998; Montrose, 1999).

As regards the manner of electronic construction, modularity is the norm. When designing a microprocessor system, for instance, the minimum number of PCBs should be used to avoid long connections between different boards. Careful thought must be given to external connections to a PCB. These should be the minimum number of boards possible, should transmit the slowest signals possible, and pass the minimum amount of current necessary. These general principles imply that using the same clock between two boards is undesirable, since a very fast signal must not be allowed to propagate the long distance. Similarly, transmitting a signal between boards to drive a large number of gates is undesirable, since this normally requires large currents. Some form of buffering can substantially reduce the magnitude of these currents. If there are unavoidable external connections, these should be shielded to reduce EMI interference (Christopoulos 1995). For instance, PCB-level shielding is used to reduce the EMI emissions from circuit components by installing small, PCB-mounted shielding enclosures over them. These small shielding cans or enclosures utilize the PCB ground plane as their bottoms, and when used in conjunction with a multilayer PCB, will effectively control EMI emissions to provide a total reduction of approximately 55 dB. They also minimize crosstalk and susceptibility at the board level.

Briefly there are two approaches that can be used to reduce the emission from the PCB and other level systems.

1. Electronic approach. The electronic approach includes circuit design, careful IC component selection and PCB layout, good cabling, and grounding, as well as insert suppression components such as filters, ferrite beads, and bypass capacitors into the circuit to reduce its bandwidth. This technique relies heavily upon circuit redesign to make the circuits less susceptible to EMI and, more importantly, to reduce the amount of EMI emissions of theses circuits. The advantage of the electronic approach is that it attacks the problem at the source, reducing the level of radiated emissions dramatically. Unfortunately, redesigning the electronic circuitry is expensive and does not protect the circuit from external interference. Furthermore, as circuitry becomes more and more complex, this approach becomes more difficult to achieve (GE Plastics, n.d.).
2. Shielding. Shielding is the only noninvasive suppression technique. Since the shielding is not inserted into the circuit, it does not affect the high-frequency operating speed of the system, nor does it affect the operation of the system should changes be made to the design in the future.

Moreover, shielding does not create timing problems and waveform distortion; it does not decrease system reliability; and it reduces crosstalk. In addition, shielding works for emission suppression as well as susceptibility (immunity) problems (Instrument Specialties, 1998). As mentioned earlier, shielding cannot only be used at PCB level, but it also can be used at many other space levels. The details will be addressed in the following chapters.

Even with the overall advantages of shielding, the most cost-effective approach is to use a combination of the electronic approach (circuit suppression/hardening) and shielding. Whenever possible, the system should be designed with a logic family that operates at low frequencies (minimum circuit bandwidth) and has low current, minimum radiating loop areas, and optimal board layout and construction, and add shielding as necessary to meet these seemingly incompatible objectives at various space levels such as IC, PCB, chassis or system enclosures, I/O, cable and connecter, and bonding and grounding.

As the importance of EMI shielding is continuously identified, designing for EMC and ensuring compliance through the entire production cycle comes to a new stage although it is still a complex matter. All aspects of the design and manufacture of equipment that practically have an impact on EMC have started to be considered, and various EMC regulations and proper EMC quality controls are being established and maintained during the whole production period. However, experience and problems with EMC should continue being carefully analyzed and the knowledge gained should be disseminated more widely within the organization to improve future design for EMC. More details about EMC design have been summarized in many references (Chatterton and Houlden, 1992; Christopoulos, 1995; Montrose, 1999).

1.3.2 Shielding Methods and Materials Selection

EMI shielding is playing a more important role in EMC design, with consideration given throughout product design, materials selection, processing method, manufacturing, and environment compliance. As a result, a variety of advanced materials and processing methods of EMI shielding have been developed for various level enclosures, gasketing, cable and connector shielding, bounding and grounding, and integral assembling.

1.3.2.1 Shielding Housing and Enclosures

The fundamental principle of shielding housing and enclosures is to place a Faraday cage around an electrical device for shielding against EMI. If PCBs and internal cabling are properly designed, the need for enclosure shielding can be minimized. However, if it is found that enclosure shielding is required, designing the enclosure to meet the EMC requirements will minimize the level of the shielding design and associated cost. A shielded enclosure should be fabricated from materials that possess the desired physical and electrical characteristics, and conform to environment compliance. Discontinuities should be minimized and designed to maintain shielding effectiveness by proper seam treatment, and use of EMI shielding gaskets, vent/window screening, and shielding/filtered leads.

Materials used to make the enclosures and shielding housing include metal sheet and foil, nonmetallic materials such as plastic materials with conductive coating, conductive plastics, and a combination of these two plastic materials through coinjection.

1.3.2.1.1 Metal Sheet

Metal offers excellent mechanical properties and inherent shielding characteristics, and can be cost effective in making simple, nonfunctional parts. Plastics offer increased design flexibility with integration of functional features and excellent aesthetics. However, plastics require secondary operations like coating or painting, or conductive fibers to achieve shielding characteristics. Hence, one combination approach is to line a plastic housing with a metal shroud. This box-in-a-box technique may not be cost effective, as the metal liner has to be attached to the plastic housing, complicating and lengthening the assembly process. However, it can be used for a functional application. The advantage of ease of assembly that is gained by using plastic is compromised with a metal shroud (GE Plastics, n.d.).

1.3.2.1.2 Nonmetallic Materials

Many commercial electronic devices are packaged in shielding housing and enclosures of plastic or other nonconductive materials. If the devices must rely on enclosure shielding for EMC compliance, these enclosures must be treated with a conductive material to provide shielding. Metallizing techniques for this application include vacuum deposition, electroless plating, arc spray, and conductive spray paint. The latter is the most frequently used technique, which is a paintlike slurry of metal particles in a carrier. These conformal coatings are loaded with very fine particles of a conductive material such as silver, nickel, copper, and carbon. Shielding effectiveness levels of 60 dB to 100 dB can be achieved (Tecknit 1998).

Conductive plastics are a viable alternative to the surface-coated plastics and metals for EMI shielding and ESD applications. They are generally made conductive by the addition of conductive fibers in the base resin. The most common fillers are carbon fibers, nickel-coated graphite (NCG) fibers, stainless steel fibers, and aluminum fibers and flakes. The amount of fiber in the resin is usually between 7% and 20% of the total volume depending upon the level of shielding required. The conductive fibers offer good EMI shielding performance (20–60 dB). In plastics, the most efficient use of conductive fibers in the finished molded part is achieved through maintaining fiber length during processing. The longer the fibers in the resin matrix, the better the fibers overlap or network electrically. This networking provides a continuous conductive path throughout the resin. As the shielding continuity must be maintained through the even distribution of conductive fibers in the plastic matrix during molding, the network of fibers must also reach the surface of the part to achieve good surface conductivity to avoid forming a slot antenna between mating surfaces. In general, using conductive plastics as a shielding method for large, material intensive parts may not be cost competitive. However, for smaller, more complex parts, conductive plastics may be a more economically favorable shielding method than a secondary shielding operation (GE Plastics, n.d.).

Electromagnetic Interference Shielding Fundamentals and Design Guide

These plastic enclosures are usually formed with injection or foam molding. A coinjection molding has been developed for forming plastic enclosures, using two different resins simultaneously. One of these can be a conductive plastic that acts as a shield, which can be injected as the core material, and an aesthetic and insulative plastic is formed as the skin material. As with other plastic enclosures, the grounding and joint design of this coinjection molded enclosure are major concerns.

1.3.2.1.3 Ventilating Panel and Windows

Generally, a window with large area openings or apertures is required for viewing the display, status lamp, device operating status, or ventilation and de-dust. When shielding of these large areas is required for EMC purposes, several options are available: (a) laminating conductive screen between optically clear plastic or glass sheets; (b) casting a wire mesh within a plastic sheet; (c) applying an optically clear conductive layer to a transparent substrate; and (d) adding a ventilating panel with waveguide apertures made with metal honeycomb or conductive foam panel, as well as integral window with multifunctions. The design guideline of vents and apertures are all described in Section 1.2.3. Some basic principles include: (a) many small holes allow less leakage than a large hole of the same total area; (b) short, wide openings are better than long, thin openings; and (c) thicker wall sections at the opening minimize leakage.

1.3.2.2 Gasketing

A gasket may be necessary to provide a continuous conductive path. There haven't been a wide variety of gaskets available for a broad range of applications. Based on their geometry and materials, shielding gaskets can be basically categorized as metal (CuBe, CuNiSn, S.S., etc.) spring fingers, knitted wire mesh, metal-filled conductive elastomers, and fiber on foam. Regardless of the gasket type, the important factors to be considered when choosing a gasket are EMI impedance, shielding effectiveness, material compatibility, compression forces, compressibility, compression range, compression set, and environmental sealing. For instance, when used for direct current (DC) or low frequency grounding/bonding, the conductivity across the seam is the most important parameter. At the higher frequencies, the gasket inductance becomes the most important parameter. The conductivity/inductance relationships are both material and shape dependant. For electric field and plane-wave shielding, EMI shielding gaskets made from CuBe provide the highest shielding effectiveness of any gaskets made from metal spring materials because of its high conductivity and low inductance. The high conductivity also permits CuBe to provide better magnetic shielding at frequencies over approximately 1 MHz. For magnetic fields below about 50 kHz, however, CuBe, like other nonmagnetic materials, suffers because of its unity permeability. As a result, magnetic field shielding is not normally provided by gasket materials regardless of their composition. Low-frequency magnetic field attenuation is usually provided by using highly permeable enclosure materials and not by the gasket. This is accomplished by designing the enclosure mating surfaces to be in contact with one another to create minimum air gaps in the path of the magnetic flux (Instrument Specialties, 1998).

When selecting gasketing materials, they should be galvanically compatible with the metallic coating to minimize corrosion. Generally, gaskets are subject to minimum and maximum pressure limits to achieve a proper EMI seal. The greater the pressure applied to the gasketed joint, the better the apparent environmental and EMI seal. However, if the pressure exceeds the maximum pressure limit of the gasket, permanent damage to the gasket can occur. This damage may decrease pressure across the seam and degrade the environmental and EMI shielding characteristics. Moreover, the gasket can mount into a groove of a tongue-in-groove joint to form near continuous electrical contact between the mating surfaces. If an enclosure is to be painted for aesthetics, the decorative paint must not extend inside and cover the gasket groove. However, the conductive coating (paint, plating, etc.) should cover the gasket groove to maintain the electrical continuity. If screws are used, the gasket should fit on the inside of the screw to protect against leakage around the screw hole. In addition, the selection and location of gasket materials for a given application must consider what damage may result from broken fingers or from tiny broken wire strands or conductive flakes, or particles that can be distributed throughout the electronics through natural or forced air-cooling systems.

Solid metal, flat, spring-type gaskets can be used in either a shear application or under direct compression. The capability of being used in shear permits the design of a seam where the solid metal compression forces will be aligned parallel with the mating surfaces of an enclosure. The knife-edge arrangement is an example of such a seam. Direct compression forces will be normal to the mating surfaces. There is a distinct advantage to parallel force alignment because it eliminates the requirement for multiple, evenly spaced fasteners to maintain the gasket compression. In either case, with a spring finger configuration, there will be shear forces developed because the shield and gasket surfaces will slide across each other. Since the surface of spring alloys like CuBe is very hard compared to the oxidation or corrosion films, the gasket and mating surfaces form a self-cleaning contact. Knitted wire mesh or conductive elastomers cannot be used in shear. They can only be used in direct compression. But when they are compressed, no shear forces are developed, hence no sliding action occurs, and they are usually not self-cleaning. In some cases, however, the wiping action of wires in the knitted wire gasket can break through the surface film to make excellent interface contact. Thus at points of high contact pressure, the wire mesh and elastomer gaskets generally create a gas tight seal, which prevents corrosion directly under the contact yet permits corrosion adjacent to the contact. So as long as the gasketed joint is not disturbed, the shielding effectiveness remains high. If the mating surfaces are moved, the gasket will generally not realign its corrosion-free contact points with those of the mating surfaces, and the shielding will be degraded (Instrument Specialties, 1998).

1.3.2.3 Integral Assembly

Integral assembly shielding must also be considered when designing for EMC. For example, electronic filters or shielded cables, connectors, and I/O shields are usually used to maintain the shielding integrity of the enclosure at cable penetrations. The waveguide principle can be applied to feedthroughs of an enclosure for nonconductive shafts or air vents. The assembly design of plastic enclosures or covers has moved toward a "snap-fit" approach to achieve better manufacturability and a higher level of subassembly integration. These fasteners can be molded into the plastic enclosure allowing for simplified assembly. A durable, uniform, metallic coating is important with snap-fit assembly, especially if parts will be separated often. The metallic coating must be tough enough to resist normal abrasion and wear in service to maintain the effectiveness of the EMI shield. In addition to snap fits, a unit can be held together mechanically by using molded-in bosses. Bosses act as position guides and can tie the device together electrically. Metal inserts can be fixed into the bosses and the device assembled with threaded fasteners. The inserts line up so that mating sections of the device are in electrical contact through boss-to-boss contact and form a continuous conductive path (GE Plastics, n.d.).

The shielding effectiveness and durability of the coating can be further optimized in the design of the plastic enclosure via (a) uniform wall sections; (b) changes to thickness progressively rather than abruptly, if needed; and (c) designing for uniform cooling after molding to minimize warpage of the plastic enclosure, allowing for a more uniform, consistent, and durable metallic coating on it.

Among the integral assembly shielding, grounding is probably the most important aspect of EMI control. If the setup of grounding connections is improper at the frequencies of interest, the performance of the enclosure shielding, cable shielding, or filtering may be degraded. A good ground scheme is to bond, or maintain a low impedance connection between mating conductive parts. This requires that mating parts of enclosures not be painted, the ground straps not be attached to painted surfaces, and, perhaps, in corrosive environments, special attention be given to the use of dissimilar metals to preclude the effects of galvanic action. The goal is to maintain, as close as practical, a single potential safety ground system (Tecknit, 1998).

1.3.2.4 Corrosion and Material Galvanic Compatibility

Corrosion is a major challenge and problem for shielding design in electronic systems because of (a) oxides, resulting from chemical corrosion and oxidation; and (b) electrochemical reactants, which can be either insulators that degrade the shielding effectiveness, or semiconductors, which create unwanted EMI signals through nonlinear mixing. For instance, electrical conductivity of pure metals may not always reflect shielding effectiveness due to surface oxidation. Pure copper is highly conductive, but its oxide exhibits relatively poor conductivity. Therefore, any shielding system that uses copper must protect its surface against oxidation. The use of a surface coating that acts as an oxidation barrier is the most common method of protection.

Electrochemical reaction is influenced by exposed surface area, materials dissimilarity, and the presence of an electrolyte, usually moisture. Stamped or formed solid metal gaskets have minimum surface area. The surface is nonporous, and thus the gaskets do not permit the infusion of moisture. These characteristics act to reduce but not eliminate corrosion. Wire mesh on the other hand, has increased surface area, thus the corrosion resistance is not quite as good as for the solid metal gaskets. Either gasket type can be easily plated with other metals to minimize galvanic corrosion. Elastomer gaskets tend to be self-sealing, which reduces moisture infiltration beyond the edges. This characteristic frequently limits the corrosion to the outer edges of the gasket contact region. Unfortunately, the porosity is a big variable among the different elastomers that are used, and many elastomers permit the infusion of moisture especially after long-term exposure, therefore, resulting in electrochemical corrosion with the form of worm holes (Instrument Specialties, 1998).

The generally accepted criterion for galvanic compatibility is MIL-STD-1250. This standard does not permit galvanic couples that exceed 0.25 volts (electrochemical potential difference). Most commercial applications in a controlled environment will allow up to 0.5–0.6 volts. For such applications where large contact voltages occur, the more reactive material will be destroyed. To prevent this problem, either the gasket material or the mating surfaces, or both, will need to be plated with a material that is compatible.

Consequently, the use of shielding to meet EMC requirements is increasing as part of electronic equipment designs, in conjunction with or after, circuit suppression, isolation, and desensitization measures have been implemented at PCB and other space levels. Shielding effectiveness of a system can be improved by following basic design guidelines, such as (Instrument Specialties, 1998):

- Always fabricate enclosures from the fewest number of pieces and utilize existing metal structures such as a disc drive base plate, as the part of shield. Do not have free floating pieces of metal.
- Use EMI gaskets as seals for removable discontinuities.
- Design removable panels, covers, doors, and so forth to provide controlled gasket compression. Choose a gasket with minimum compression force, maximum working height range, and provide a positive stop.
- Whenever possible, use a knife-edge or pan-edge panel arrangement for doors and access panels. This arrangement eliminates the need for periodic threaded fasteners to preserve the shielding effectiveness of the enclosure
- Shield gaskets should be installed to provide the highest surface conductivity, metal-to-metal condition. If the mating materials must be environmentally protected, the protective coating should be electrically conductive. If the gasket includes a separate environmental seal, that seal should be facing the environment.
- In all cases, the gasket material should be selected or plated to achieve galvanic compatibility.

There are many other considerations, including selection of an appropriate gasket type, enclosure material, surface finish, and closure mechanism.

1.3.3 Environmental Compliance

Environmental compliance must be evaluated during EMI shielding design because it has become a key advantage of competitive shielding products. The requirements are regulated by the Restriction of the Use of Certain Hazardous Substances in Electrical and Electronic Equipment (RoHS) and Waste Electrical and Electronic Equipment (WEEE) Directives promulgated by the European Union (EU). The directives make selections of shielding materials critical. These environmental compliance regulations, together with existing requirements for EMC, provide forceful guidelines for designing products with EMC and environmentally compliant.

TABLE 1.1
Restricted Substances

Material/Substance Category	Maximum Concentration in Weight
Asbestos	Not detectable
Certain Azo colorants and Azo dyes	Not detectable
Cadmium/cadmium compounds	0.01%
Hexavalent chromium/hexavalent chromium compounds	0.1%
Lead/lead compounds	0.1%, 0.03% (PVC cables only)
Mercury/mercury compounds	0.1%
Ozone depleting substances (CFCs, HCFCs, HBFCs, carbon tetrachloride, etc.)	Not detectable
Polybrominated biphenyls (PBBs)	0.1%
Polybrominated diphenylethers (PBDEs)	0.1%
Polychlorinated biphenyls (PCBs)	Not detectable
Polychlorinated naphthalenes (more than 3 chlorine atoms)	Not detectable
Radioactive substances	Not detectable
Certain short-chain chlorinated paraffins	Not detectable
Tributyl tin (TBT) and triphenyl tin (TPT)	Not detectable
Tributyl tin oxide (TBTO)	Not detectable

1.3.3.1 RoHS Directive

RoHS Directive 2002/95/EC, together with WEEE Directive 2002/96/EC, became European law in February 2003, setting collection, recycling, and recovery targets for all types of electrical goods. Producers have had to comply with all provisions of the WEEE Directive since August 13, 2005, while the RoHs Directive required that producers prohibit any of the banned substances on the market after July 1, 2006. Producers failing to comply with the WEEE and RoHS Directives' requirements face legal penalties and potential restriction from selling products in the EU.

Table 1.1 shows restricted materials currently in the RoHS Directive. These substances are restricted to the threshold level in all products. All homogeneous material in purchased articles (i.e., materials, components, subassemblies, or products) must be free from the substances or cannot contain higher concentrations than the defined

TABLE 1.2
Banned Substances in Product Packaging

Material/Substance Category	Maximum concentration in weight
Cadmium, mercury, lead, chromium VI	Total amount 0.01%
Chlorofluorocarbons (CFCs)	Not detectable
Hydrochlorofluorocarbons (HCFCs)	Not detectable

TABLE 1.3
Reported Substances

Material/Substance Category	Maximum Concentration in Weight
Antimony/antimony compounds	0.1%
Arsenic/arsenic compounds	0.1%
Beryllium/beryllium compounds	0.1%
Bismuth/bismuth compounds	0.1%
Brominated flame retardants (other than PBBs or PBDEs)	0.1%
Nickel (external applications only)	0.1%
Certain phthalates	0.1%
Selenium/selenium compounds	0.1%
Polyvinyl chloride (PVC; disclosure is limited to "is present"/"is not present" in amounts that exceed threshold)	0.1%

percent threshold levels as listed in the table. Table 1.2 gives the banned substances in product packaging. The substances must be reported when concentration exceeds the indicated threshold level shown in Table 1.3. Exemptions to the maximum allowed concentrations of restricted materials are identified for cases where technology does not yet allow for substitutions or where alternatives may have a worse impact on human health and the environment. Some exemptions include mercury in several kinds of fluorescent lamps; lead in steel, copper, and aluminum alloys; lead in some types of solder; and military applications.

RoHS Article 3(a) states that RoHS covers electrical and electronic equipment "which is dependent on electric currents or electromagnetic fields in order to work properly and equipment for the generation, transfer and measurement of such currents and fields falling under the categories set out in Annex IA to Directive 2002/96/EC (WEEE) and designed for use with a voltage rating not exceeding 1000 volts for alternating current and 1500 volts for direct current." With that said, a microwave oven would not be covered by RoHS because it cannot perform its intended function with the power switched off. On the other hand, a talking doll can still be used as a doll even when the batteries are removed, so it isn't covered by RoHS. RoHS does allow for noncompliant components after the July 1, 2006, deadline, but only as spare parts for equipment on the market before July 1, 2006.

1.3.3.2 WEEE Directive

The WEEE Directive imposes the responsibility for the disposal of waste electrical and electronic equipment on the manufacturers of such equipment. Those companies should establish an infrastructure for collecting WEEE, in such a way that "users of electrical and electronic equipment from private households should have the possibility of returning WEEE at least free of charge." Also, the companies are compelled to use the collected waste in an ecological-friendly manner, either by ecological disposal or by reuse/refurbishment of the collected WEEE. The WEEE Directive identifies producers as any company that sells electronic and electrical equipment directly or indirectly under its own brand name. The intent of the WEEE Directive is to require producers to design products and manufacturing processes that prevent the creation of WEEE and, barring that, reuse, recycle, dispose of, or incinerate WEEE. The WEEE Directive calls for set percentages of IT and telecommunications equipment (category 3 includes PCs, wireless devices, and similar devices) to be recovered and reused, or recycled (minimum 65%), incinerated (maximum 10%), or safely disposed of (maximum 25%).

Therefore, the goal of the EMI shield design and production should be compliant to the RoHS and WEEE Directives in a minimized cost manner. Meanwhile, designing for EMC and environmental compliance must assure that time to market is minimized to preserve the product's competitive advantages.

REFERENCES

Bjorklof, D. 1999. Shielding for EMC. *Compliance Engineering*. http://www.ce mag.com/99APG/Bjorklof137.html

Bridges, J. E. 1988. An update on the circuit approach to calculate shielding effectiveness. *IEEE Transactions on Electromagnetic Compatibility* 30: 289–293.

Carter, D. 1999. *Metal Barriers with Discontinuities*. England: Tecknit Europe Ltd.

Chatterton, P. A., and M. A. Houlden. 1992. *EMC Electromagnetic Theory to Practical Design*. Chichester: John Wiley & Sons.

Chomerics. 2000. *EMI shielding theory*. http://www.chomerics.com/product/documents/emicat/pg192theory_of_emi.pdf

Christopoulos, C. 1995. *Principles and Techniques of Electromagnetic Compatibility*. Boca Raton, FL: CRC Press.

CISPR11. 2004. *Industrial, scientific and medical (ISM) radio-frequency equipment electromagnetic disturbance characteristics: Limits and methods of measurement*. International Special Committee on Radio Interference.

Duffin, W. J. 1968. *Advanced Electricity and Magnetism*. London: McGraw-Hill.

Evans, R. W. 1997. *Design guidelines for shielding effectiveness, current carrying capability, and the enhancement of conductivity of composite materials*, NASA contractor report 4784.

GE Plastics. n.d. *EMI/RFI Shielding Guide*. Pittsfield, MA: GE Plastics.

Holmes, D. G., and T. A. Lipo. 2003. *Pulse Width Modulation for Power Converters*. New York: John Wiley & Sons.

Instrument Specialties. 1998. *Engineering Design and Shielding Product Selection Guide* (acquired by Laird Technologies). Delaware Water Gap, PA.

Mardiguian, M. 2001. *Controlling Radiated Emissions by Design*, 2nd ed. Boston: Kluwer Academic.

Molyneux-Child, J. W. 1997. *EMC Shielding Materials*. Oxford: Newnes.

Montrose, M. I. 1999. *EMC and the Printed Circuit Board*. New York: Institute of Electrical and Electronics Engineers.

Nave, M. J. 1991. *Power Line Filter Design for Switched-mode Power Supplies*. New York: Van Nostrand Reinhold.

Ott, H. 1988. *Noise Reduction Techniques in Electronic Systems*, 2nd ed. New York: John Wiley & Sons.

Pozar, D. M. 1988. *Microwave Engineering*, 2nd ed. New York: John Wiley & Sons.

Tecknit. 1998. *Electromagnetic Compatibility Design Guide*. http://www.tecknit.com/REFA_I/EMIShieldingDesign.pdf

Vasaka, G. J. 1956. *Theory, design and engineering evaluation of radio-Frequency shielded rooms* (Report NADC-EL-54129). Johnsville, PA: U.S. Naval Development Center.

White, D. R. J., and M. Mardiguian. 1988. A Handbook Series on Electromagnetic Interference and Compatibility *Electromagnetic Shielding* (Vol. 3). Gainesville, VA: Interference Control Technologies, Inc.

2 Characterization Methodology of EMI Shielding Materials

The purpose of this chapter is to introduce and review characterization methodology of electromagnetic interference (EMI) shielding and used materials. The main topics include shielding effectiveness measurement, electrical and thermal properties, electrical permittivity and magnetic permeability, mechanical property, surface finish and interface compatibility, formability and manufacturability, and environmental performance evaluation.

2.1 SHIELDING EFFECTIVENESS MEASUREMENT

The techniques to measure the shielding effectiveness of an enclosure or a conductive gasket or other EMI shielding components include a variety of standard methods and variations on standard methods. The standardized methods, however, cannot always be used because they need well-defined samples that are sometimes impossible to make due to the properties of the object samples to be measured. In many cases, therefore, it is necessary to design tailored measurement equipment based on standard methods, which can be used for measurements on specific gaskets or other shielding materials with a particular shape and size.

Most commercial and industrial shielding measurement documents used are modifications of American military standard MIL-STD-285 which was already withdrawn. MIL-G-83528 is one of the examples. It is allowed to interchange the location of the test antennas and equipment and repeat the test for the alternative configuration. This will provide with attenuation data for both emission and susceptibility, but requires that two sets of measurements be made.

A shielding effectiveness value can generally be obtained using radiated measurement methods, transfer impedance testing methods, and other methods. Each measurement method is producing a performance value that is usually hard to correlate with the value obtained using another measurement method. For instance, a measurement method where a shielding material sample is mounted inside a coaxial transmission line fixture will not necessarily expose the sample for the same electromagnetic field impedance as a method where an incident plane wave is radiated from an antenna (Wilson and Ma 1985). In general, performance of EMI shielding gaskets below 1 GHz is determined by measuring the gasket's transfer impedance based on SAE-ARP-1705A. This method allows minor variation in performance to be easily evaluated and is much more repeatable than the direct radiated method called out in MIL-G-83528. Because of correlation problems, transfer impedance data does not lend itself to full-scale enclosure evaluations. Some typical test methods usually used for shielding effectiveness are described next.

2.1.1 Testing Methods Based on MIL-STD-285

Established in 1956, the American military standard MIL-STD-285 is the foundation of shielding effectiveness measurements. The standard was withdrawn; however, improved measurement methods have made new standards evolve. The method uses two shielded rooms with one common wall as shown in Figure 2.1 (Lundgren 2004). The wall has an aperture where test samples are mounted. In one room is the transmitting antenna and in the other room is the receiving antenna. The two antennas are directed toward each other and are at a fixed distance from each other. The transmitter transmits at constant power and the receiver measures the transferred power with and without the test sample mounted in the aperture. The shielding effectiveness of the test sample can be obtained from the difference between these two measurements. For instance, a modified method according to MIL-STD-285 has been used to examine shielding effectiveness of conductive composites and performance of conductive gaskets made of the composite (Bodnar et al. 1979). In this measurement, gasket material is used to seal an opening in a shielded enclosure. A transmitting antenna is located outside of the enclosure, and a receiving antenna is located inside. The shielding effectiveness is basically the difference in measured signal strength inside the enclosure before and after the gasket is installed. Drawbacks with the method are that measured attenuation is dependent on the antenna placement and the reflections of the electromagnetic wave inside the shielded enclosure. Poor repeatability resulting from enclosure resonance, reflections, and antenna loading, as well as a lack of standardization in mechanical dimensions, severely limit the use of this kind of radiated measurement in determining gasket performance.

Improved versions of the method in MIL-STD-285 have been developed to minimize the problems with reflections by the use of absorbing material in the shielded rooms or chambers (Bodnar et al. 1979). New versions, like MIL-G-83528, can also be found in the standard IEEE-STD-299 (1997), with a frequency range of a few megahertz to 18 GHz.

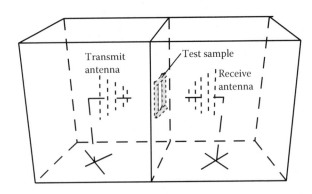

FIGURE 2.1 Shielding effectiveness measurement according to the standard MIL-STD-285.

Characterization Methodology of EMI Shielding Materials

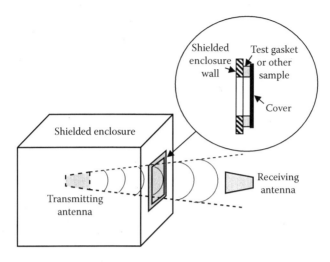

FIGURE 2.2 Radiated shielding effectiveness testing method (MIL-G-83528).

2.1.2 Modified Radiation Method for Shielding Effectiveness Testing Based on MIL-G-83528

The radiated energy approach of the shielding effectiveness test described in MIL-G-83528 is a modified MIL-STD-285 test method. MIL-STD-285 stipulates that the wave generated by the transmitting antenna be directed at discontinuities or joints between the test sample and the common wall in the enclosure, and that the receiving antenna be positioned to receive the maximum field strength emanating from the joint. MIL-G-83528 regulates that the electromagnetic wave be directed at the center of a large 28 in × 28 in (711.2 mm × 711.2 mm) cover plate and that the receiving antenna be located directly behind the plate, as shown in Figure 2.2. Prior to installing the cover, the open reference is set up by placing the transmitting and receiving antenna one meter from the wall of the shielded room and the test fixture. The appropriate signal generator and radio frequency (RF) amplifier are attached to the transmitting antenna. A spectrum analyzer is attached to the receiving antenna. After measuring the open reference, the closed reference is set up with the test equipment in the exact place of the open reference and the premier test sample is placed on the test fixture. The test sample is held in place with the plastic spacer and compression plate. The compression plate is used to hold the test sample in place and apply an even load.

The shielding effectiveness is obtained by taking the difference between the power level from the open reference and the power level recorded in the closed reference.

2.1.3 Dual Mode Stirred Chamber

The dual mode stirred chamber method was developed by National Institute of Standards and Technology (NIST) of the United States and other organizations in

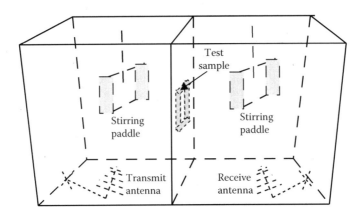

FIGURE 2.3 Shielding effectiveness measurement using a dual mode stirred chamber facility.

the world using two mode stirred chambers next to each other with a common wall having an aperture for mounting the test samples (Spiegelarr and VanderHeyden, 1995). A mode stirred chamber is a shielded room or enclosure with a mode stirring arrangement, to create electromagnetic fields with a large number of modes within the chamber, as shown in Figure 2.3 (Lundgren, 2004). The mode stirring arrangement can be a mechanical paddle wheel or some other highly conductive structure that can be rotated stepwise or continuously. The mode stirring can also be implemented by modulating the signal source to create frequency mode stirring. The multimode electromagnetic field in the chamber can be of high amplitude using a small power amplifier (Quine, 1993). The facility can be used for mode tuning when the stirring mechanism is stepped between measurements so that several measurements are obtained in different electromagnetic environments. A mode stirring action can also be used when the stirring mechanism is causing a continuously changing electromagnetic environment in the chamber (Lundgren 2004). This makes the measurements considerably easier. The placement of antennas, for instance, is not critical here. The main drawback of this method is that the equipment needed to control the mode stirring can be expensive. The low frequency limit for the method is about 500 MHz and it is dependent on the smallest size of the mode stirred chamber (the smallest distance inside a chamber should be seven times the wavelength of the lowest frequency [Hatfield et al. 1998], i.e., 500 MHz gives a distance of more than 4 meters). Other ways to derive the lowest useful frequency gives different lower frequency limits and measurement of the field uniformity in the completed chamber as the best way to get an exact realistic value (Lundgren, 2004).

2.1.4 Transverse Electromagnetic (TEM) Cell

An apertured transverse electromagnetic (TEM) cell in a reverberating chamber is a simplified method that makes use of one mode stirred chamber. Inside the chamber is a transmitting antenna and an apertured TEM cell, which is the receiver (Rahman et al., 1992). The location of the transmitting antenna and the TEM cell

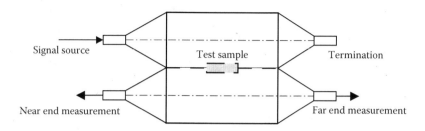

FIGURE 2.4 Shielding effectiveness measurement using a dual transverse electromagnetic (TEM) cell.

is not critical, however, the antenna should not be aiming toward the TEM cell and the TEM cell should not be close to the walls or other reflecting objects (Lundgren, 2004). A TEM cell is an expanded section of a rectangular co-axial transmission line. The sample is mounted over an aperture in the TEM cell. An electromagnetic field is then generated with the antenna and a receiving instrument connected to the TEM cell is used to measure the leakage through the sample. The usable frequency range is 200 MHz to 1 GHz and the dynamic range is about 100 dB (Rahman et al., 1992). A related and even cheaper method is the dual TEM cell method. The two TEM cells are connected together in a piggy-back manner, as shown in Figure 2.4 (Lundgren, 2004). The TEM cells are coupled through an aperture in the common wall. An important feature with this measurement method is that near-field shielding effectiveness measurements can be obtained, including the E-field and H-field shielding effectiveness. One TEM cell is connected to a signal source and terminated in the other end. The second TEM cell has two outputs where electric-field coupling and magnetic-field coupling can be measured. The aperture is covered with the sample to be measured. A drawback with this method is that the polarization of the electric field is normal to the sample (Rahman et al., 1992). A nice feature with this method is that electric- and magnetic-field shielding can be simultaneously investigated by measuring the received power in both ends of the secondary TEM cell. The frequency range for this method is 1 MHz to 200 MHz and the dynamic range is about 60 dB (Lundgren, 2004).

Another kind of TEM cell is the split TEM cell, also called a rectangular split transmission line holder. The split TEM cell is made as an ordinary TEM cell in two halves. The shielding effectiveness of the test sample is calculated from a measurement of attenuation through the empty cell with the halves joined and a measurement of attenuation through the shorted cell with the halves joined with the sample in between. Both the center conductor and the outer conductor must make good contact with the sample on both sides so that the cell is shorted by the sample.

The receiving half of the TEM cell can be modified to measure the magnetic-field shielding efficiency. Then a loop antenna is combined with a box equipped with a 90-degree angle reflector on one wall. The loop antenna is mounted through the reflector such that three quarters of the loop is inside the box and one quarter outside. When performing a measurement, the wall with the reflector and quarter loop antenna is joined with the half TEM cell and the sample in between. The frequency range for this method is 1 MHz to 1 GHz (1 MHz to 400 MHz for H-field) and the dynamic range is about 70 to 80 dB (Lundgren, 2004; Rahman et al., 1992).

2.1.5 Circular Coaxial Holder

The test fixture of circular coaxial holder has two different versions (Rahman et al. 1992). The continuous conductor version is an expanded 50 Ω coaxial connector. The test sample has to have an annular washer shape to fit between the inner and outer conductor and thereby short the transmission line. The continuous conductor test fixture has an operating frequency range of direct current (DC) to 1 GHz and a dynamic range of 90 to 100 dB (Rahman et al., 1992). The other version, called a split conductor, is described in the standard ASTM D4935 and has a similar design but is split in two halves. This simplifies the mounting of the test samples. Test samples are needed for the reference measurement and for the actual measurement of the material. For the reference measurement, a small disc-shaped piece is fitted between the center conductors and a large washer-shaped piece is fitted between the two halves of the fixture. This method keeps distance and material between the two parts of the fixture constant except for the region between the center and the other conductor. The outer conductor is equipped with flanges to offer good capacitive coupling between the two halves of the fixture (Lundgren, 2004).

The shielding effectiveness is calculated as the difference between the measurement of the large disc sample and the reference measurement. The split conductor fixture has a frequency range of 1 MHz to 1.8 GHz and a dynamic range of 90 to 100 dB (Lundgren, 2004; Rahman et al., 1992).

2.1.6 Dual Chamber Test Fixture Based on ASTM ES7-83

A dual chamber test fixture, based on ASTM ES7-83, has an enclosure or chamber split into two sections. Each section has an antenna fixed on the inside. A sheet of the sample material is sandwiched between the two sections and the transmission through the material is measured. As a reference, the transmission between the antennas is measured without the sample present. The shielding effectiveness is calculated as the difference between these measurements. This test method has been used for frequencies from 100 kHz to 1 GHz and gives a dynamic range of 80 dB (Lundgren, 2004).

2.1.7 Transfer Impedance Methods

The transfer impedance is usually used to describe the high-frequency characteristics of an EMI shield in lumped or distributed circuit elements. Parameters influencing the shielding performance of an EMI shield are skin depth, geometrical shape, conductivity, and permeability. In a fixture for transfer impedance measurement, the purpose of the fixture design is usually to establish a uniform current distribution over the test sample. Several measurement methods for shielding effectiveness have been developed and standardized through different fixture designs. Transfer impedance methods are typically used to characterize the shielding performance of shielded cables, shielded connectors, and conductive gaskets; however, they can also be used for shielded enclosures or other EMI barriers in general.

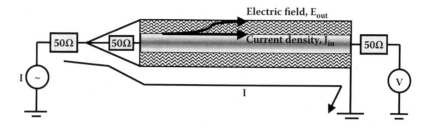

FIGURE 2.5 Schematic measurement setup example for the triaxial transfer impedance fixture used in standard IEC 96-1A.

2.1.7.1 Transfer Impedance of Coaxial Cables

The transfer impedance (Z_t) represents the diffusion through the thickness of the shield and the magnetic-field coupling through imperfections in the shield. It can be obtained from measurements as the ratio of a potential difference across an electrical field (E_{out}) on the outside of a shield due to a current density (J_{in}) on the inside. That is, $Z_t = E_{out}/J_{in}$. For characterizing the shielding effectiveness of a coaxial cable shield, its transfer impedance is commonly measured. The measurement procedure and the design of its test fixture are described in the standard IEC 96-1A (1976). Figure 2.5 illustrates the triaxial fixture setup used in this standard for a 50 Ω characterization. This fixture is usually used with a network analyzer, and transmission through the fixture (S_{21}) is recorded. The transfer impedance of the cable shield can then be calculated by a formula given in the standard (Lundgren, 2004):

$$Z_t = abs\,(S_{21} \cdot 2 \cdot 1.4 \cdot 60 \cdot \ln\,(60/d)) \qquad (2.1)$$

where d is the outer diameter of the cable shield being tested; S_{21} is the transmission coefficient; S is the scattering parameter; and a and b are incident and transmitted wave strengths, respectively.

The useful frequency range for the IEC 96-1A fixture is direct current (DC) to 30 MHz. More advanced fixtures using quadraxial and quintaxial setups are suitable for a frequency range up to 1 GHz (Hoeft and Hofstra, 1992; Lundgren, 2004).

For shielded connectors, the transfer impedance characterization is also commonly used. Different fixtures are used as described in the literature (Dunwoody and VanderHeyden, 1990).

2.1.7.2 Transfer Impedance Method for Shielding Effectiveness Measurement of Conductive Gaskets

In the standard SAE ARP 1705 (1981) and the revised version SAE ARP 1705A (1997), transfer impedance measurements are described to determine the shielding effectiveness of EMI gaskets, as shown in Figure 2.6. In the transfer impedance test, a signal is applied through a 50 Ω load to a gasket specimen clamped between two plates. The testing measures the impedance of the gasketed joint and normalizes the impedance in terms of a meter length of gasket. Transfer impedance relates a

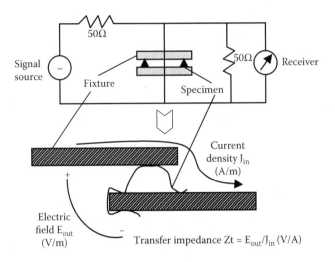

FIGURE 2.6 Transfer impedance measurement on a conductive gasket.

current on one surface of the gasket to the voltage drop generated by this current on the opposite surface of the gasket. This value can be calculated by $Z_t = E_{out}/J_{in}$. Here J_{in} is a longitudinal disturbing current generated on one surface of the gasketed joint and E_{out} is the longitudinal voltage per unit length, generated by J_{in}, appearing on the opposite surface of the gasketed joint. This method has excellent repeatability and is sensitive enough to detect small differences in performance between different materials, finishes, or loads.

Figure 2.7 illustrates a typical test fixture for transfer impedance measurement (Stickney, 1984). It is a heavy wall welded steel structure to prevent mechanical distortion under heavy loads. Housing components are tin plated to assure they are corrosion free with low-impedance electrical contact. An acting pneumatic cylinder system is used to apply loads to the gasket material. Force is applied through the test plates (bottom plate and top plate) at three distinct points, and three load cells are used to measure the force. If a gasket material is to be tested at a fixed height rather a fixed load, nonconductive coaxial spacers are installed in the gap between the test plates along with the test specimen. The bottom plate is a disk with a threaded hole in the center. The top plate is rectangular with a threaded hole in the center and six bolt holes around the perimeter. The plate material can be any metal or conductive composite capable of withstanding the desired test load. Electromagnetic continuity is maintained with the movable bottom plate by means of two silver-plated CuBe finger rings sliding within the center bore of a silver-plated center conductor. This conductor is screwed into the center hole of the bottom test plate and is located by a phenolic insulation plate that is supported by the air cylinders. The fixture is closed by bolting down the top plate. All instrumentation systems for transfer impedance measurement include a signal source and some type of receiver. Both instruments are capable of the measurement of a frequency range of approximately 10 kHz to 1 GHz. The measurement accuracy is limited by fixture resonance, which usually

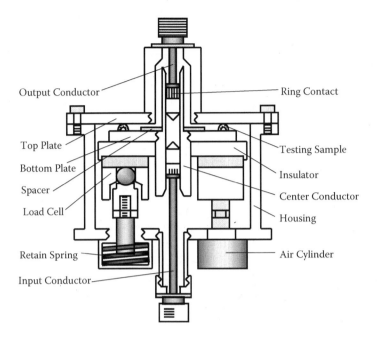

FIGURE 2.7 A typical test fixture for transfer impedance measurement.

occurs at approximately 700 MHz. Interconnections are made with 50 Ω coaxial cable using type *N* connectors, and they should be as short as possible to minimize signal losses.

2.2 ELECTRICAL AND THERMAL CONDUCTIVITIES

2.2.1 Electrical Conductivity and Contact Resistance

Electrical conductivity and interface contact resistance are important parameters in ultimately determining the shielding effectiveness that the EMI gaskets or other shields would provide. They are especially becoming more critical factors at higher frequencies where skin effect or surface resistivity is predominant. Surface resistivity and volume resistivity are commonly tested for characterizing electrical conductivity of EMI shielding materials as a quick and practical quality check.

The test method for measuring surface resistivity is defined by ASTM D257. Surface resistivity is measured in units of ohm/sq, which indicates the resistance of a material to moving or distributing a charge over its surface. The lower the surface resistivity value, the more easily a material will redistribute an electrical charge over its entire surface area. It is also important to have a low enough interface contact resistance to allow a continuous conductive path between two mating parts. This contact is especially critical across seams and joints.

Volume resistivity measures the ability of a material to resist moving a charge through its volume or bulk. The lower the volume resistivity of a material, the more conductive the material. The measurement of volume resistivity is defined by test method ASTM D257. The volume electrical resistivity, ρ, is expressed as the electrical resistance of a material per unit length and multiply by cross-section area:

$$\rho = (RA)/L \qquad (2.2)$$

where R is the resistance (ohm); A is the cross-section area (in^2 or cm^2); and L is the length (in or cm).

Electrical conductivity is the reciprocal of the material's volume or bulk resistivity. Electrical conductivity of copper alloys is usually in units of % IACS, which is the acronym for percentage of International Annealed Copper Standard. The conductivity of the annealed copper 5.81×10^7 S/m (P = 1.72×10^{-7} ohm-cm) is defined to be 100% IACS at 20°C. All other material's conductivity values are related back to this conductivity of annealed copper. For example, iron, with a conductivity value of 1.044×10^7 S/m, has a conductivity of approximately 18% of that of annealed copper and this is reported as 18% IACS. Because processing techniques have improved since the adoption of this standard in 1913, and more impurities can now be removed from the metal, commercially pure copper products now often have IACS conductivity values greater than 100%. Electrical conductivity itself is a very useful property since its values are usually affected by the material's chemical composition and the stress state of crystalline structures. Therefore, electrical conductivity information can be used for measuring the purity of a substance, sorting materials, checking for proper heat treatment of metals, and inspecting for heat damage in some materials.

The electrical contact resistance of the separable or nonseparable interfaces plays an important role in providing a highly conductive path between two mating parts of an EMI shielding system. Contact resistance consists of constriction resistance and film resistance. It is usually dominated by the constriction resistance of tiny apparent areas (A-spots), as shown in Figure 2.8 (Brush Wellman, 1995). The film resistance is due to thin layers (as low as 20 angstroms) of insulating material between the contacts, caused by (a) intentional surface coating and (b) unintentional oxidation of the contact material or other contamination. These thin films conduct electrons by means of a tunneling effect. Both constriction and film resistance depend on the normal force, and physical properties of the contacting surfaces and hardness of the contacting materials. The relationship between contact force and contact resistance is also shown in Figure 2.8. The magnitude of the separable contact interface resistance is a few milliohms.

2.2.2 THERMAL CONDUCTIVITY

Thermal conductivity is another important parameter especially for elevated temperature application. It is the ease with which a material dissipates heat. In a homogeneous material under steady-state conditions, thermal conductivity is the rate of heat flow per unit time per unit area, per unit temperature gradient in a direction perpendicular to that area. The temperature rise is a function of the bulk resistance of the material, which depends on the thermal dissipation of heat generated, which in turn depends on the thermal conductivity, the magnitude of the electrical current flow, and the heat sinking or convection of the shielding material.

Characterization Methodology of EMI Shielding Materials

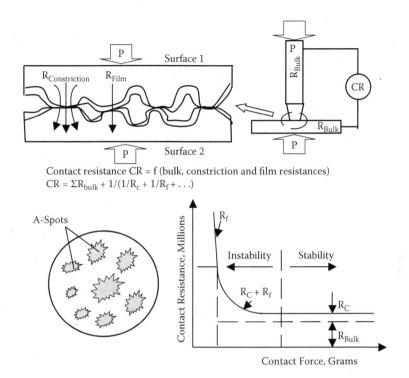

FIGURE 2.8 Contact resistance and the relationship between contact force and contact resistance.

Both magnitude and stability of electrical/thermal contact resistance are critical for EMI shielding requirements. Conventional copper strip material as a contact shielding part is effective up to 1 GHz operating frequency when proper considerations are given to electrical and mechanical designs.

2.3 PERMEABILITY AND PERMITTIVITY

2.3.1 PERMEABILITY

A material's permeability, μ, also called magnetic permeability, is a constant of proportionality that exists between magnetic induction (B) and magnetic field intensity (H). That is, $B = \mu H$. This constant is equal to approximately 1.257×10^{-6} H/m in free space (a vacuum), which is symbolized μ_o. The relative permeability of materials is defined as $\mu_r = \mu/\mu_o$. Materials with a $\mu_r < 1$, are called diamagnetic. When $1 < \mu_r < 10$, the materials are called paramagnetic; when $\mu_r > 10$, the materials are called ferromagnetic. The permeability of some materials changes with variation of temperature, intensity, and frequency of the applied magnetic field. Certain ferromagnetics, especially powdered or laminated iron, steel, or nickel alloys, have μ_r that can range up to about 1,000,000. When a paramagnetic or ferromagnetic core is inserted into a coil, the inductance is multiplied by μ_r compared with the inductance of the same coil with an air core. This effect is useful in the design of transformers and chokes for alternating currents (AC), audio frequencies (AF), and radio frequencies (RF).

2.3.2 Permittivity

Permittivity, ε, also called electric permittivity, is a constant of proportionality that exists between electric displacement (D) and electric field intensity (E): $D = \varepsilon E$. The vacuum permittivity is $\varepsilon_0 = 1/(c^2\mu_0) \approx 8.85 \times 10^{-12}$ farad per meter (F/m), where c is the speed of light and μ_0 is the permeability of vacuum. The linear permittivity of a homogeneous material is usually given relative to that of vacuum, as a relative permittivity ε_r: $\varepsilon_r = \varepsilon/\varepsilon_0$. When ε_r is greater than 1, these substances are generally called dielectric materials, or dielectrics, such as glass, paper, mica, various ceramics, polyethylene, and certain metal oxides. A high permittivity tends to reduce any electric field present. For instance, the capacitance of a capacitor can be raised by increasing the permittivity of the dielectric material. Dielectrics are generally used in capacitors and transmission lines in filtering EMI by AC, AF, and RF.

2.3.3 Characterization of Permeability and Permittivity

Common methods used for the measurement of permeability and permittivity include the loaded resonant waveguide cavity, the open-ended coaxial line, and the loaded coaxial transmission line. Loaded resonant waveguide cavity methods offer good measurement accuracy, particularly for measurement of imaginary parts of the permittivity for determining the energy losses in a material (Baker-Jarvis et al., 2001). The sample material is mounted inside a resonant cavity to fit well with certain dimensions and thereby loading the cavity. This method only covers a narrow frequency band to maintain the resonance status of a cavity. However, this can be improved to some extent by including higher-order resonant modes.

The open-ended coaxial line is a nondestructive test method. The coaxial probe is pressed against a test sample to measure the reflection coefficient. And then the permittivity can be determined from the reflection coefficient. Because this method is sensitive to air gaps that disturb the electric field, the probe, and the sample, its measurement accuracy is not as good as the other two methods.

In the loaded coaxial transmission line method the material being tested is placed to fill the volume between the inner and outer conductor in a section along the transmission line. The dimensions of the sample piece are critical to ensure a precise fit. The material may load the line and cause a change of characteristic impedance. Both reflection and transmission through the fixture is used when calculating the test material data (Baker-Jarvis et al., 2001). The measurement accuracy of this method is not as good as that of loaded cavity resonance measurement method. An advantage of the loaded coaxial transmission line method is that it offers the possibility of performing measurements in a broad frequency band (Lundgren, 2004).

The selection of the measurement method depends on restrictions on sample preparation and the desired frequency range and accuracy of the data. The loaded cavity resonance measurement method gives high accuracy data for narrowband measurements at frequencies over 100 GHz. The open-ended coaxial line offers a

Characterization Methodology of EMI Shielding Materials

method where only a flat surface of sample material is needed. The loaded coaxial transmission line method makes it possible to measure over wide frequency range with better accuracy than what the open-ended coaxial line method offers (Baker-Jarvis et al., 1998, 2001; Lundgren, 2004).

2.4 MECHANICAL PROPERTIES

EMI shielding materials must have reliable and endurable mechanical properties to meet the shield performance requirements. For instance, the shielding gaskets should provide a reliable contact over wide deflection ranges, while withstanding multiple mechanical flexures. Except for excellent conductivity, the shielding contact materials must have high resilience (ratio of yield strength over modulus of elasticity), high fatigue strength, excellent corrosion resistance, and acceptable manufacturability.

2.4.1 UNIAXIAL TENSILE TESTING

The uniaxial tensile testing generates a stress–strain curve, as shown in Figure 2.9, which characterizes a material's mechanical performance. The test is used to directly or indirectly measure the most mechanical properties of the test material, such as yield strength, tensile strength, elastic modulus, Poisson's ratio, elongation, resilience, and toughness.

From Figure 2.9, the normal stress, σ, is defined as the ratio of applied load (P) to the original cross-sectional area (A) in the tension test expressed in kg/mm^2. That is, $\sigma = P/A$. A measure of the deformation of the material that is dimensionless, called strain, is ε = change in length/original length = $\delta L/L$. The stress-strain curve can be

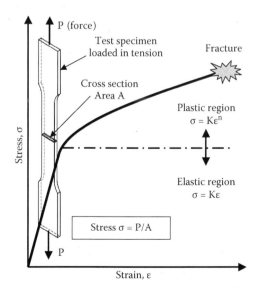

FIGURE 2.9 Uniaxial tensile testing.

divided into the elastic range and plastic range. In the elastic range, $\sigma = E\varepsilon$, where E is the elastic modulus; in the plastic range, $\sigma = K\varepsilon^n$, where n is strain hardening exponent and K is constant.

According to ASTM E8, tensile testing is usually conducted with a standard dog-bone sample. The surface finish and the size of the tensile sample are just as important as the shape in determining the results. For example, any nicks or burrs along the edge of the test piece will concentrate stress, which can cause the sample to yield and fracture prematurely. A narrow strip with a width to thickness ratio of less than 8:1 will show lower stress than a wider material (Gedeon, 2001). Samples of a small cross-section will show a small change in force for a given elongation. This decreases the accuracy of the force measurement, which in turn reduces the accuracy of stress measurement. If the material to be tested is plated, there is potential for inaccuracy as well.

Therefore, to compare one material to another, it is best to use properly prepared, unplated tensile test specimens of identical size and shape, tested at the same strain rate. This will help ensure a more accurate test.

2.4.2 Hardness

Hardness is the resistance of a material to plastic deformation. A hardness test does not measure a well-defined material mechanical property but provides a useful approximation. Hardness testing is usually used to monitor process operations, such as cold working, solution annealing, and age hardening. All test methods do not employ the same type of measurements for basis therefore, hardness values measured using one test method do not always correlate well with those from another. In other words, there are no universal hardness-conversion relationships between the results obtained by different methods. Generally, diamond pyramid (DPH) and Vickers hardness scales have the advantage that a continuous set of numbers covers the entire metallic hardness spectrum. The test allows direct hardness comparison of different gauge products (Brush Wellman, 1995).

2.4.3 Poisson's Ratio

For elastic deformation, Poisson's ratio is commonly designed as the ratio of transverse (across the width) to longitudinal (down the length) engineering strain. For metals, Poisson's ratio usually falls somewhere between 0.28 and 0.32. This means that for a 100% elongation, the width would contract by about 30%.

2.4.4 Stress Relaxation

Stress relaxation is the gradual decay of stress at constant strain. Figure 2.10 shows a stress relaxation test apparatus and uses cantilever beam surface stress to demonstrate stress relaxation conditions. One common method of stress relaxation is to utilize the simple bend loads of the test samples. Measurement of the permanent set of the beam occurs at various times after continuous exposure to elevated temperatures. Stressing samples to a constant percentage of their individual yield strengths is preferable to a common numerical stress value for all samples. The measured permanent set of the test beam reflects a loss of spring force, or a relaxation of the initial stress, as a result

Characterization Methodology of EMI Shielding Materials

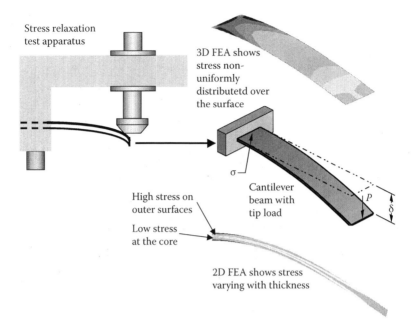

FIGURE 2.10 Stress relaxation test apparatus and stress relaxation conditions demonstrated with cantilever beam surface stress.

of the test conditions of time and temperature. The percentage loss at an established temperature and time is constant for any initial stress below the alloy's yield strength. In general, temperature is more important than time for stress relaxation. Higher temperature means more rapid relaxation (Brush Wellman, 1995). Materials with higher conductivity will therefore reach a lower maximum temperature that is advantageous from a stress relaxation point of view.

2.4.5 Contact Force

Contact or normal force is the most important attribute of a spring contact system directly related to contact reliability. Contact force is generated by surfaces contacting each other and perpendicular to those surfaces. As shown in Figure 2.10, for a cantilever beam-type spring component, the contact force, P, can be expressed as:

$$P = dE\,wt^3/(4\,L^3) = \sigma_{max}\,wt^2/(6\,L) \tag{2.3}$$

where d is the deflection of beam; E is the modulus of elasticity; w is the beam width; t is the beam thickness; L is the beam length; and σ_{max} is the maximum surface stress. At small deflection, for a given contact configuration and a constant deflection, the contact force is a function of the modulus of elasticity only. At large deflections, above the yield strength, the rate of increase in contact force with deflection decreases. This leveling off in force occurs as a result of plastic deformation, or permanent set, along the flexed portion of the beam.

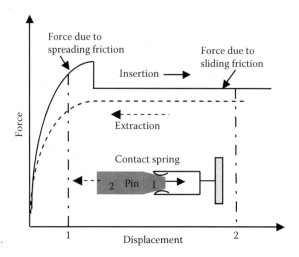

FIGURE 2.11 Variation of force vs. displacement during sliding insertion and extraction of a contact spring.

2.4.6 Friction Force

The friction force, F, depends on the contact or normal force, P, and is very sensitive to conditions at the interface. The related relationship can be expressed as $F = \mu P$, where μ is the coefficient of friction. Figure 2.11 illustrates a variation of insertion and extraction forces vs. displacement for a contact spring system (Brush Wellman, 1995). Here the insertion and extraction forces are dependent on the lead-in angle of the mating parts as well as the friction force.

2.4.7 Permanent Set

Permanent set is the plastic deformation that remains after releasing the deformation producing stress. This is a result of repeated deflections during component mating, oblique insertions, or insertions of oversized test probes. The contact performance degradation from permanent set reduces the contact forces exerted on the pin by reducing the amount of available deflection.

2.5 SURFACE FINISH AND CONTACT INTERFACE COMPATIBILITY

In the EMI shielding, a proper material surface finish is usually required for the following reasons: (a) as corrosion and oxidation protection for the base metal of a shield; (b) to establish a metallic contact interface between two mating parts; (c) to improve surface solderability; and (d) to maintain contact interface compatibility to eliminate galvanic corrosion. The surface finish is usually applied by electroplating; other processes such as painting, cladding, or spraying also have been chosen for some specific applications.

2.5.1 CORROSION AND OXIDATION PROTECTION

In many cases, EMI shields are subject to corrosion in typical operating environments because of oxidation, sulfidation, galvanic incompatibility, and so forth. Application of a surface finish seals off the shielding components from the environment and therefore prevents corrosion. For this reason, the contact finish itself must be corrosion resistant, usually provided by noble metal finishes or by forming a passivating surface via tin or nickel plating (nonnoble metal finishes).

2.5.1.1 Noble Finish Selection

Noble finishes, such as gold (Au), palladium (Pd), and some alloys of these metals, are intrinsically corrosion or oxidation resistant. When using this kind of finish, however, the extrinsic factors (e.g., contamination, base metal diffusion or intermetallic compound formation, and contact wear) would degrade their functions. Nickel underplating is generally applied in providing such protection. Typical surface finish platings are 0.4 to 0.8 μm thick for Au, or in some cases Au over Pd alloy, and 1.25 to 2.5 μm for nickel (Ni) underplating (Mroczkowski, 1997). These finishes vary in their degree of nobility and will be considered separately.

2.5.1.1.1 Gold (Au)

Gold provides an ideal contact surface finish due to its excellent electrical and thermal characteristics, as well as corrosion and oxidation resistance in virtually all environments. Pure gold, however, is relatively soft, therefore alloyed gold plating is usually applied, with cobalt being the most common, at levels of a few tenths of a percent to increase the hardness and wear resistance (Mroczkowski, 1997). Cost reduction objectives have led to the use of (a) selective plating practices such as reductions of the plating area and thickness; and (b) alternative noble metals, most commonly gold-flashed palladium or palladium alloy.

2.5.1.1.2 Palladium (Pd)

Palladium is not as good as gold in corrosion resistance or electrical/thermal conductivity. It is, however, significantly harder than gold, which improves durability performance. Palladium can catalyze the polymerization of organic deposits and result in increased contact resistance under fretting motions. Palladium is, therefore, not as noble as gold, although the effect of these factors on contact performance depends on the operating environment. In most applications, palladium is used with a gold flash (around 0.1 μm thick) to provide a gold contact interface (Mroczkowski, 1997).

2.5.1.1.3 Noble Metal Alloys

Noble metal alloys mainly include gold alloys and palladium alloys. The major gold alloy used is 69 wt% Au–25 wt% Ag–6 wt% Pt. There are two palladium alloys in use: 80 wt.% Pd–20 wt.% Ni and 60 wt% Pd–40 wt% Ag. The Pd–Ni alloy is electroplated and the Pd–Ag finish is primarily an inlay (Mroczkowski, 1997). In general these finishes include a gold flash to counter the lower corrosion resistance of these alloys compared to gold.

When applying these noble metal finishes onto EMI shields, nickel underplating provides the following benefits (Mroczkowski, 1997): (a) nickel, through the formation of a passive oxide surface, seals off the base of porosities and surface scratches sometimes presented in the noble metal finishes; (b) nickel provides an effective barrier against the diffusion of base metal constituents to the surface plating where the intermetallic compounds could be formed; (c) nickel provides a hard supporting layer under the noble surface, which improves contact durability; and (d) nickel provides a barrier against the migration of base metal corrosion products, reducing the potential of their contaminating the contact interface.

These benefits allow for equivalent or improved performance at reduced noble metal thickness. The effects of discontinuities are moderated, and the durability is improved. Nickel underplating also allows a reduction in the size of the contact area that must be covered with the noble metal finish. All these functions serve to maintain the nobility of the contact surface finish at a reduced cost.

2.5.1.2 Nonnoble Finishes

Nonnoble contact finishes differ from noble finishes in that they always have a surface oxidation film, which can be disrupted during contact surface mating and the potential for recurrence of the films during the application lifetime of the contact EMI shields. Tin (Sn), silver (Ag), and nickel (Ni) are typical nonnoble finishes usually used. Tin is the most commonly used nonnoble finish; silver offers advantages for high conductive contacts; and nickel is used in high-temperature applications.

2.5.1.2.1 Tin (Sn)

The utilization of a tin surface finish derives from the fact that the oxide film is easily disrupted on mating, and metallic contact areas are readily established. Generation of a tin contact interface for tin finishes results from cracks in the oxide under an applied load. The load transfers to the soft ductile tin, which flows easily, opening the cracks in the oxide, and tin then extrudes through the cracks to form the desired metallic contact regions. Here normal force alone may be sufficient; however, the wiping action that occurs on mating of contact spring gaskets, for instance, virtually ensures oxide disruption and creation of a metallic interface. The potential problem with tin is the tendency for reoxidation of the tin at the contact interface if it is disturbed. This process is called fretting corrosion. The tin finish surface would be reoxidized continuously when it is repeatedly exposed. The end result is a buildup of oxide debris at the contact interface leading to an increase in contact resistance. The driving force for fretting motions include mechanical (vibration and disturbances/shock) and thermal expansion mismatch stresses. Two approaches to mitigating fretting corrosion in tin and tin alloy finishes are high normal force (to reduce the potential for motion) and contact lubricants (to prevent oxidation). Each has been used successfully, and each has its limitations. High normal forces limit the durability capability of the finish, which is already low due to tin being very soft, and result in increased mating forces. Contact lubricants require secondary operations for application, have limited temperature capability, and may also result in dust retention (Mroczkowski, 1997).

Another potential issue for tin plating finish is tin whisker growth. The details will be discussed in Chapter 7.

2.5.1.2.2 Silver (Ag)

Silver can react with sulfur and chlorine to form silver sulfide and silver chloride films on its surface. Silver sulfide films tend to be soft and readily disrupted, and do not result in fretting corrosion. Silver chloride films, however, are harder, more adherent, and more likely to have detrimental effects on contact performance. In addition, silver is susceptible to electromigration, which can be a problem in some applications (Mroczkowski, 1997). Therefore, silver has limited use in the contact surface of EMI shields. Because of its high electrical and thermal conductivity and resistance to welding, however, silver finish still is a candidate for high-conductive contacts. The thickness of silver finish is generally in the range of 2 to 4 μm.

2.5.1.2.3 Nickel (Ni)

Nickel plating forms a passivating oxide film that reduces its susceptibility to further corrosion. This passive film also has a significant effect on the contact resistance of nickel. Because both nickel oxide and base nickel are hard, nickel finishes usually require higher contact normal forces to ensure the oxide film disruption. The self-limiting oxide on nickel, however, makes it a candidate for high temperature applications. With the same mechanism as tin finishes, nickel contact finishes are also susceptible to fretting corrosion. When nickel underplating is covered with thin noble metal finishes, for instance, wear-through of the finish due to mating cycles or fretting action, which exposes the nickel underplating, can result in fretting corrosion (Mroczkowski, 1997).

2.5.2 Solderability of Surface Finishes

Solderability of some EMI shield surface finishes is required when the shield needs to be soldered with other components. Solder bonds depend on whether the solder is either soluble in or capable of forming a metallic bond (a thin layer of intermetallic layer) with the material being soldered. Solderability demands that the metal surface finish be clean and remain clean and wettable by the solder with or without the aid of flux. Different metals have different affinity to a particular solder. Table 2.1 describes the general solderability of some typical metal surface finishes of EMI shields. Among them, tin plating is the most common surface finish and has an excellent solderability. However, the solderability can degrade over time. Loss of solderability in tin and tin-alloy plating, for instance, usually stems from three conditions: (1) excessive thickness of the intermetallic layer; (2) excessive amounts of oxides and other surface contaminants on the plating; and (3) high levels of codeposited carbon from the organic brighters in bright tin coatings. Each of these conditions can become worse over time; storage conditions, especially temperature, can influence the rate of degradation and therefore shelf-life of the plating. Therefore, the solderability of the surface finish needs to be inspected before soldering.

TABLE 2.1
Solderability of Some Typical Surface Finishes of EMI Shields

Surface Finish/Base Metal	Solderability	Remarks
Gold, palladium, silver, tin, cadmium, rhodium	Excellent	Noble metals dissolve easily in solders and mostly form intermetallic compounds, resulting in brittle joints in many cases. Nonnoble metals like tin may lose solderability in some conditions.
Copper, bronze, brass, lead, nickel silver (CuNiZn), CuNiSn, copper beryllium (CuBe), CuTi	Good	Some base metals like copper usually require high heat input during soldering due to their high thermal conductivity. Proper flux must be used because they oxidize quickly.
Carbon steels, low-alloy steels, zinc, nickel	Fair	The surface needs to be cleaned with proper flux before soldering. Solder joints become brittle in sulfur-rich environments.
Aluminum, aluminum bronze, aluminum alloys	Poor	Tough oxides on the surface prevent wetting (formation of solder bonding). Solders have to be specially selected to avoid galvanic incompatibility.
High-alloy steels, stainless steels	Poor	The surface needs to be cleaned with an aggressive flux to remove chromium oxides.
Cast iron, chromium, titanium, tantalum, magnesium	Very difficult	Require preplating with a solderable metal finish.

Two most common methods used to predict and measure solderability of a surface finish are the "dip and look" method and the wetting beam balance test. The dip and look test is performed according to JEDEC J-STD002A and a similar test using lead-free solder. Parts are dipped in flux for 5 seconds and then immersed in the liquid solder for 5 seconds. The part is examined for the extent of solder wetting on the surface. If the wetting area has greater than 95% solder coverage, the solderability of the surface finish is deemed acceptable. This is the test most commonly performed in the manufacturing environment to ensure ongoing product solderability. The wetting beam balance test is another method used for solderability evaluation. In this test, the part is first fluxed and then partially submerged into a solder pot. During the submersion process, the force required to submerge the part is measured using a delicate balance. The insertion process produces a characteristic diagram of force vs. displacement with several key features related to solderability. During insertion, as shown in Figure 2.12, the surface tension of the solder initially resists insertion by the terminal. This appears as a downward dip in the curve. At some point, the solder begins to wet the surface of the terminal, drawing the terminal into the solder pot. The wetting force produces an upswing in the force curve. The first key metric is the zero-cross time. This indicates the time required for the wetting force to overcome the surface tension of the solder. Acceptability conditions vary, but generally times of less than 1 second are desired (Hilty and Myers, 2004).

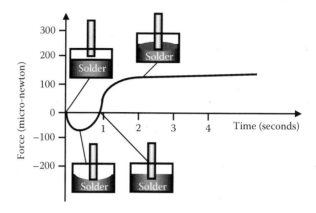

FIGURE 2.12 Schematic of wetting beam balance curve for solderability testing.

2.5.3 Effects of Mating Cycles and Operating Environments on Contact Finishes

The contact durability of a spring EMI shield-like gasket depends on the contact finish, the normal force, contact geometry, and the material's stress-relaxation resistance. With respect to contact finish, the dominant factors of its durability performance include the hardness of the finish, contact geometry, normal force, surface roughness, and state of lubrication of the interface. Gold-flashed palladium or palladium alloy finishes provide the highest durability capability. Silver and tin finishes are severely limited in the number of mating cycles they can support. Contact lubricants are usually used to improve the mating-cycle capability of noble finishes. An appropriate contact lubricant can improve the durability of mating finishes by upto an order of magnitude under favorable conditions (Mroczkowski, 1997). In addition, finish thickness largely affects the durability, with a roughly linear relationship between mating cycles and thickness for a given contact force/geometry configuration. The effect of normal force on mating life has two aspects: (1) wear rates increase with mating force and (2) the wear mechanism may change from burnishing to adhesive wear as normal force increases. The effect of contact geometry on mating cycles results from localization of the contact areas. High curvatures result in narrow contact wear tracks and increased wear for a given value of normal force (Mroczkowski, 1997).

Operating environments including temperature and corrosion severity will impact finish durability. The corrosion resistance of noble metal finishes decreases in the following order: gold, palladium, and palladium-nickel (80 wt% Pd–20 wt% Ni) alloy (Mroczkowski, 1997). The corrosion of the noble metal finishes are mainly caused by chlorine and sulfur. As mentioned earlier, nickel underplating reduces the corrosion susceptibility of the noble metal-coated parts. Tin finishes usually exhibit good stability with respect to corrosion, although fretting corrosion must always be taken into consideration for tin finishes.

Temperature limitations for noble metal finishes exhibit a similar pattern to corrosion resistance. In both cases, the order is determined by the presence of the nonnoble constituent in the palladium alloys, because it is alloying elements that are susceptible to corrosion. Hard gold finishes, even though the alloy element is of the order of tenths of a percent, are also subject to oxidation as operating temperature increases. In general, soft, or pure, gold finishes are recommended for temperatures above 125°C (Mroczkowski, 1997).

Tin has a temperature limitation due to an increasing rate of intermetallic compound formation, a reduction in the already low mechanical strength, and an enhanced oxidation rate. The interaction of these factors results in a recommendation that tin not be used above 100°C in conventional contact spring gasket design (Mroczkowski, 1997).

Thermal cycling is another important environmental consideration and is arguably the major driving force for fretting corrosion of tin systems. In addition, thermal cycling accelerates the effects of humidity on contact performance degradation (Mroczkowski, 1997).

2.5.4 Galvanic Corrosion and Contact Interface Compatibility

EMI shielding design requires an understanding of galvanic corrosion and contact interface compatibility. Galvanic corrosion is the process by which the materials corrode in contact with each other. Three conditions must exist for galvanic corrosion to occur (Engineers Edge, 2008): (1) two electrochemically dissimilar metals are in contact with each other; (2) an electrically conductive path is present between the two metals; and (3) a conductive pathlike electrolyte must be present for the metal ions to move from the more anodic metal to the more cathodic metal. If any one of these three conditions does not exist, galvanic corrosion will not occur. Usually when a design requires that dissimilar metals come in contact, the galvanic compatibility is managed by plating or coating with other finishes. The finishing and plating selected facilitate the dissimilar materials being in contact and protect the base materials from corrosion. The typical anodic index of some EMI shielding materials is shown in Table 2.2. For harsh environments, such as outdoor, high humidity, and salt environments, there should typically be not more than 0.15 V difference in the anodic index. For example, gold/silver would have a difference of 0.15 V being acceptable in these kinds of environments. For normal environments, such as storage in warehouses or nontemperature and humidity-controlled environments, there should typically be not more than a 0.25 V difference in the anodic index. For controlled environments, that are temperature and humidity-controlled, 0.50 V can be tolerated. Caution should be maintained when considering this application, as humidity and temperature vary by region (Engineers Edge, 2008).

2.6 FORMABILITY AND MANUFACTURABILITY

Metal strips or sheets have been widely used for EMI shield manufacturing, such as enclosures, gaskets, board-level shields, and connecting and grounding shields. These EMI shields are usually manufactured by a mechanical forming process. Various shapes and geometries are formed using a variety of thicknesses of thin sheet metals. When selecting metal strips, therefore, their formability and manufacturability are critical for the mechanical design and production of EMI shields.

TABLE 2.2
Typical Anodic Index of Some EMI Materials

Surface Finishes/EMI Shielding Materials	Anodic Index
Gold (solid and plated), gold platinum alloys, wrought platinum, graphite carbon	0.00
Rhodium plated on silver-plated copper	0.05
Rhodium plating	0.10
Silver (solid or plated) high-silver alloys	0.15
Nickel (solid and plated), nickel copper alloys, titanium and its alloys, Monel metal	0.30
Copper (solid and plated), low brass or bronzes, silver solder, nickel silver (CuNiZn), NiCr alloys, CuBe alloys, CuNiSn alloys, CuTi alloys, austenitic stainless steels	0.35
Yellow brass and bronzes	0.40
High brasses and bronzes, naval brass, Muntz metal	0.45
18%-Cr type corrosion-resistant steels, 300 series stainless steels	0.50
Tin plating, chromium plating, 12% Cr-type corrosion-resistant metal, 400 series stainless steels	0.60
Terne plate, tin–lead solder	0.65
Lead (solid or plated), high-lead alloys	0.70
Aluminum; 2000 series wrought aluminum alloys	0.75
Iron (wrought, gray, or malleable), plain carbon and low-alloy steels, Armco iron, cold-rolled steel	0.85
Aluminum alloys other than 2000 series, cast Al–Si alloys, 6000 series aluminum alloys	0.90
Cast aluminum alloys other than Al–Si, cadmium plating	0.95
Hot-dip galvanized or electrogalvanized steels	1.20
Wrought zinc, zinc die casting alloys, zinc plating	1.25
Magnesium and its alloys (cast or wrought)	1.75
Beryllium	1.85

2.6.1 Bending Formability

As shown in Figure 2.13, the bending formability ratio, r/t, is the ratio of the bend radius (r) to the strip thickness (t). This value defines the sharpest forming radius allowable prior to failure. Here the failure is identified as orange peel surface or surface cracks after bending. Generally, only a smooth surface is acceptable for mechanically formed EMI shields. Usually a V block or W block with a 90-degree bend is used for a bending formability test. An r/t value of zero defines a material that can form a very sharp bend without failure. Larger r/t values indicate reduced formability. Formability is related to a material's yield strength, elongation, and toughness, and formability decreases for higher strength and harder tempers (hardness). However, the bending formability ability improves when the strip width decreases at a constant thickness. For a thin strip (usually when the thickness over width ratio is less than 8), the formability is more sensitive to edge conditions and generally damage associated with stamping. Moreover, there is a directionality or anisotropy of strip properties as a result of cold rolling. Formability is the function of the bend orientation relative

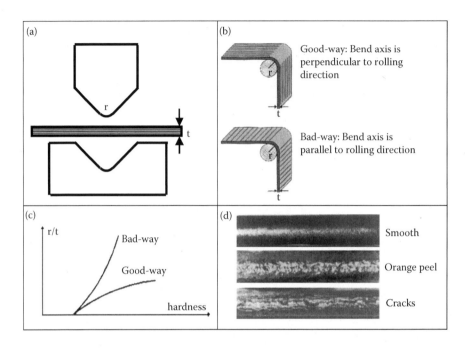

FIGURE 2.13 Schematic of metal strip bending formability: (a) V block die; (b) good-way and bad-way bending; (c) r/t variation trends with different tempers (hardness); and (d) surface status after bending.

to the rolling direction in a heavily cold rolled metal strip. The formability in the longitudinal direction, good-way bending, is usually better than that present in the transverse (bad-way) direction (Brush Wellman, 1995).

2.6.2 Effect of Strain Hardening on Formability

A material with a high strain hardening exponent, or n value, has a greater formability (a capacity for being stretched or bent in a forming die) than a material with a low n value. Once strain is imparted by forming in a die or by cold rolling in the mill, the material has less capacity for additional straining. This is why materials that have been cold rolled to higher hardness have less formability than those with less cold work. In general, the alloys at different tempers with the best formability (lowest r/t ratio) usually have much higher strain hardening exponent than the other alloys.

For designing EMI shields like metal gaskets, contacts, and connectors, where the important material properties are typically strength and conductivity, the 90-degree V-block formability data is usually adequate to ensure that the parts can be formed without a problem. For some EMI shields like deep draw board-level shields or complex-shaped stamping gaskets, the design concerns include strength, ductility, springback, and formability. In this case, the r/t ratio is not enough; the n value would help immensely in choosing the proper punches and dies. The n value is good for determining how the material will bend and stretch, however it is not so useful in determining how the material can be drawn. There are other material parameters that come into play in this case.

2.6.3 ANISOTROPY COEFFICIENT

Deep drawing is an operation that involves a substantial amount of material flow and stretching. The flat strip must flow around the base of the draw into the cup and must then elongate as the cup is drawn out. This uses a substantial amount of strain hardening. For severe drawing operations, the material must be annealed in between draws to soften the material and give it more capacity for work hardening. In effect, annealing is resetting the original n value. The annealed material would then be ready for additional drawing operations.

The anisotropy coefficient, r, provides a measure of how well strip material can be drawn. Typically, this data would be available only for annealed or very lightly strengthened alloys suitable for deep drawing. For strip material in tension, it is the ratio of the true transverse strain to the true through-thickness strain, $r = \varepsilon_w/\varepsilon_t$. Since it is difficult to mount a strain gauge on the edge of a piece of a thin strip, the strain is usually measured in the longitudinal and transverse directions. Since the volume and the density of the material do not change in tension, the longitudinal, transverse, and thickness strains must all sum up to zero. The through-thickness strain is thus equal to the inverse of the sum of the longitudinal and transverse strains, $\varepsilon_t = -\varepsilon_w - \varepsilon_L$. Therefore, $r = -\varepsilon_w/(\varepsilon_w + \varepsilon_L)$. If the material is perfectly isotropic, then r is equal to 1. If r is more than 1, the material stretches more than it thins, and will thus flow easily into a drawing die without excessive thinning. If r is less than 1, the material will thin more than it stretches which may lead to tearing when the material is drawn. Strip materials with high r values are suitable for deep drawing operations, whereas those with low r values can only be lightly drawn before they crack.

Generally, longitudinal and transverse strain response within the plane of the strip are not necessarily the same; the r value also depends on which direction the sample was stretched. Therefore, the stretched r value is usually stretched in three directions: longitudinal, transverse, and at 45 degrees to the rolling direction. An average of the r values in all three directions is determined by using the following equation (Davis, 1996):

$$r_{Ave} = (r_L + 2 r_{45°} + r_T)/4 \tag{2.4}$$

The directionality of the r value accounts for the phenomenon of earring when material is drawn into a cup shape.

Points along the rim of the cup at which the r value is higher will have more excess material than those where the r value is lower. The size of ears will be dependent on the parameter Δr (Davis, 1996):

$$\Delta r = (r_L - 2 r_{45°} + r_T)/2 \tag{2.5}$$

A large value of Δr implies that there will be substantial leftover material in some directions, while there will be very little leftover material, or even not enough material, in other directions. Therefore, it is desirable to have a low Δr value along with a high r value.

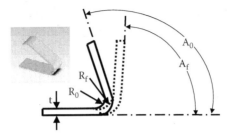

FIGURE 2.14 Schematic of bend springback.

Consequently, while high strain hardening exponents (n values) are important for ensuring that materials can be bent and stretched in a stamping die, high anisotropy coefficients (r values) are important for ensuring that materials can be drawn in a stamping die.

2.6.4 SPRINGBACK DURING METAL STRIP FORMING

As shown in Figure 2.14, springback causes dimensional change in formed components after release of the forming tool pressure. Overformed bends compensate for springback to maintain the desired part geometry. Springback increases with increasing yield strength and punch radius, and with decreasing elastic modulus and strip thickness. The differences in springback do not allow a temper change once tooling is built (Brush Wellman, 1995).

For all punch radius/strip thickness ratios, a 90-degree plane strain bend springback, K, can be empirically expressed as (Brush Wellman, 1995):

$$K = A_f/A_0 = -25.54\ X^3 + 17.91\ X^2 - 5.85\ X + 1.08 \tag{2.6}$$

where $X = (Ys/E)(R/t)$, A_f is the formed angle, A_0 is the bend or die angle (90°), Ys is the 0.2% offset yield strength, E is the modulus of elasticity, R is the punch radius, and t is the strip thickness.

2.7 ENVIRONMENTAL PERFORMANCE EVALUATION

Appropriate environmental qualification tests are generally required for EMI shields. For instance, packaging and cleanliness requirements protect the EMI shields from mechanical damage and contamination during shipping and handling. Shipping and storage conditions are the conditions prior to installation. The operating environment consists of corrosive gas, particulate, and psychometric (temperature and humidity) conditions. Environmental tests usually consist of accelerated conditions to simulate the lifetime of an EMI shielding system and the test results are a measure of the reliability of the system. The assembly process for the test vehicles should be representative of the application manufacturing processes. Environmental stress tests include thermal cycling, thermal aging, gaseous testing, temperature and humidity testing, dust sensitivity, and vibration and shock testing (Brush Wellman, 1995).

2.7.1 Thermal Cycling

Thermal cycling provides a short-term simulation of the long-term effects of temperature cycling. The Coffin-Manson relationship for metal fatigue determines the number of cycles required at test temperatures to equate to the number of anticipated cycles at field temperatures (Nelson, 1990):

$$N = C/(\Delta T)^\gamma \quad (2.7)$$

where N is the number of cycles to failure, C and γ are the characteristic constants of the metal, and ΔT is the temperature range of the thermal cycle. Here, the thermal cycling fatigue life mainly depends on the material's thermal mismatch (CTE).

For example, the typical conditions of simulated shipping and operational environment tests can be designed as 5 thermal cycles under $-40°C$ to $+65°C$ for shipping, and 100 thermal cycles under $0°C$ to $+75°C$ for operation (Brush Wellman, 1995).

2.7.2 Thermal Aging

Thermal aging provides a short-term simulation of the long-term effects of exposure to elevated temperatures. Dry oxidation of base materials and thermal stress relaxation are related key material properties. Typical conditions are 700 hours at $100°C$ (Brush Wellman, 1995).

2.7.3 Gaseous Testing

The purpose of gaseous testing is to provide a short-term simulation of exposed contact base metal corrosion. Gold finishes degrade through the ingress of corrosion products from various sources. Corrosion migration and pore corrosion are the two degradation mechanisms. The gaseous test involves exposing contacts to corrosive gases during a thermal cycling or aging. Typical conditions are twenty-five cycles of thermal cycling with gas concentrations in the part per billion ranges (H_2S, NO_2, Cl_2; Brush Wellman, 1995).

2.7.4 Pore Corrosion and Fretting Corrosion

Figure 2.15 illustrates the pore corrosion and fretting corrosion processes. The pore corrosion occurs when the defect site is a pore, which is a small discontinuity in the contact finish from which corrosion growth occurs. The presence of pores will cause corrosion migration that refers to the movement of corrosion products into the contact area from sites away from the contact interface. Such sites include contact edges and defects in the contact finish. Corrosion migration is of concern predominantly in environments where sulfur and chlorine are present. Nickel underplating, for instance, is usually applied to prevent pore corrosion in gold-plating surface finish.

Fretting corrosion presents in tin plating, nickel plating, or other surface finishes. Tin is the most common plating subject to fretting corrosion. Tin is resistant to surface corrosion because a protective oxide film forms on the tin surface limiting further corrosion. This oxide film does not affect contact resistance since mating of the contact part disrupts it. The reason is that the thin, hard, and brittle tin oxide fractures under the application of contact normal force; and the soft, ductile tin

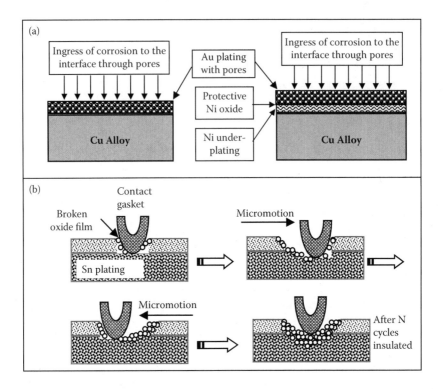

FIGURE 2.15 Comparison of pore corrosion and fretting corrosion.

flows to enlarge the cracks in the oxide and extrudes through the cracks to establish the desired metallic interface. However, the oxide forming tendencies of tin remain active, and the contact interface moves, due to thermal cycling, and contact resistance increases. This degradation mechanism is so-called fretting corrosion. Fretting refers to the small motions, hundredths to tenths of a millimeter, which occur randomly due to mechanical disturbances or thermal expansion mismatches. The tin surface re-oxidizes during the fretting exposure and is a degradation mechanism causing increases in contact resistance (Brush Wellman, 1995). Contact lubricants and high normal forces can be used to minimize fretting corrosion.

2.7.5 Temperature and Humidity Testing

Temperature and humidity testing provides a short-term simulation of the long-term effects of high temperature and humidity. Humidity affects galvanic corrosion. Fretting corrosion is the degradation mechanism. Typical conditions are 300 hours at 50°C and 80% RH (relative humidity; Brush Wellman, 1995).

2.7.6 Dust Sensitivity Test

The dust sensitivity test tracks the response of contact parts exposed to increasing amounts of dust to determine the point at which performance declines. Contact resistance change or the number of openings that occur is a monitor of performance. (Brush Wellman, 1995). Typical conditions and testing procedures are described in MIL-STD810F method 510.4.

2.7.7 Vibration and Shock Test

EMI shield vibration and shock testing are performed at the next level assembly since the input level is very dependent on the mounting structure. The four vibration input types are random, sine-on-random, sine dwells, and sine sweep. The plane and amplitude of expected shock and vibration affect connector design. A resonantly vibrating member can destroy electrical continuity, causing intermittent failures. Vibration tests determine contact resonance at their natural frequencies. A drop table simulates impact during shipping. Here a shock load determines the level that will cause the contacts to separate. (Brush Wellman 1995). The detail testing conditions and procedures are described in MIL-STD 810F method 514.5 (Vibration) and method 516.5 (Shock).

REFERENCES

Baker-Jarvis, J., R. G. Geyer, J. H. Grosvenor, Jr., M. D. Janezic, C. A. Jones, B. Riddle, C. M. Weil, and J. Krupka. 1998. Dielectric characterization of low-loss materials a comparison of techniques. *IEEE Transactions on Dielectrics and Electrical Insulation*, 5(4): 571–577.

Baker-Jarvis, J., M. D. Janezic, B. Riddle, C. L. Holloway, N. G. Paulter, and J. E. Blendell. 2001. *Dielectric and conductor-loss characterization and measurements on electronic packaging materials* (Technical note 1520). Boulder, CO: National Institute of Standards and Technology.

Bodnar D. G., H. W. Denny, and B. M. Jenkins. 1979. Shielding effectiveness measurements on conductive plastics. *IEEE International Symposium on Electromagnetic Compatibility*, 27–33.

Brush Wellman.1995. *Connector engineering design guide: Materials selection in the design of spring contacts and interconnections.* Cleveland, OH: Brush Wellman.

Davis, J. R. 1996. *Carbon and Ally Steels.* Cleveland, OH: ASM International.

Dunwoody, S., and E. VanderHeyden. 1990. Transfer impedance testing of multiconductor shielded connectors of arbitrary cross-section. *1990 IEEE International Symposium on Electromagnetic Compatibility*, 581–585.

Engineers Edge. 2008. Galvanic compatibility. http://www.engineersedge.com/galvanic_capability.htm.

Gedeon M. 2001. Tensile testing. *Technical Tidbits*, 3(9). http://www.brushwellman.com/alloy/tech_lit/September 01.pdf.

Hatfield, M. O., E. A. Godfrey, and G. J. Freyer. 1998. Investigations to extend the lower frequency limit of reverberation chambers. *IEEE International Symposium on Electromagnetic Compatibility*,1, 20–23.

Hilty, R. D., and M. K. Myers. 2004. Solderability of lead free electrodeposits in tin/lead solder (Tyco document, 503-1001). http://www.tycoelectronics.com/customersupport/rohssupportcenter/pdf/503-1001_Solderability_backwards_compatable.pdf

Hoeft, L. O., and J. S. Hofstra. 1992. Measurement of surface transfer impedance of multi-wire cables, connectors and cable assemblies. *IEEE 1992 International Symposium on Electromagnetic Compatibility*, 308–314.

IEC 96-1A. 1976. *Radio frequency cables, part I: General requirements and measuring methods.* International Electrotechnical Commission, IEC Standard Publication 96-1A.

IEEE-STD-299. 1997. IEEE standard method for measuring the effectiveness of electromagnetic shielding enclosures.

Lundgren, U. 2004. Characterization of components and materials for EMC barriers. PhD diss., Lulea University of Technology, Lulea, Sweden.

MIL-STD-285. 1956. Attenuation measurements for enclosures, electromagnetic shielding.

Mroczkowski, R. S. 1997. Connector and interconnection technology. In *Electronic Packaging and Interconnection Handbook*, 2nd ed., edited by C. A. Harper, 3.1–3.82. New York: McGraw-Hill.

Nelson, W. 1990. *Accelerated Testing: Statistical Models, Test Plans and Data Analyses*. New York: John Wiley & Sons.

Quine J. P. 1993. Characterization and testing of shielding gaskets at microwave frequencies. *IEEE 1993 International Symposium on Electromagnetic Compatibility*, 306–308.

Rahman, H., P. K Saha, J. Dowling, and T. Curran. 1992. Shielding effectiveness measurement techniques for various materials used for EMI shielding. *IEE Colloquium on Screening of Connectors, Cables and Enclosures*, 9/1–9/6.

SAE ARP 1705. 1981. Coaxial test procedure to measure the RF shielding characteristics of EMI gasket materials. Society of Automotive Engineers, Committee AE-4G.

SAE ARP 1705A. 1997. Coaxial test procedure to measure the RF shielding characteristics of EMI gasket materials. Society of Automotive Engineers, Committee AE-4G.

Spiegelaar, H., and E. VanderHeyden. 1995. The mode stirred chamber: A cost effective EMC testing alternative. *1995 IEEE International Symposium on Electromagnetic Compatibility*, 368–373.

Stickney, W. 1984. Improved testing for EMC gasket performance. *R. F. Design*, 7(1), 40–44.

Wilson, P. F., and M. T. Ma. 1985. Factors influencing material shielding effectiveness measurement. *IEEE International Symposium on Electromagnetic Compatibility*, 29–33.

3 EMI Shielding Enclosure and Access

As electronic systems become more complex, the need for electromagnetic interference (EMI) barriers becomes more urgent. These barriers include physical barriers like filters and geometrical barriers. As one of most common geometrical barriers, conductive shielding enclosures are usually used to reduce radiated EMI emissions by enclosing and isolating the radiating source and the victim device or system. The design and material selection for a shielding enclosure can be a comprehensive task as the enclosure often plays many roles and must fulfill requirements other than only those of EMI shielding. When designing shielding enclosures, the following multifunctions must be offered: EMI shielding barrier with discontinuities; allowing for the installation of EMI gaskets and access feedthroughs; ventilation or cooling for interior components; moisture and dust seal; mechanical protection for interior components; communication path with user and/or other devices; aesthetics and environmental compliance; optimal selection of materials; and production techniques to make the enclosure have low production cost and last for the required lifetime.

The choice of material will influence the production technique selection to meet the overall requirements on an enclosure. To achieve EMI shielding, for instance, the enclosure material can be metal, magnetic material, plastics coated with conductive layer, or conductive composites. However, an offset should be reached among the material and production cost, shielding performance, and user friendliness.

This chapter will focus on EMI shielding enclosure design methodologies, materials selection, magnetic shielding, enclosure, and access integration for achieving required EMI shielding performance.

3.1 ENCLOSURE DESIGN AND MATERIALS SELECTION

A shielding enclosure is a box or housing that provides isolation to the EMI emitter or receiver. Shielding reduces emissions and improves immunity, but does not slow the system as filters do. However, shielding enclosures or other components can increase the cost of the system, as shown in Table 3.1.

Proper shielding enclosure design can eliminate EMI, and such a design process consists of following EMC regulations and EMI standards, defining goals and EMI parameters, choosing the proper materials, and the judicial use of gaskets. Shielded enclosures are normally designed for emission suppression or to limit susceptibility to electromagnetic radiations. Depending on the equipment that goes into the enclosure and the range of frequencies of the electromagnetic waves, the enclosure needs to meet some performance objectives. There are many possible design and performance objectives in different kinds of environments, such as shielding effectiveness,

TABLE 3.1
Relative Price to Use the Shielding

Individual ICs	$0.25
Segregated areas of PCB circuitry	$1
Whole PCBs	$10
Subassemblies and modules	$15
Complete products	$100
Assemblies (e.g., industrial control and instrumentation cubicles)	$1,000
Rooms	$10,000
Buildings	$100,000

galvanic compatibility, longevity of gaskets, easy and less expensive maintenance, and easy accessibility to the equipment inside the enclosure. General principles of the shielding enclosure design and material selection include:

1. High permeability materials should be used for low impedance magnetic fields to provide low-reluctance path shielding.
2. High-conductivity materials are effective for electric fields to provide Faraday's cage shielding; they also can be used for high frequency magnetic fields to provide induced eddy current shielding.
3. Enclosures containing emitters should be designed for maximum absorption loss of the emitted field.
4. Enclosures containing receptors should be designed for maximum reflection loss of the received field.
5. Discontinuities should be minimized and designed to maintain shielding effectiveness by proper seam treatment, and use of EMI gaskets, vent/window screening, and shielding/filtered leads. Connectors must be regarded as discontinuities and shielded.
6. Before getting involved in design and evaluation, it is good to write down the goals, and to identify the parameters that need limits or values to be assigned. Parameters might include design for emission suppression, design for limiting susceptibility, shielding standard to be met, frequency range of the equipment inside the enclosure, attenuation levels to be met, and type of environment where the enclosure will be used.

There are many factors in selecting materials for shielding enclosures. Table 3.2 shows the conductivity and permeability of common shielding enclosure materials for some commercial applications (Beaty, 1990; English, 2007; Evans, 1997). First of all, enclosure materials must meet electrical and mechanical requirements. For an enclosure where a high level of EMI shielding is desired, a metallic material is the best choice since metals offer the best conductivity. The enclosure will thereby offer high shielding effectiveness for both electric and magnetic fields at reasonable high frequencies. However, sometimes marketing decisions concerning appearance

TABLE 3.2
Typical Conductivity and Permeability of Enclosure Materials

Materials	Approximate Composition (Wt%)	Conductivity (% IACS)	Initial Relative Permeability
Annealed copper	100 Cu	100	1
Beryllium copper	98 Cu, 2Be	22	1
CuNiSn	9 Ni, 6 Sn, 85 Cu	12	1
Phosphor bronze	8 Sn, 0.2 P, 91.8 Cu	13	1
Brass	63 Cu, 37 Zn	26	1
Nickel silver	18 Ni, 27 Zn, 55 Cu	6	1
Aluminum	100 Al	61	1
Al 6061	1.2 Mg, 0.8 Si, 98 Al	46	1
Commercial iron	99.8 Fe	16	250
Purified iron	99.95 Fe	18	10,000
Cold-rolled steel	0.5 Mn, 0.06C	12	200
Stainless steel	17 Cr, 7 Ni	2.4	1.02
Mg AZ91D	9 Al, 0.7 Zn	12	1
Silver plating	100 Ag	105	1
Gold plating	100 Au	70	1
Zinc plating	100 Zn	29	1
Electroless Ni plating	88-95 Ni, 5–12 P	15–25	1
Electrolytic Ni plating	~100 Ni	20–25	~100
Tin plating	100 Sn	15	1
EMI Paint	—	<0.1	1
Plastic composite	Steel filaments	8.62E-04	1
Plastic composite	10% S.S. filaments	2.10E-04	1
Plastic composite	Glass fiber	9.58E-02	1
Plastic composite	40% carbon fiber	1.72E-06	1
Polycarbonate composite	5% Ni-coated graphite	5.07E-09	1
Polycarbonate composite	10% Ni-coated graphite	3.32E-06	1
Polycarbonate composite	15% Ni-coated graphite	1.08E-03	1
Polycarbonate composite	20% Ni-coated graphite	1.57E-03	1
78 Permalloy	78.5 Ni, 21.5 Fe	10.8	8,000
MoPermalloy	79 Ni, 4.0 Mo, 17 Fe	3.13	20,000
Supermalloy	79 Ni, 5.0 Mo, 15 Fe	2.87	100,000
48% nickel–iron	48 Ni, 52 Fe	3.82	5,000
Monimax	48 Ni, 3 Mo, 49 Fe	2.15	2,000
Sinimax	43 Ni, 3 Si, 54 Fe	2.02	3,000
Mumetal	77 Ni, 5 Cu, 2 Mo, 15 Fe	2.87	20,000

or cost, for example, may dictate material selection rather than good engineering practice. Therefore, it is important to carefully balance system requirements, performance, EMI protection, and overall cost. For instance, all metallic materials corrode and oxidize, and corrosion or oxidation is poorly conductive; conductive paint prevents corrosion and improves appearance, but is not very conductive; metal plating prevents corrosion and oxidation and is very conductive, but costs more.

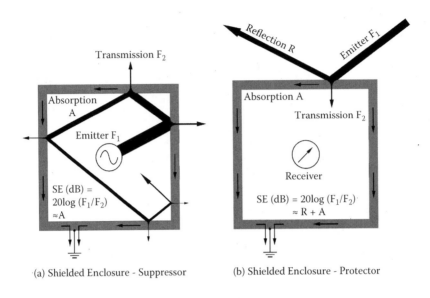

FIGURE 3.1 Illustration of EMI shielding enclosure acting as (a) suppressor and (b) protector.

Figure 3.1 illustrates the EMI shielding enclosure that acts as a suppressor or a protector. As shown in Figure 3.1a, if the goal is emission suppression, the enclosure material should have a maximum absorption loss of the emitted field; if the components in the enclosure are to be protected from outside fields (Figure 3.1b), the enclosure material should have a maximum combination of absorption and reflection losses. As discussed in Chapter 1, the absorption loss is the same for all electromagnetic waves. It mainly depends upon the material thickness, relative permeability of material to free space, relative conductivity of the material with reference to copper, and the frequency of the incident wave. On the other hand, the reflection loss is different for electric, magnetic, and plane waves. It depends on the distance of the source to the material, relative conductivity of the material with reference to copper, and the frequency of the incident wave. Shielding effectiveness of the enclosure material used for emission suppressions approximately equals the absorption loss. Shielding effectiveness of the enclosure material used for limiting susceptibility is approximately equal to the sum of absorption and the appropriate reflection loss depending on the type of incident wave.

Another important factor in selecting material for a shielded enclosure is the continuity of the conductive enclosure. A perfect welded closed box would be ideal. However, the user needs to have easy access inside the enclosure to mount equipment. So it is common to have doors, side panels, and input/output (I/O) panel openings. Most of the time, shielding effectiveness depends on how the electromagnetic waves are transmitted across the boundaries of these openings. These openings cause a major problem in the design of shielded enclosures because a commercial enclosure cannot be completely sealed. The most common openings in an enclosure are doors, side panels, I/O panels, ventilation ports, and cables that ingress and egress. All openings in an

EMI Shielding Enclosure and Access

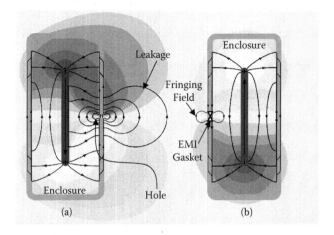

FIGURE 3.2 EMI shielding enclosure encloses a radiated field with (a) leaky shielding and (b) gasketed shielding.

enclosure will act as slot radiators of electromagnetic waves unless continuity is restored, as shown in Figure 3.2. Any time two surfaces meet in a shielded enclosure the possibility of emissions exists. The doors and panels of an enclosure do not have perfectly flat surfaces, and gaps created by them will act as slot radiators. To eliminate this, mating surfaces should be extremely rigid so when compressed against each other they will not leave any gaps. The gaps can also be filled with conductive material to maintain continuity. If metallic materials are used to make an enclosure offering a high level of shielding, for instance, it is common practice to include bolts, rivets, and gaskets in the enclosure design to ensure high conductivity between the contact parts.

Using a highly conductive material for the enclosure, gaskets have apparent advantages to seal discontinuity and help increase shielding effectiveness. To deal with imperfections inherent in commercial enclosure design, as shown in Figure 3.2b, gaskets can be used to help maintain electrical continuity across enclosure openings. The gasket materials should be galvanically compatible with the material of the enclosure. When the gasket material is dissimilar from the enclosure metal, suitable protection (painting or plating) against galvanic corrosion must be applied. One example is the chassis application, which involves two types of enclosure forces on the gasket: compression and shear. When gaskets are installed under a flat cover panel in a compression configuration, pressure is used to preserve the shielding effectiveness of the seam. The alternative is a shear application where a flange or a channel arrangement shears against a gasket before the enclosure is closed to accomplish shielding.

The purpose of the gasket is to improve conductivity to reduce emissions; however, all porous spots in the gasket can act as slot radiators at high frequencies. It is important to calculate the shielding effectiveness of the gasket given its porosity. In addition, most gaskets have a limited lifetime, normally defined in cycles for a particular compression limit. Based on the usage of the particular opening, the longevity of the gasket can be determined and an appropriate maintenance or replacement schedule can be planned.

Some enclosures have natural or forced convention cooling that require an inlet and an exhaust for air. In this kind of situation, an arrangement needs to be made allowing air to come in and out, but not the electromagnetic waves. The answer is a waveguide. A waveguide is nothing more than a tube made long enough such that it cuts off electromagnetic waves and acts like a brick wall. The length of the waveguide should be at least five times its diameter, as discussed in Chapter 1.

Sometimes cables and connectors are considered part of the enclosure and must also be shielded. The cables carrying power in and out of the enclosure radiate some electromagnetic waves. Proper insulating covers for the cables and their associated connectors should be used to eliminate any chance of cable-based EMI radiation.

After a prototype enclosure is built, one way to find any discontinuities prior to testing to a particular standard is by using the principle of skin effect. This requires using special equipment to determine the surface resistivity at different points, as discussed in Chapter 2. (Resistivity is the inverse of conductivity; good conductivity should be maintained to minimize EMI.) Figure 3.3 illustrates the shielding impedance of an enclosure. Accordingly, the relationship between the combined shielding effectiveness of the enclosure and its impedance can be expressed as:

$$SE_{enclosure} = 20 \log_{10} [I_{Shield}/I_{Radiated}] = 20 \log_{10} [Z_{Radiated}/Z_{Shield}] \quad (3.1)$$

where $Z_{Shield} = Z_{Door} + Z_{Gasket} + Z_{Enclosure}$.

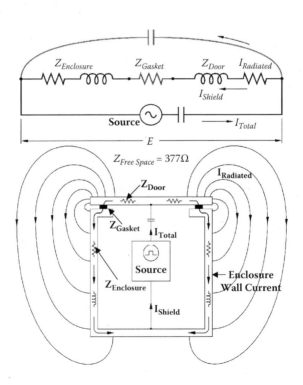

FIGURE 3.3 Shielding impedance of an enclosure.

TABLE 3.3
Calculated Values Using Equation 3.2

Shielding (dB)	Z_{shield} milliohms	Comments
40	3800	Average shielding
50	1200	Average shielding
60	380	Good shielding
70	120	Good shielding
80	38	Good shielding
90	12	Good shielding
100	3.8	Excellent shielding
110	1.2	Excellent shielding
120	.38	Excellent shielding

If the $Z_{Radiated} = 377\ \Omega$, then Equation 3.1 expressing the relationship between the shielding effectiveness and impedance of the enclosure would become:

$$SE_{enclosure} = 20 \log_{10} [377\ \Omega\ /Z_{Shield}] \quad (3.2)$$

Table 3.3 shows the calculated values of SE and Z_{Shield}. In order to obtain a combined effectiveness as good as possible, the gasket impedance must be selected that is less than the enclosure shield impedance. In general, apertures and cables are dominated by the shield impedance when the combined shielding effectiveness is below 60 dB. On the other hand, it is difficult to achieve a shielding effectiveness more than 105 dB.

3.2 MAGNETIC FIELD SHIELDING ENCLOSURE

3.2.1 Basics of Magnetic Shielding

A magnetic field results from a source of magnetic flux, usually with low frequencies ranging from direct current (DC) to 100 Hz. Direct current (DC) fields are static. A DC field might come from Earth, a permanent magnet, or a coil carrying direct current. Alternating current (AC) magnetic fields oscillate in direction that reverses cyclically at a frequency. The most common AC magnetic fields are the 60 Hz (USA) fields emitted by the electric power total amount of a magnetic field. Other common examples of magnetic fields are those emitted by motors, transformers, electric-arc furnaces, and electric-power generating and transfer stations (Green, 2001).

Unlike electricity, magnetic fields cannot be blocked or insulated. No materials have been found to be able to block magnetic fields without themselves being attracted to the magnetic force. In other words, magnetic fields can only be redirected, not created or removed. In general, high permeability shielding alloys are used for doing this. The magnetic field lines are strongly attracted into the shielding material. All shielding materials work by diverting the magnetic flux to them;

therefore, although the field from a magnet will be greatly reduced by a shield material, the shield material will itself be attracted to the magnet. Shielding factor (*SF*) can be expressed as the ratio between the unperturbed magnetic field (*Bo*) and the shielded magnetic field (*B*). For a spherical shell of magnetic shielding placed in a uniform magnetic field *Bo*, the *SF* can be formulated as (Fitzpatrick, 2002):

$$SF = Bo/B \approx 2/3 \, \mu \, (d/r) \quad (3.3)$$

where μ is the relative permeability of shell material, *r* is the inter radius of the sphere shell, and *d* is the thickness of the shell. Thus, if $\mu \approx 10^5$ for magnetic material, then the magnetic field strength inside the shell can be reduced by almost a factor of 1000 using a shell whose thickness is only 1/100th of its radius. However, as the external field strength, Bo, is increased, the magnetic shell material eventually saturates, and μ gradually falls to unity. Therefore, extremely strong magnetic fields (typically, $Bo \geq 1$ tesla) are hardly shielded at all by common magnetic materials. In general, the magnetic shielding design depends on several critical factors: maximum predicted worst-case magnetic field intensity (magnitude and polarization) or the geomagnetic (DC static) field at that location, whichever is greater; shield geometry and volumetric area; type of materials, permeability, induction, and saturation; and number of shield layers.

There are two basic types of magnetic shields: lossy shields and flux-entrapment shields. As discussed in Chapter 1, lossy magnetic shielding depends on the eddy current losses that occur within highly conductive materials, such as copper, aluminum, iron, steel, and silicon-iron. When a conductive material is subjected to a time-varying (AC) magnetic field, currents are induced within the material that flow in closed circular paths, which are perpendicular to the inducing field. According to Lenz's law, these eddy currents oppose the changes in the inducing field, therefore, the magnetic fields produced by the circulating eddy currents attempt to cancel the larger external inducing magnetic fields near the conductive surface, thereby generating a shielding effect (Chatterton, 1992; VitaTech Engineering, 2000).

A flux-entrapment shield is constructed with soft ferromagnetic, highly permeable magnetic shielding materials (they themselves do not develop permanent magnetization), such as nickel iron alloys, Hipernom alloy, Aumetal, and Mumetal, which either surrounds (cylinder or rectangular box) or separates (U shaped or flat plate) the area from the magnetic source. Closed shapes—cylinders with caps, boxes with covers, and similar shapes—are the most efficient for magnetic shielding. Ideally, when magnetic flux lines encounter the flux entrapment shield, they prefer to enter the highly permeable material, traveling inside the material via the path of least magnetic reluctance, rather than passing into the protected (shielded) space (VitaTech Engineering, 2000).

3.2.2 Magnetic Shielding Materials Selection

As the absorption loss is the primary shielding mechanism to shield low frequency magnetic fields, a magnetic material has the ability to increase the absorption loss. It is worth noting that using a magnetic material would decrease the shielding in the case of low-frequency electric fields or plane waves, since the primary shielding mechanism is reflection. When magnetic materials are considered as a shield, some overlooked properties must be taken into account. Such as (Ott, 1988):

- The permeability decreases as frequency increases. The permeability values given for magnetic materials in Table 3.2 are static or DC permeabilities. For example, the permeability of Mumetal is no better than cold-rolled steel at 100 kHz, even though the DC permeability of Mumetal is thirteen times that of cold-rolled steel. High permeability materials are most useful as magnetic field shields at frequencies below 10 kHz.
- Permeability depends on field strength. Maximum permeability, and therefore magnetic shielding, occurs at a medium level of field strength. The effect at high field strengths is due to saturation, which varies depending on the type of material and its thickness. To overcome the saturation phenomenon, multilayer magnetic shields are usually used.
- Machining, working, or heat treating high permeability materials, such as Mumetal, may change their magnetic properties.

Magnetic shielding materials are, therefore, chosen mainly based on their discrete characteristics, especially with respect to permeability and saturation. Because of their effectiveness in redirecting low-frequency magnetic fields, high permeability materials such as 80 wt% nickel–iron alloys like Mumetal, are the most commonly employed shielding materials. These alloys conform to MIL-N-14411C, composition 1; and ASTM A753-97, Type4. They are usually fabricated as foils or sheets and are annealed at around 1093°C in a dry hydrogen-rich atmosphere to improve the material's shielding effectiveness. These high permeability materials are typically employed when maximum magnetic field reduction is required in the minimum amount of space. While these materials are often the appropriate choice, there are cases when they may offer more shielding than is required or when the strength of the magnetic fields (typically in the case of elevated DC fields) requires a material with higher saturation values (Green, 2001; Maltin and Kamens, 2001).

When the frequency is low (50 Hz or below), it is difficult to achieve good shielding effectiveness with any reasonable thickness of ordinary metals. High permeability materials such as Mumetal and Radiometal generally have relative permeability over 10,000. Their skin depth is correspondingly very small, but they are only effective up to a few tens of kilohertz.

When the shielding objective requires only a small reduction in field strength (one to four times reduction) or when the field strength is intense enough to saturate a high permeability shield, ultra low carbon steels (ULCS) can be the ideal choice. These lower cost materials typically have a carbon content of less than 0.01%, giving them a relatively high permeability (close to 1000) compared to other steels, and excellent saturation characteristics. This material is less brittle and easier to form than silicon steels, allowing for easy installation in large-area shielding projects and for fabrication into small components. ULCS can be used together with high permeability materials to create optimum shields for applications requiring both high saturation protection and high levels of attenuation (Maltin and Kamens, 2001).

All metallic shielding materials with relative permeability greater than one can saturate in intense magnetic fields, but then they don't work well as shields often heat up. A steel or Mumetal shield box over a main transformer to reduce its hum fields, for example, can saturate and fail to achieve the desired effect. It is usually necessary to make the box larger so it does not experience such intense local fields.

For shielding applications at cryogenic temperatures, Cryoperm 10 (a registered trademark of Vacuumschmelze GmbH, Hanau, Germany), like Mumetal, is a high permeability nickel–iron alloy that is specially processed to provide increasing permeabilities with decreasing temperatures. Standard magnetic shielding alloys, like Mumetal, lose a significant percentage of their permeability at cryogenic temperatures. However, the permeability of Cryoperm 10 increases in the range of 77.3 to 4.2 degree Kelvin (Maltin and Kamens, 2001).

As mentioned earlier, typical magnetic shielding alloys such as Mumetal, can reach their optimum permeability only after they undergo a special heat-treatment cycle called hydrogen annealing. The annealing should always be done once all fabrication has been completed, because any shock or vibration to the shield after annealing will degrade the materials' performance. Strict adherence to the special annealing cycle insures optimum magnetic shielding characteristics, as the material's permeability can be increased by forty times on average compared with that of material before annealing. In addition, since the high permeability material costs can account for 50% of a shield's price, it is an advantage to use the thinnest gauge that is still able to yield the required shielding and structural performance (Maltin and Kamens, 2001).

To improve performance of current magnetic materials, many novel materials have been developed. For instance, a meterial is designed based on the concept that the magnetic field is actually "deflected" away from the shielded area. This material is made with a mixture of coal slag, silica, silver, calcium, magnesium and zinc powders. As all constituents of the material are nonferrous, one of its advantages is that it has no saturation limitation when using it in high strength magnetic fields (Gordon and May, 2007).

3.2.3 Magnetic Shielding Enclosure Design Consideration

Magnetic shielding materials offer a very high permeability path for magnetic field lines, directing the electromagnetic energy through the thickness of the alloy, and keeping the electromagnetic field from going where it is not wanted. It is important that the shield offers a complete path for the field lines, so that they do not exit the material in a place where they will cause unintended interference.

However, most magnetic shielding formulas and models were developed based on the theoretical geometry of a sphere or an infinitely long cylinder. In real-world applications, the actual geometry of a given shield will be driven by device configuration and the space available for the shield itself. Figure 3.4 shows some schematic examples of poor design and good design. One of the most important constructs to consider when designing a magnetic shield is that it is difficult for magnetic flux lines to turn 90 degrees. Therefore, rounded shields, such as cylinders or boxes with round corners, are better at redirecting lines of flux than shields with square corners. Similarly, gentle radii are better than sharp corners to contain and redirect flux that

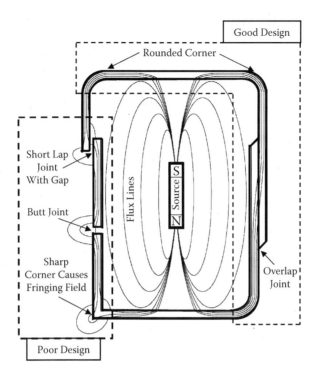

FIGURE 3.4 Examples of poor design and good design for magnetic shielding design.

is already entrapped in the shielding material. It is important to keep the shape of a shield simple, focusing on providing a low-reluctance path, or path of least resistance for the magnetic fields to travel. Wherever possible, a shield should be closed on all sides to avoid field leakage. This configuration, even if rectangular, most closely approximates a sphere and creates a closed magnetic circuit. In addition, complete closure provides shielding in all axes and guarantees the highest possible shielding performance. Removable covers, lids, and doors can be incorporated into the shield design when both exceptional performance and access are required.

If a shield cannot be closed on one or both ends, the ratio of the shield's length to diameter of the open end and it is protected should be at least 4:1 to avoid "end effect" or flux penetration into the critical shielding region. A common rule is that a shield needs to overhang the device by an amount equal to the radius of the opening. Wherever possible pro-flanges and overlapped joints should be used to provide a high level of magnetic continuity at a significant cost savings as compares to heliarc welding (Maltin and Kamens, 2001).

Size plays an important role in the effectiveness and the cost of a shield. The smaller the effective radius of a shield, the better its overall performance. Therefore, the objective is to design a shield that will envelop the component or space to shield as closely as possible. Because high permeability material is a major cost component in shield design, a smaller shield has the benefit of yielding better performance at a lower cost. In many applications, a single-layer shield cannot provide either the level of attenuation or saturation protection required. In these cases, multilayer or nest-typed shields are employed. Nesting two or more high permeability shields within one another, utilizing air gaps provided by the spacing between them, results in excellent shielding factors. The more demanding the shielding objective, the more layers that may be required (Maltin and Kamens, 2001; Ott, 1988).

When shields need to operate in very high intensity magnetic field environments, such as those in close proximity to electric arc furnaces or large superconducting magnets, the saturation of materials is a concern. In these cases, an operation is to use nested shields constructed from different composition materials. The layer closest to the highest field levels should be fabricated from a low permeability, high saturation material. This buffering shield and the added attenuation due to the spacing of the air gap should reduce the source field to a level where it is safe to use high permeability materials without the fear of saturation.

3.3 SHIELDING ENCLOSURE INTEGRITY

An ideal shielding enclosure would be a large, thick, hollow sphere of iron, with gold plating on the outside and no seams or openings, as shown in Figure 3.5. Unfortunately this precludes the use of power cords, data cables, displays, ventilation openings, or peripheral devices. Therefore, a perfect shield is not practical. Using the optimal design techniques and carefully selected materials, however, it is not difficult to produce an enclosure that provides around 100 dB of shielding effectiveness across a wide frequency spectrum from DC to light for electric and electromagnetic fields. Moreover, the low-frequency magnetic field screening should be specially considered. In fact, theoretical shielding specifications can never be achieved because of a loss of shielding integrity due to a number of items that need access to the enclosure. Taking the desktop computer system as an example from Figure 3.5, the following

FIGURE 3.5 An ideal sphere shell enclosure and a practical enclosure integrity of a desktop computer system.

EMI Shielding Enclosure and Access

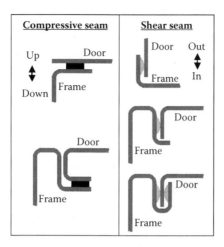

FIGURE 3.6 Typical gasketed seams.

shielding integrity should be given proper consideration in each shielding design and assembly: cover plates, ventilation and cooling apertures, indicator lamps, switches, cathode ray tube (CRT) and display widows, power controls, Internet cables and connectors, keyboard openings, etc.

From a performance perspective, shielded enclosures should be rigid and made of thick metal. However, regarding the factors of cost, weight, size, convenience, and aesthetics, enclosures are usually made from thin sheet metal or conductive-coated plastic and have seams. And in many cases, these materials are flexible, and the seams can leak EMI. The best seams are made by welding, brazing, soldering and riveting. Gaskets are used to create the same level of shielding in a separable joint. For instance, the cover plates for access to equipment for maintenance, adjustment, or alignment of equipment, need to be removed from time to time and the most usual technique, when metal cover plates are involved, is using a resilient gasket system. Alternatively, knitted wire mesh or a metal particle loaded silicone gasket configuration can be used. Some typical types of such gasket seams are shown in Figure 3.6. Cycle life is important parameter for a resilient gasket product. How many times is a gasket expected to be compressed and then return to its original position? High cycles could be 100,000 times or more; low cycles could be once. Gasket failure or damage can occur from excessive cycling. Damage can include breakage, fraying, flaking, or wear of protective surface. Moreover, improper design or installation can cause gasket failure, and compression set of elastomer or foam core products may cause a decrease in gasket performance. Where very occasional access is required, a conductive caulking compound can be considered. When the cover or access plate is removed, the existing conductive caulk needs to be scraped off to reveal a clean mating surface and fresh calk carefully applied to preserve the integrity of the joint when the cover is replaced (Molyneux-Child, 1997).

However, anything conductive that penetrates a shielded enclosure can create leaks. Typical latch or hinge hardware is conductive and makes a good antenna. Moreover,

Panels / Characters	Thin Film Reflective Surface	Knitted Mesh	Woven Screen	Perforated Metal	Honeycomb	Solid Metal
Shielding	10MHz	100MHz	200MHz	400MHz	18GHz	Maximum
Viewing	Excellent	Poor	Good	Fair	Poor	None
Rigidity	Excellent	None	None	Fair	Excellent	Excellent
Ventilation	None	Good	Good	Better	Excellent	None

FIGURE 3.7 Typical shielding panels used for EMI enclosure.

no matter how well an enclosure is designed, its shielding effectiveness can be compromised by using the wrong hardware or installation techniques. For instance, while shielding at board level or enclosing a signal generator will reduce the amount of shielding required on external covers, the electrical conductivity, construction, aperture, and type of gasket affect shielding effectiveness. The higher the frequency, the smaller the wavelength. Therefore, the high frequency signal will leak through small openings. Shielding products that have openings (apertures) include metal finger stock, wire mesh, honeycomb vents, and board-level shields (BLS). Some shielding products, such as fabric-over-foam (FOF), electrically conductive elastomer (ECE), and form-in-place (FIP) have very small or no openings, which would result in very small or no leaking for high frequency signals and would provide better environmental sealing.

The need for viewing panel meters, scopes, and digital displays, together with other types of monitoring equipment and information presentations can also be parts of the shielding integrity of an enclosure. Figure 3.7 shows typical shielding panels used for EMI enclosure integrity. For example, cooling apertures primarily include honeycomb vent systems and the cheaper wire mesh arrangements. Frequently, shielded housings or enclosures require either convection or forced-air cooling, and since associated openings compromise the integrity of the basic shield material, a suitable electromagnetic mask must be sought that will provide substantial EMI attenuation while not significantly impeding the mechanical flow of air. A honeycomb vent is unable to remove dust or even produce a reduction in dust levels in the process of ventilation; as an alternative, the screen is made from either woven wire mesh or corrugated metal mesh. This type of construction can even incorporate an oil coating to improve its dust-removal effectiveness. The EMI filter and dust shield perform three functions: attenuate EMI, provide for a passage of cooling air, and filter dust from the airflow. It is designed for use in industrial environments at low to medium airflow rates and can be cleaned by either inversion or agitation in a suitable solvent or by air blasting. The filter itself can be formed from multiple layers of corrugated aluminum mesh aligned to provide maximum dust trapping effectiveness and minimum resistance to airflow. The filter material is contained in an aluminum channel frame of either all welded construction or made from lightweight formed channel (Molyneux-Child, 1997). For some shielding applications it may be possible to stretch a simple, knitted wire mesh tape over the ventilation aperture and secure this in place with a conductive epoxy adhesive, thus directly electrically bonding the mesh to the clean conductive enclosure wall. A problem with this approach is that the fairly fragile mesh layer may at some stage become damaged or deformed. To obviate this occurrence an external mechanical protection in the form of a metal grill may have to be fitted in addition.

Another approach is use of thin, sputtered metallic coatings on shielding enclosure or panels. Coatings can be deposited on flexible polymer film substrates, such as polyster, polyimide, or polycarbonate. Positively charged ions, attracted by the strong negative voltage of the targeted material are physically dislodged. The ejected high-energy atoms bond to the substrate, forming coating layers with precisely controlled properties. In a typical vacuum roll coaster, extremely precise thin layers of pure elements, alloys, and compounds can be applied at multiple coating stations. The entire process takes place in a tightly sealed chamber under high vacuum. The process can be used, in addition to produce conductive optical film, to coat fabrics, metal foils, meshes, and various flexible polymers. Low-temperature substrate vacuum processing affords coatings at relatively low cost and with an excellent quality, as continuous monitoring and feedback control is integral to the process. Virtually any metal, alloy, or a variety of compounds can be deposited using these vacuum deposition techniques, although gold and silver have been the preferred metals and alloys to date. The high energy of the deposited particles provides excellent adhesion to the substrates, together with good conductive film uniformity. When large areas need to be provided with a semitransparent shield, multilayer sputtered coatings may be used. These multilayer metal-sandwich coatings are considerably more economical than the gold coatings and, in many cases, provide comparable shielding. EMI shielding efficiency of sputtered thin film coatings is achieved mainly by the process of reflection, rather than by absorption, because of their highly conductive, metallic nature. As the frequency of the radiation increases, the shielding efficiency decreases (Molyneux-Child, 1997).

Shielding of displays is another specific situation, as the displays require apertures in enclosures. Mounting a display outside a shielded enclosure can avoid the aperture, but removes the benefit of the shielding from the display and creates the new problem of what to do with the display's data and power cables and their penetration of the enclosure. Therefore, shielding windows are needed where a display needs shielding by an enclosure although some high-grade CRTs can provide a good shield when the metal frame around the front of their tube is electrically bonded to the front panel all around the aperture. A variety of shielded windows are available: thin metal films on plastic sheets, usually indium-tin-oxide (ITO); embedded metal meshes, usually a fine mesh of blanketed copper wires; and honeycomb metal display screens for the requirement of very high shielding performance. A vital issue for screened windows is that their conducting layers (mesh, film, or honeycomb metal) must be bonded directly to the enclosure shield surface around all the edges of their cutout. Conductive sealants/glues can be used to avoid the need for mechanical fixings. The use of UV-curable conductive adhesives can make assembly times equal or better than mechanical fixing methods.

In addition, wherever metal control shafts or spindles penetrate the shielded enclosure integrity, they should be fitted with a conductive gland seal, frequently of compacted knitted wire mesh, and the spindle should also pass through a tightly fitting metal bush or tube.

3.4 SPECIALIZED MATERIALS USED FOR SHIELDING ENCLOSURE

Following is a summary of typical commercial materials applied to the EMI shielding enclosure integral system, which will be described in detail in following chapters.

3.4.1 EMI Gaskets

Table 3.4 lists the primary performance of typical commercial EMI gaskets used for shielded enclosure integral systems (Fenical, 2005), and Figure 3.8 and Figure 3.9 illustrate the shielding effectiveness ranges of EMI gaskets (Instrument Specialties, 1998).

3.4.1.1 Metal Strip Gaskets

This is a traditional established gasketing technique, using spring metal strips such as CuBe, CuNiSn, stainless steels, and phosphor bronzes stamped into a wide range of configurations. The design is such that each spring finger of the gasket is capable of being flexed to a different degree and hence accommodates a considerable amount of joint unevenness. The surface finish of the spring gaskets are provided either as untreated, bright clean, or coated with zinc, nickel, silver, gold, or chromated aluminum to conform the contact surfaces with the galvanic compatibility. Depending on the

TABLE 3.4
Typical EMI Gaskets Used for EMI Shielding Enclosures

Material Parameter	CuBe Finger	CuBe Mesh	Monel Mesh	FOF	ECE
Shielding effectiveness 10k–10GHz	100–120 dB	60–115 dB	40–95 dB	75–100 dB	35–120 dB
Compression range	20%–80%	15%–75%	5%–15%	20%–60%	5%–20%
Typical closure pressure	0.3 kg/cm^2	0.15 kg/cm^2	1.41 kg/cm^2	0.15 kg/cm^2	3.5–7 kg/cm^2
Usable in shear	Yes	No	No	Limited	No
Compression set	Very low	Low	Medium	Low	Highest
Flame retardant	Yes	Yes	Yes	Yes	Yes
Fluid seal	No	No	No	No	Excellent
Surface area	Minimum	Medium	Maximum	Medium	Medium
Weight	Light	Light	Medium	Light	Heaviest
Gas permeable	No	No	No	Yes	Yes

Source: Modified from Fenical 2005.

EMI Shielding Enclosure and Access

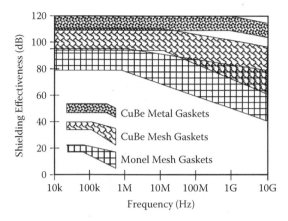

FIGURE 3.8 The shielding effectiveness of CuBe strip, CuBe mesh, and Monel mesh gaskets.

mounting techniques used, there is an enormous number of profiles with clip-on styles, self-adhesive strip mounted types, riveting finger in place, and twisted configurations. Parts can be silver soldered, or other soldering techniques can be used. Metal finger stock is best for frequent cycling applications with high shielding effectiveness, useable in wiping closures, large deflection ranges, and a variety of finishes to provide good solderability and galvanic compatibility.

3.4.1.2 Knitted Metal Wire Gaskets

The commercial knitted wire gaskets mainly include CuBe and Monel wire knitted mesh gaskets. Gaskets can be knitted using a variety of wire materials with low costs, environmental gaskets are available, and there are a variety of mounting styles.

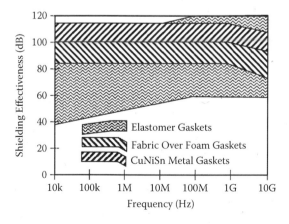

FIGURE 3.9 The shielding effectiveness of elastomer, fabric over foam, and CuNiSn strip gaskets.

3.4.1.3 Metal-Plated Fabric-over-Foam (FOF)

Metal-plated fabric-over-foam (FOF) is a resilient urethane foam wrapped in coated or metallized nylon fabric. The flexible outer coating conforms to the tiniest irregularities in a metal mating surface, with a claimed 60% to 70% more surface contact than wire mesh (even with an elastomeric core) and 40% more contact than metal strip gaskets. Useful environmental sealing performance is assured by the core of urethane foam. Its multidimensional cellular structure slips into gaps and spaces. The dense molecular mass of an elastomer, on the contrary, moves away from a compressive load, and unlike loaded elastomers, the soft foam makes cabinet doors and hatches easy to open, close, and latch. Another development of this kind of material is the conductive fabric over a closed-cell thermoplastic elastomer (TPE; also called thermoplastic rubber) extruded foam. EMI shielding is provided by a woven polyester fabric plated with a copper layer with an overcoating of nickel for corrosion and abrasion resistance. The complete gasket offers excellent surface conductivity, shielding effectiveness, and resistance to moisture and salt spray. The closing force is a function of density and geometry of the profile. Open-cell urethane foam and closed-cell neoprene foam are available as alternative core fabrics and can be changed into a nonwoven material. There is almost no limitation in custom design profiles including full or partial wrap of EMI fabric over the foam core. The advantages of FOF gaskets include low compression force, good appearance, good shielding effectiveness, and snag resistance (Molyneux-Child, 1997; Instrument Specialties, 1998).

3.4.1.4 Electrically Conductive Elastomer (ECE) Gaskets

Electrically conductive elastomer systems are normally loaded with metal particles and used as extruded profiles, thin sheet, die-cut sheet, screen printed gaskets, molding in place, and forming in place. Other types in use are elastomers loaded with oriented wires and metal meshes. ECE provides an extensive range of gasketing solutions. In general, ECE is best for miniature devices or where environmental sealing is required. ECE provides high levels of shielding effectiveness, excellent environmental sealing, a wide variety of fillers for optimum galvanic corrosion and electrical performance, and a wide variety of polymer systems for optimum mechanical and fluid properties. ECE also qualifies for nuclear, biological, and chemical (NBC) environments, and is flame retardant (Instrument Specialties, 1998).

3.4.2 MAGNETIC SCREENING MATERIALS

Magnetic screens are fabricated from high permeability alloys, either protecting the susceptor from stray magnetic fields, for example cathode ray tubes, or preventing the EMI emission from products such as transformers. Screening usually utilizes nickel-based alloys, such as Mumetal (80% Ni alloy), Radiometal 4550 (50% Ni alloy), and Radiometal 36 (36% Ni). Shields are normally formed from blanks stamped from rolled metal sheet. Any holes or cutouts are made in the blank, and the shield is then formed around a tool. The formed blank may then be overlapped or seam welded. An alternative method of fabrication is to manufacture deep drawn cans, which provide the best type of shield since there are no joints. However, sizes of cans are limited since nickel–iron alloys work-harden rapidly during the drawing process (Molyneux-Child, 1997).

3.4.3 Shielding Tapes

For small components or for small quantity runs where shielding is not very critical, wraparound techniques may be used with thin shielding tapes. The method to be employed is for the component to be wrapped and actual shielding performance measured. Then the modification is easily made if necessary, however this type of approach is not likely to achieve particularly consistent shielding results. There have been several types of shielding tapes (such as plan, plated, or laminated metal tapes), die-cuttapes, metallized cloth tape, painted tape, and wire mesh tape. Gasketing-type wire alloys are usually used for mesh tapes and these include Monel, tinned copper, stainless steel, copper, silver-plated copper, aluminum, and tinned copper clad steel.

3.4.4 Thermoformable Alloys

Thermoformable alloys typically have a very high proportion of zinc, normally 78% zinc and 22% aluminum. At room temperature, the alloy is hard and strong, and yet when heated to 220°C–270°C, it becomes very pliable and can be drawn to more than 1500% of its original size. After molding, it displays the intrinsic hardness and strength that it had prior to thermoforming. After thermoforming, the parts can be plated, painted, buffed, or left as the formed finish. Parts made by this process are readily solderable and weldable. These alloys are ideal for providing shielded covers, enclosures, and other EMI shielded or screened parts in relatively small volume runs, entailing low tooling costs and affording engineering design flexibility. Thermoformable alloys can be formed in two ways. One, the highly formable alloy is usually thermoformed on aluminum molds. Using this process the shielding components can be made by plastic injection molding and is cost effective for low volume runs. Or when using vacuum forming processes, there are a number of design limitations to be considered and most parts for economic production necessitate a minimum draft angle of about 1%. Inside radii should be designed to coincide with sheet material thickness selected, and minimum outside radii should be twice the sheet thickness, as should corner radii (Molyneux-Child, 1997).

3.4.5 Honeycomb Materials

A honeycomb material is generally used because it provides high shielding effectiveness and streamlined airflow for an enclosure with large openings. In the typical honeycomb construction, the hexagonal elements use the waveguide effect to provide the desired shielding effectiveness. Meanwhile, the honeycomb shield openings can be used for heat dissipation, ventilation, and lighting from the passage of undesirable EMI signals. Other advantages of the material include light weight, low resistance to airflow, high shielding efficiency, and reduced turbulence. In addition, in situations where aluminum honeycomb panels may be vulnerable to accidental damage, panels can be fitted with either flattened expanded metal or heavy woven mesh built into one, or both, faces. A variety of gasket types can be used to provide an EMI seal between housing and the aluminum frame of the vent; a gasket of knitted wire mesh and silicone sponge is quite usual (Molyneux-Child, 1997).

3.4.6 Painted or Plated Plastic Enclosures

Painted or plated plastic enclosures have been used for a pleasing appearance with reasonable shielding effectiveness, which is achieved by surface coating with conductive paint or metal plating. The weak points for effective shielding are usually the seams between the plastic parts. They often cannot ensure a leak-tight fit and are usually hard to be gasketed. In addition, paint or plating thickness on plastic is generally limited; therefore, the number of skin depths achieved can be quite small. Nickel and other permeable metals have been developed to take advantage of their reasonably high permeability to reduce skin depth and achieve better shielding effectiveness. Other practical problems with painting and plating include making them stick to the plastic substrate over the life of the product in its intended environment. Conductive paint or flaking off inside a product can short out conductors, causing unreliable operation and risking fire and electrocution. A special problem with painting or plating plastics is voltage isolation. For any plastic-cased product, adding a conductive layer to the internal surface of the case can encourage potential electrostatic discharge (ESD) to occur through seams and joints, possibly replacing the EMI problem with one of susceptibility to ESD. For commercial reasons it is important that careful design of the plastic enclosure occurs from the beginning of the design process if there is any possibility that shielding might eventually be required (Molyneux-Child, 1997).

3.4.7 Conductive Composite Shielding Materials

Volume-conductive plastics or resins generally use distributed conductive particles or threads in an insulating binder, which provide the mechanical strength and shielding effectiveness for an enclosure design. Sometimes insulating the skin of the basic plastic or resin makes it difficult to achieve good conductive continuity, to prevent long apertures being created at joints, and to provide good bonds to the bodies of connectors, glands, and filters. Problems with the consistency of mixing conductive particles and polymers can make enclosures weak in some areas and lacking in shielding in others. However, some novel conductive composite materials have been developed, showing promising performance for shielding enclosure applications.

3.5 SUMMARY

Proper system enclosure design can eliminate EMI, and such a design process consists of following EMI standards, defining goals and EMI parameters, choosing the proper materials, and the judicial use of gaskets. Depending on the equipment that goes into the enclosure and the range of frequencies of the electromagnetic waves, the enclosure needs to meet specified performance objectives. There are many possible design and performance objectives in different kinds of environments, such as shielding effectiveness, galvanic compatibility, longevity of gaskets, easy and less expensive maintenance, and easy accessibility to the equipment inside the enclosure.

There are many factors in selecting material for the enclosure. First, enclosure materials must meet electrical and mechanical requirements. However, sometimes marketing decisions concerning appearance rather than good engineering practice may dictate material selection. Therefore, it is important to carefully balance system requirements, performance, EMI protection, and overall cost.

Magnetic shielding materials offer a very high permeability path for magnetic field lines to travel through, directing them through the thickness of the shielding alloy and keeping them from going where they are not wanted. It is important that the shield should offer a complete path for the field lines, so that they do not exit the material in a place where they will cause unintended interference.

Shielded enclosures are normally designed for emission suppression or to limit susceptibility to electromagnetic radiations. Enclosures containing emitters should be designed for maximum absorption loss of the emitted field. Enclosures containing receptors should be designed for maximum reflection loss of the received field.

Discontinuities should be minimized and designed to maintain shielding effectiveness by proper seam treatment, and use of shielding gaskets, vent/window screening, and shielding/filtered leads. A variety of shielding materials and processing technologies have been selected and used for EMI shielding enclosure integral systems.

REFERENCES

Beaty, H. W. 1990. *Electrical Engineering Materials Reference Guide.* New York: McGraw-Hill.
Chatterton P. A., and M. A. Houlden. 1992. *EMC electromagnetic theory to practical design.* Chichester, UK: John Wiley & Sons.
English. J. 2007. Addressing EMI shielding problems with specialty engineered materials. *Compliance Engineering.* http://www.ce-mag.com/archive/02/07/English.html
Evans, R. W. 1997. *Design guidelines for shielding effectiveness, current carrying capability, and the enhancement of conductivity of composite materials.* NASA Contractor Report 4784.
Fenical, G. 2005. Ten considerations when specifying an RF gasket. *Interference Technology.* Emc Test & Design Guide, pp. 64–72. http://www.interferencetechnology.com/ArchivedArticles/shielding_aids/01_tdg_05.pdf
Fitzpatrick, R. 2002. *Magnetic shielding.* http://farside.ph.utexas.edu/teaching/jk1/lectures/hode52.html
Gordon, W., and W. May. 2007. Deflecting magnetic field shield. US Patent 7220488. http://www.freepatentsonline.com/7220488.html
Green, A. H. 2001. *Shielding magnetic fields.* Magnetic Shield Corp. http://www.tipmagazine.com/tip/INPHFA/vol-7/iss-5/p24.pdf
Instrument Specialties. 1998. *Engineering Design and Shielding Product Selection Guide* (acquired by Laird Technologies). Delaware Water Gap, PA: Instrument Specialties.
Maltin, L., and A. Kamens. 2001. *Magnetic shielding theory and practice: A guide to cost-effective magnetic shield design.* http://www.interferencetechnology.com/ArchivedArticles/magnetic_shielding/Magnetic%20shielding%20theory%20and%20practice.pdf?regid=.
Molyneux-Child, J. W. 1997. *EMC Shielding Materials.* Oxford: Newnes.
Ott, H. 1988. *Noise Reduction Techniques in Electronic Systems,* 2nd ed. New York: John Wiley & Sons.
VitaTech Engineering. 2000. *Magnetic shielding designs.* http://www.vitatech.net/magnetic-shielding.php4

4 Metal-Formed EMI Gaskets and Connectors

This chapter will introduce metal strip selection; typical metal and alloy for electromagnetic interference (EMI) gasket and connector production; a design guideline for gaskets and connectors; and the fabrication process and mounting methods.

4.1 INTRODUCTION

Mechanically formed metal gaskets can be dated back to the 1940s when spring clips were used to ground chassis in electronic equipment (Hudak, 2000). They have been widely used for EMI applications with high-traffic door opening and closing under compression or shear forces, wiping applications, and elevated temperature environments. The appeal of these gaskets results from their robust construction with various configurations and high level of shielding effectiveness. Metal-formed gaskets are usually made from high strength and conductivity spring alloys such as copper beryllium (CuBe), stainless steels, and copper–nickel–tin (CuNiSn) spinodal alloys. Gaskets made from CuBe alloys are most preferred due to their elastic resilience and highest electrical conductivity of all the conductive spring materials. Comparably, stainless steel provides a low-cost alternative to CuBe gaskets, which is particularly attractive in the commercial electronic industry. However, stainless steel gaskets are not recommended for high-cycling applications because stainless steel lacks the resilient spring properties of CuBe. They are suited for use in static joints and limited-access panels. In addition, spinodal CuNiSn alloys have been developed for making EMI gaskets and provide much better shielding effectiveness and spring performance than stainless steel.

The shielding effectiveness of metal gaskets typically ranges from a high level of 120 dB for CuBe parts, to 100–110 dB for CuNiSn parts, to 60–70 dB for stainless steel parts. Shielding performance of metal gaskets tends to fall off at high frequencies due to the slots or openings intrinsic to their construction. The lower levels of conductivity will affect the shielding of stainless steel gaskets to a higher degree than other gasket types. Depending on metal strip thickness and gasket profiles, compression forces applied on the gaskets generally range from 15 to 65 kg per linear meter during application. Usable compression height for CuBe gaskets ranges from 20% to 80% of the free height dimension. Compression set values of metal gaskets range from an industry-leading figure of less than 1% for CuBe up to 25%–40% for some grades of stainless steel when used within the gaskets' specified ranges of compression. Overcompressing any metal-formed gasket greatly increases its compression set (Hudak, 2000).

4.2 METAL STRIP SELECTION AND PERFORMANCE REQUIREMENT

Most metal strips available for EMI gasket production have a thickness range of 0.05 to 0.3 mm. When choosing a metal strip, major factors to be considered include property/performance, manufacturability, and commercial attributes, as shown in Table 4.1. Gasket spring materials require various properties, such as responsiveness to formation of complex configurations peculiar to various assemblies; initial strength to reach contact pressures above certain levels in contact areas; ability to maintain these pressures in the long term; and fatigue resistance for high-cycling applications. Specifically, these properties include yield strength and fatigue strength; elastic resilience (ratio of yield strength over elastic modulus); electrical conductivity; stress-relaxation resistance; and bending workability.

High spring strength is required to obtain high contact pressures; the index of this is yield strength. In addition, spring-bending elastic limit and Vickers hardness can be used for evaluating spring-strength properties (Takano et al. 2006). Relatively high fatigue strength is also required to obtain a high-cycling life. When a spring gasket is downsized, the cross-sectional area of the current-carrying portion becomes smaller and the temperature rises due to heat generation. Temperature rise has the potential to cause a reduction in the contact pressure due to stress relaxation. Therefore, electrical conductivity becomes a more important property with downsizing of spring gaskets.

As mentioned in Chapter 2, stress relaxation is the phenomenon of gradual reduction in the stress generated in a material with time under a fixed strain (deformation). A major accelerating factor for stress relaxation is temperature. When in use, a spring gasket is always subject to a constant strain under a contact pressure. When its stress-relaxation resistance property is poor, the contact pressure decreases with time, sometimes leading to malconduction and loss of shielding effectiveness. Spring gaskets can be exposed to elevated temperatures, in automotive applications in

TABLE 4.1
Major Factors for Metal Strip Selection

Properties/Performance	Manufacturability	Commercial Attribute
Elastic resilience	Mechanical formability	Price and maintaining cost
Toughness and ductility	Heat treatment	Temper availability
Shielding	Chemical thinning	Gauge and width availability
Plating capability	Plating ability	Supply and delivery
Stress relaxation resistance	Dimension tolerance	Operating environment
Fatigue life	Oxide cleaning	Environmental compliance
Wear life	Photoetching	Recyclability
Conductivity	Solderability	Econometrics
Galvanic compatibility	Weldability	Aesthetics
Tarnishing & surface appearance	Polishing & surface finishing	
Corrosion resistance		

particular, and careful selection of materials is required in such severe environments (Takano et al., 2006).

Because of the complexity of their shapes, spring gaskets usually need to experience extreme bending during manufacturing; therefore, it is necessary for metal strips to have excellent bending workability in various bending directions. Consequently, improved bending workability and suppression of anisotropy to low levels are generally required for spring gasket materials.

Spring metal strips must be strong, but strong materials are usually difficult to form, especially for mill-hard or non-heat-treatable alloys. In most cases, miniature gasket parts must be formable, but formable materials are usually too weak to provide high enough elastic resilience. Finding a material for a miniature spring gasket usually requires a compromise between strength and formability. Strength versus formability gives the designer a way to visualize mechanical properties and to identify a material with the correct mix of properties.

In addition, a metal strip's surface appearance, platability to improve its solderability and galvanic contact compatibility, and environmental compliance and recyclability to conform the RoHs and WEEE regulations have become more important factors in metal strip selection.

4.3 COPPER BERYLLIUM ALLOYS

Copper beryllium (CuBe) alloys are among the strongest and most conductive of all conductive spring alloys. This is attributable to the optimally synthesized microstructure, and properties via interaction of composition, cold work, and thermal processes of CuBe alloys. Small beryllium additions to copper, activated by mechanical and thermal processing, result in strength exceeding that of most copper-base alloys and many hardened steels. Furthermore, controlled mechanical working and specific thermal processing allow properties to be tailored to meet a broad range of requirements (Brush Wellman, 2002).

4.3.1 Phase Constitution and Primary Processing of CuBe Strips

CuBe alloys can be basically categorized as two types: high strength and high conductivity. With increasing beryllium content, the strength increases, while the electrical and thermal conductivity decreases. A binary equilibrium phase diagram is shown in Figure 4.1, approximating the behavior of high-strength alloys (Brush Wellman 2002). These high-strength alloys usually contain up to 0.6 wt% total cobalt and nickel in addition to beryllium; therefore, the binary phase diagram here is not rigorously appropriate but it helps in understanding the alloy behavior. At concentrations of 1.6 to 2.0 wt% beryllium, a beryllium rich γ phase is present below 593°C (1100°F), which precipitates from a solid solution of beryllium (α-phase). In these alloys precipitation hardening is the primary contributor to their strengthening. Before metal forming, the alloy can be tempered by a solution annealing process, that is, heating above 704°C (1300°F) causes beryllium to dissolve in the alpha solid solution phase, and then a rapid quench to room temperature maintains beryllium in the solid solution. This process makes the alloy soft, ductile, and good for metal forming; helps regulate grain size; and prepares the alloy for age hardening. After metal forming, the alloy can be

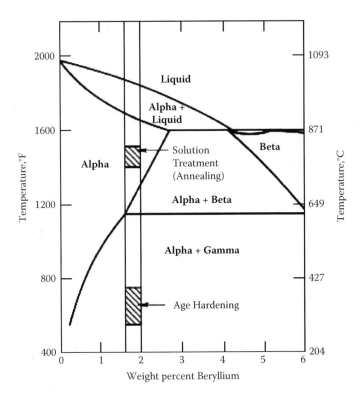

FIGURE 4.1 CuBe binary equilibrium phase diagram approximating the behavior of high strength CuBe alloys. (Source: Brush Wellman, 2002.)

hardened by heating the supersaturated solution to 316°C (600°F) and holding for two to three hours to precipitate the strengthening γ phase. In addition, the phase diagram shows that the solubility limit is 2.7 wt% (the cobalt addition reduces this to about 2.3 wt%) at 866°C (1590°F). And less than 0.25 wt% beryllium is soluble at room temperature. This difference in solubility is the driving force for effective hardening. In some high-strength alloys, the addition of cobalt is used to promote a fine grain size by limiting grain growth during solution annealing, and makes precipitation hardening less time sensitive, therefore increasing the maximum strength slightly.

The high-conductivity CuBe alloys contain much less beryllium, spanning the range from 0.15 to 0.7 wt% beryllium. As an example, Figure 4.2 shows a pseudobinary phase diagram (nickel beryllide in copper) for high-conductivity CuBe alloy 3 (Brush Wellman, 2002). In these alloys most of the beryllium is portioned to beryllide intermetallics. Coarse beryllides are formed during solidification and limit grain growth during annealing; fine beryllides are formed during precipitation hardening and impart strength. The temperature ranges for solution annealing and age hardening are higher for these alloys than for the high-strength alloys. In these high-conductivity alloys, the stability of the strengthening phase at elevated temperatures results in high resistance to creep and stress relaxation.

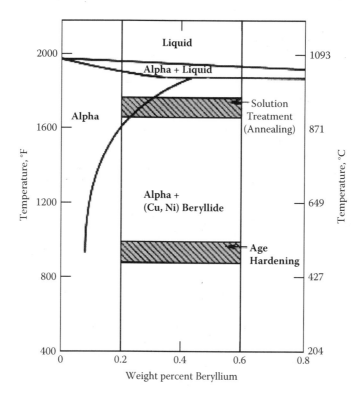

FIGURE 4.2 CuBe binary equilibrium phase diagram approximating the behavior of high conductivity CuBe alloys. (Source: Brush Wellman, 2002.)

The primary processing for CuBe alloy strips usually includes metal working, age hardening, surface cleaning, electroplating, and soldering.

Metal working is usually used to harden and change the dimensions of the metal strip, with rolling, drawing, bending, and upsetting being prime examples of cold-working operations. In contrast to hot working, cold working elongates grains in the working direction, and deformation effects accumulate.

Age hardening response depends on time, temperature, and amount of cold work because strengthening is governed by precipitate size and distribution. For each alloy there is a temperature–time combination that is designated as standard practice because it produces maximum strength. Departures from standard practice, either higher or lower temperatures for example, may be used to meet requirements that permit less than maximum strength or hardness. A temperature higher than the standard causes more rapid precipitation and thus faster strengthening, while a lower temperature results in a slower strengthening rate. Cessation of aging at times shorter than the time needed to achieve peak strength is known as underaging. Toughness, fatigue strength, and, in some cases, corrosion resistance benefit from the underaged microstructure. Overaging involves heating for a time longer than needed to achieve peak strength. This results in precipitate coarsening, and consequently causes hardness and strength below peak values. Overaging improves ductility, but there is evidence that toughness is reduced. Electrical and thermal conductivity and dimensional stability are maximized by overaging. Caution is required to avoid severe overaging. Age hardening of CuBe alloys normally does not require either controlled cooling or a special furnace atmosphere. A protective atmosphere is useful, however, especially when it is recirculated to reduce furnace thermal gradients. A forming gas atmosphere of 5% hydrogen in nitrogen is an example of one that economically aids heat transfer while minimizing posthardening cleaning requirements. Vacuum age hardening is difficult because of the non-uniform nature of radiant heating. Age hardening increases the density of the high strength CuBe alloys slightly as a result of the precipitation reaction. This density change is accompanied by a decrease in linear dimensions of approximately 0.15%–0.2%. The dimensional change in high-conductivity alloys is negligible for most applications. Fixtures may be used for age hardening to prevent distortion. A salt bath provides the precise control needed for short-time, high-temperature aging, which results in minimum distortion and the short production cycle. Residual stress, which may arise from certain types of deformation after age hardening, may be thermally relieved without loss of hardness. Heating at temperatures of 149°C–204°C (300°F–400°F) for up to two hours is generally adequate to provide moderate stress relief (Brush Wellman, 2002).

Surface cleaning is usually taken as the first step in preparation of a CuBe strip for subsequent plating or soldering, including removal of surface oxidation and all soils, particularly oils and greases. Sulfur-bearing lubricants used during forming, if not moved quickly, can stain CuBe. Conventional cleaners, such as organic solvents and alkaline solutions, are normally adequate for removing oily residues. Vapor degreasing is especially effective for removing oils and greases. Trisodium phosphate and similar alkaline solutions, and ultrasonic or electrolytic agitation can supplement these media for best results. Cleaning solutions should be thoroughly rinsed from all surfaces.

Like all Cu alloys, CuBe can form a thin surface oxide, or tarnish, when exposed to air. Tarnish formation is accelerated by the presence of moisture and elevated temperatures. Oxidation normally results from heat treatment. Even when protective atmospheres are used, the formation of sufficient surface oxides could cause plating or soldering problems. Surface oxides take two forms: beryllium oxide, present on surfaces exposed to the high temperatures needed for solution annealing; and combinations of beryllium and copper oxides, present on parts after precipitation hardening. The surface of CuBe can be prepared for plating or soldering, or restored to its original, lustrous appearance, with the following procedure (Brush Wellman, 2002):

1. Immerse parts in a 75°C alkaline solution for up to five minutes to remove soils.
2. Immerse parts in a 50°C–55°C aqueous solution of 20–25 volume percent sulfuric acid plus 2–3 volume percent hydrogen peroxide. Immersion time should be long enough to remove dark coloration and provide the desired response or appearance.
3. Rinse thoroughly and dry.

Metal-Formed EMI Gaskets and Connectors

Electroplating, coloring, and polishing are common practices on CuBe alloys. Nickel, gold, silver, tin, chromium, copper, and other metals are commonly electroplated on CuBe. CuBe products can also be colored by all conventional techniques used for copper alloys. Satin black oxide to an artificial patina is an example. Wet brushing, buffing, and electropolishing are used to produce extremely fine surface finishes on CuBe. The best electropolishing results are obtained with a nitric acid/methanol electrolyte at −57°C (−70°F). A phosphoric acid, chromate electrolyte, can be used at room temperature, but this may leave certain intermetallic particles in relief. Phosphoric, nitric, and acetic acids can be mixed to produce a chemical polishing solution for use at 71°C (160°F).

Soldering is normally specified when the anticipated service temperature is below 149°C (300°F), and electrical and thermal continuity are difficult to insure with a mechanical joint. CuBe can be soft soldered after age hardening without detriment to mechanical properties. It is good practice to join the parts as soon as possible after surface preparation. If delays are unavoidable, parts should be stored in clean, dry locations free from acidic, sulfurous, or ammoniacal fumes (Brush Wellman, 2002).

4.3.2 Performance and Availability of CuBe Alloys

Varying the beryllium content from about 0.15 to 2.0 wt%, CuBe alloys offer physical and mechanical properties differing considerably from those of other copper alloys, principally because of the nature and action of the alloying element beryllium. Whether a high-strength or a high-conductivity alloy, some physical properties remain similar. For example, the elastic modulus of the high-strength alloys is 30 GPa; for the high-conductivity alloys, 37 GPa. Poisson's ratio is 0.3 for all compositions. However, the thermal and electrical conductivity differs significantly: about 105 W/mK, 15–28% IACS for high-strength alloys; 245 W/mK, 20–60% IACS for the high-conductivity grades. The good thermal and electrical conductivities of CuBe alloys promote their use in applications requiring heat dissipation and current carrying capacity, in addition to shielding effectiveness. The thermal expansion coefficient of CuBe, 17.5×10^{-6}/K, is independent of alloy composition over the temperature range in which these alloys are used, which closely matches that of steels, including the stainless steels. This insures that CuBe and steel are compatible in the same assembly. High-strength CuBe alloys have smaller densities than other conventional copper alloys, often providing more pieces per unit weight of input material. Excellent strength, conductivity, resilience, and other elastic properties make CuBe the best choice of spring gasket application. Table 4.2 shows the typical commercial CuBe alloys available (Brush Wellman, 2002).

4.3.2.1 High-Strength Alloys

Alloy 25 is the highest strength and hardness heat-treatable copper alloy available. Intricate forms with small radii can be made in the soft temper condition, yet after heat treatment, very high strength and resilience can be achieved. Heat treatment also relieves internal stress, which allows bidirectional loading and provides excellent stability. In the fully aged condition, the electrical conductivity is around 22% IACS. Alloy 25 is the most common material used for metal-formed gasket production.

TABLE 4.2
Commercial CuBe Alloys Available

CuBe Alloys			Composition (Weight Percent)						
Type	Brush Designation	UNS Number	Beryllium	Cobalt	Nickel	Cobalt + Nickel	Co + Ni + Fe	Lead	Copper
High Strength									
	25	C17200	1.8–2.00	—	—	0.20 min	0.6 max	0.02 max	Balance
	190								
	290								
	M25	C17300	1.80–2.00	—	—	0.20 min	0.6 max	0.20–0.60	Balance
	165	C17000	1.60–1.79	—	—	0.20 min	0.6 max	—	Balance
High Conductivity									
	3	C17510	0.20–0.60	—	1.40–2.20	—	—	—	Balance
	10	C17500	0.40–0.70	2.40–2.70	—	—	—	—	Balance
	174	C17410	0.15–0.50	0.35–0.60	—	—	—	—	Balance
	60	C17460	0.15–0.51	—	1.0–1.4	—	—	—	Balance

Source: Brush Wellman 2002.

Metal-Formed EMI Gaskets and Connectors 97

The gaskets can provide maximum spring performance, have a high cycle life, have excellent shielding effectiveness (80–120 dB up to 3 GHz), can conform to large gap variations (0.25–10 mm), have a wide deflection operating range (20%–80% deflection), have adjustable compression force, have exceptional resistance to stress relaxation, and can be applied in a wide temperature variation (up to 121°C). This alloy is also used for springs, switch blades, grounding strips, connector parts, and battery contacts in commercial, aerospace, and military industries.

Alloy 190 is a high-strength, mill-hardened alloy that contains the same beryllium as heat-treatable alloy 25. Alloy 190 is used in high-performance applications where heat treatment is unnecessary or undesirable. Alloy 190 is well suited to applications where larger radius bends or lower strength can be tolerated, including EMI shielding, switch springs, connector parts, and battery contacts. *Alloy 290* is similar to 190 with improved formability.

4.3.2.2 High-Conductivity CuBe Alloys

Alloys 3 and *10* combine moderate yield strength, up to 965 MPa, with electrical and thermal conductivity from 45% to 60% of pure copper. *Alloy 174* is a high-conductivity, mill-hardened alloy with excellent formability. It is most commonly used for automotive connectors, battery contacts, EMI shielding, and lighting fixtures. *Alloy 60* has lower strength, but its higher conductivity allows it to be used primarily in high-temperature spring applications at a lower cost than *Alloy 25*. *Alloy 390* combines the strength of *Alloy 25* and the conductivity of *Alloy 3* and *174*.

4.3.2.3 Nickel Beryllium

Nickel beryllium (BeNi) is a heat-treatable, high-performance alloy (1.85–2.05 wt % Be, 0.4–0.6 wt % Ti, balance Ni) used primarily in high temperature applications. It can withstand continuous exposure to 400°F (204°C) and short-term exposure to 700°F (370°C). Its good formability in the soft temper condition allows intricate, miniature parts to be fabricated. After heat treatment, strength is in the range of 200 ksi (1380 MPa).

4.4 COPPER–NICKEL–TIN SPINODAL ALLOYS

As mentioned earlier, CuBe metal-formed gaskets have long set the benchmark for EMI shielding and elastic performance. Other alloys come with compromises in terms of shielding effectiveness and elastic performance. Offering attenuation performance nearly comparable to CuBe and stress relaxation similar to that of CuBe, copper–nickel–tin (CuNiSn) spinodal alloy offers an appealing alternative to traditional CuBe gaskets (Fenical et al., 2006).

4.4.1 COMPOSITION AND PHYSICAL PROPERTIES

CuNiSn alloy used for forming EMI gaskets is one of spinodally hardening alloys. This kind of alloy hardening results from the spinodal decomposition process, which exhibits spontaneous solute clustering (nanoscale size) without a crystal structure change after phase transformation. Spinodal decomposition was thermodynamically predicted one hundred years ago, however, it was not observed in the copper–nickel–tin system until the 1960s, where it was found to greatly increase strength. In the 1990s, spinodally hardened CuNiSn continuously casting alloys and hot-worked wrought products were developed and produced. Since the early 2000s, cold-worked and spinodally hardened strip coils have been used in the electronic industry (Cribb 2006). EMI gaskets made from the CuNiSn spinodal alloy have been developed and commercialized, such as Recyclable Clean Copper™ (Laird Technologies, St. Louis, Missouri; Fenical et al., 2006; Laird Technologies, 2006a).

Typical composition of CuNiSn spinodal alloy used to replace CuBe for gasketing application is Cu9Ni6Sn. It is beryllium free, which alleviates environmental, safety, and segregation concerns associated with the traditional use and recycling of beryllium-based copper alloys.

The physical and mechanical properties of the CuNiSn alloy differ considerably from CuBe and other copper alloys, primarily because of the nature and action of alloying elements accompanied with the use of spinodal decomposition. Typical values of some of these properties are presented in Table 4.3 (Mitsubishi Electric, 2006).

Some physical properties of CuNiSn are similar to those of CuBe. For example, the elastic modulus of CuNiSn is the same as that of CuBe. Poisson's ratio is close to 0.3 for all four alloys in Table 4.3. The properties that differ significantly between these alloys are thermal and electrical conductivities. The thermal and electrical conductivities of CuNiSn are lower than those of CuBe, but similar to phosphor bronze. The thermal expansion coefficient of CuNiSn is close to that of CuBe and stainless steel, which insures that CuNiSn, CuBe, and steel are thermally compatible in the same assembly. The specific heat of CuNiSn is relatively lower than CuBe and similar to phosphor bronze. CuNiSn is a little bit denser than CuBe, and is less deformed (distorted) after age-hardening treatment, which will be explained later. Due to relative low copper content compared with CuBe, CuNiSn has lower electrical conductivity (12% IACS) than that of CuBe alloy 25 (22% IACS), but similar to that of phosphor bronze.

TABLE 4.3
Comparison of Electrical and Physical Properties of CuNiSn and Other Alloys

Properties	CuNiSn C72700	CuBe C17200	Phosphor Bronze C52100	Stainless Steel S30100
Density	8.88 g/cm^3	8.36 g/cm^3	8.80 g/cm^3	8.03 g/cm^3
Electrical conductivity	12% IACS	22% IACS	13% IACS	2.4% IACS
Thermal expansion coefficient	17.1 × 10^{-6}/K	17.5 × 10^{-6}/K	18.2 × 10^{-6}/K	16.6 × 10^{-6}/°K
Thermal conductivity	50 W/m/°K	105 W/m/K	62 W/m/°K	16 W/m/K
Specific heat	0.375 J/g·K	0.42 J/g·K	0.38 J/g·K	0.50 J/g·K
Elastic modulus	130 GPa	130 GPa	110 GPa	190 GPa
Poisson's ratio	0.30	0.27	0.33	0.30

TABLE 4.4
CuNiSn Alloy Temper Designations

Vendor Designation	ASTM Designation	Description	Cold Rolled Thickness Reduction (%)
O	TB00	Solution annealed	0
1/4H	TD01	Quarter hard	10.9
1/2H	TD02	Half hard	20.7
H	TD04	Full hard	37.1
EH	TD06	Extreme hard	50.1
OT	TF00		
1/4HT	TH01	The suffix "T" added to vendor rolled temper designations indicates that the material has been age hardened by the standard heat treatment.	
1/2HT	TH02		
HT	TH04		
EHT	TH06		
OM	TM00		
1/4HM	TM01		
1/2HM	TM02	Mill hardened to specific property ranges; no further heat treatment is required.	
HM	TM04		
EHM	TM06		
XHM	TM08		

4.4.2 Strip Gauges and Temper Designations

CuNiSn alloy properties are determined in part by composition, but cold work, age hardening, and spinodal decomposition are also important. The combined effect of noncompositional factors is defined in the alloy's *temper*. The temper designations of the CuNiSn alloy are shown in Table 4.4 (Mason, 2005; Mitsubishi Electric, 2006). When alloy temper is specified on drawing or order, a specific set of properties would be assured. When replacing CuBe with CuNiSn in gasket design, it should be noticed that there is no equivalence between two different materials with the same temper because different materials develop different properties with the same percentage of rolling reduction. Similar to CuBe, in many tempers, strength and forming characteristics of CuNiSn strips vary little with direction, that is, they are close to isotropic; for some hard tempers, the variation could be apparent. This should be carefully considered when designing CuNiSn gaskets. In a gasketing assembly, for instance, CuNiSn gasketing regulates compression force to encourage contact wiping action, provides a high normal force to minimize contact resistance, and maintains the contact force to ensure conducting path integrity. This accomplishment often requires an intricate stamped or formed gasket that combines flexing and stiffening members in the same part. Among the many CuNiSn tempers, there is one to be chosen with compliance and formability to meet requirements of gasket design.

Thickness is another critical factor in gasket design, which strongly influences force-deflection characteristics. For this reason, the strip thickness should be uniform within tolerance limits. Strip curvature, either edgewise (camber) or in the plane of strip (coilset or crossbow), also needs to be carefully controlled. In press working, excellent strip shape aids proper feed, particularly with progressive die.

In addition, CuNiSn strips for gasketing and grounding have demanding forming requirements, but also they also require strength and endurance. For instance, high fatigue strength offers resistance to the action of repeatedly applied loads, sometimes at moderately elevated temperatures.

4.4.3 Mechanical Properties and Bending Formability

CuNiSn alloy is heat treatable. Before aging treatment, this alloy has good formability similar to conventional age-hardening alloys, such as CuBe. After aging treatment, CuNiSn exhibits high mechanical strength and excellent relaxation-stress resistance. These characteristics make its formability and elastic performance close to that of CuBe.

Table 4.5 shows the typical mechanical properties and the bending formability of CuNiSn alloy (Mitsubishi Electric 2006) compared to CuBe and other alloys, with a guide to temper selection based on forming requirements. CuNiSn has superior yielding strength to many other spring alloys, and only slightly lower than that of CuBe. The r/t ratios in the table indicate the allowable punch radius (r) for a 90-degree bend as a function of strip thickness (t). Low r/t implies high formability. The formability of CuNiSn is similar to CuBe with both longitudinal and transverse bends posing no forming problem. Certain mill-hardened tempers also have low directionality.

TABLE 4.5
Typical Mechanical Properties and Bending Formability of CuNiSn Compared with Other Alloys

Properties	CuNiSn C72700	CuBe C17200	Phosphor Bronze C52100	Stainless Steel S30100
Minimum bend ratio r/t, 1/4 H	1 good way, 1 bad way	0 good way, 0 bad way	0 good way, 0 bad way	0.5 good way, 1.5 bad way
0.2% offset yield strength, 1/4 HT	685–800 MPa	1034–1276 MPa	241–428 MPa (1/4 H)	517–759 MPa (1/4 H)
Minimum bend ratio r/t, 1/2 H	1 good way, 1 bad way	0.5 good way, 1 bad way	0 good way, 0.5 bad way	1 good way, 2 bad way
0.2% offset yield strength, 1/2 HT	780–910 MPa	1103–1345 MPa	352–517 MPa (1/2 H)	760–931 MPa (1/2 H)
Minimum bend ratio r/t, H	1 good way, 1.5 bad way	1 good way, 2.9 bad way	0.5 good way, 3.0 bad way	2.5 good way, 4.0 bad way
0.2% offset yield strength, HT	890–1020 MPa	1138–1413 MPa	538–655 MPa (H)	966–1483 MPa (H)
Minimum bend ratio r/t, EH	1 good way, 4 bad way	—	1 good way, 4.3 bad way	—
0.2% offset yield strength, EHT	950–1080 MPa	—	634–738 MPa (EH)	1483–1551 MPa (EH)

Because of this isotropy, special considerations do not have to be given to the manner in which parts are stamped relative to the rolling direction. In many instances this will permit nesting of parts to allow efficient material utilization. Besides, CuNiSn is highly resilient compared with many other sheet metal materials, such as phosphor bronze and stainless steel. Similar to CuBe, springback of CuNiSn alloy during forming becomes more pronounced as temper and strength increases. For a given punch radius, its springback decreases with increasing strip thickness.

4.4.4 STRESS-RELAXATION RESISTANCE

Table 4.6 shows the stress-relaxation resistance of CuNiSn compared to CuBe and CuTi alloys under different testing temperatures (Mason, 2005). The stress relaxation is evaluated on samples cut from the strips and subjected to constant elastic bending strain under each temperature. A sample is secured in a fixture and deflected as a cantilever beam to an initial stress that is a predetermined 75% of the alloy's 0.2% yield strength. The fixture and deflected sample are exposed to each temperature for extended time, up to 1000 hours. After testing, stress relaxation of the sample causes its permanent set. The ratio of the permanent set to initial deflection defines the degree of loss of initial bending stress caused by relaxation. As the tests reveal (Table 4.6), stress-relaxation resistance of CuNiSn is comparable to CuBe and close to CuTi alloys. Therefore, the metal-formed gaskets, electronic contacts, and other spring elements made from these alloys have the capability to remain stable longer when operating at higher temperatures. This is more important as miniaturization in computer hardware, automotive interconnections, and aerospace systems has accented the importance of high thermal stability.

4.4.5 HEAT TREATMENT AND SPINODAL DECOMPOSITION

Most copper-base alloys develop high strength from solid solution hardening, cold working, and precipitation hardening, or by a combination of these strengthening mechanisms. However, in the ternary copper–nickel–tin alloys including Cu9Ni6Sn (wt%), the high mechanical strength can be produced by a controlled heat treatment with spinodal decomposition. Spinodal decomposition is generally a clustering reaction in a homogeneous solution that is unstable against infinitesimal fluctuations in density

TABLE 4.6
Stress Relaxation Resistance % of CuNiSn Compared with CuTi and CuBe Alloys

Stress-Relaxation Resistance	CuNiSn C17200	CuTi C19900	CuBe C17200
100°C 1000 h	99.8	100	95
150°C 1000 h	98	99	85
175°C 1000 h	95	97	79
200°C 1000 h	95	96	60

or composition. The solution therefore separates spontaneously into two phases, starting with small fluctuations and proceeding with a decrease in the Gibbs energy without a nucleation barrier. In other words, spinodal decomposition takes place spontaneously and needs no incubation period. It is a continuous diffusion process, in which there is no nucleation step, that produces two chemically different phases with an identical crystal structure. The two-phase structure in the spinodally hardened alloy is very fine (nanoscale), and continuous throughout the grains and up to the grain boundaries. The high strength of the Cu9Ni6Sn alloy resulting from spinodal decomposition can be attributed to the coherency strains produced by the uniform dispersions of Sn-rich phases in the Cu matrix (Cribb, 2006).

Certain conditions need to be fulfilled for hardening to occur by spinodal decomposition. As shown in Figure 4.3, the phase diagram of a spinodal system in the solid state must contain a miscibility gap, a region where the single phase (α) alloy separates into two phases ($\alpha_1 + \alpha_2$). The alloying elements must have sufficient mobility in the parent matrix at the heat treating temperature to allow interdiffusion. Therefore, the heat treatment steps of the CuNiSn alloy with spinodal decomposition should include (Cribb, 2006): (a) homogenization at a temperature above the miscibility gap to obtain a uniform solid solution of a single phase (around 800°C for Cu9Ni6Sn); (b) rapid quenching to room temperature; and (c) aging–reheating to a temperature within the spinodal region to initiate the reaction, and holding for sufficient time to complete the spinodal decomposition. For example, 371°C for two hours is the empirical process for the CuNiSn alloy.

As mentioned earlier, age hardening usually increases the density of high-strength alloys slightly, such as CuBe alloy 25, as a result of the precipitation reaction. This density change is accompanied by a decrease in linear dimensions of approximately 0.2% for CuBe alloy 25. However, the spinodal decomposition will minimize the

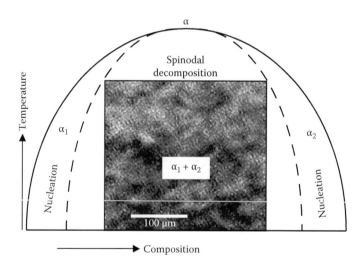

FIGURE 4.3 A part of phase diagram with a typical miscibility gap and formed spinodal microstructure by spinodal decomposition: $\alpha \rightarrow \alpha_1 + \alpha_2$.

Metal-Formed EMI Gaskets and Connectors

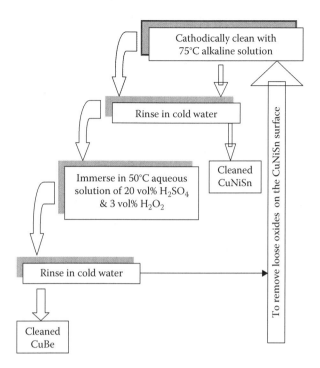

FIGURE 4.4 Precleaning procedure of CuNiSn strip compared with that of CuBe.

density and dimension changes after aging; therefore, the linear shrinkage of the Cu9Ni6Sn is only around 0.05%, and the angle movement after aging is also significantly less that that of CuBe.

4.4.6 Surface Cleaning, Plating, and Soldering

As shown in Figure 4.4, surface cleaning of CuNiSn is similar to the cleaning process of CuBe, with the exception of repeating the cathodical clean step with alkaline water solution washing at the end. This is for cleaning off the black oxides loosely attached on the CuNiSn surface. Furthermore, the platability and solderability of CuNiSn are superior or comparable to CuBe.

4.4.7 Elastic Performance

CuNiSn alloy is highly resilient when compared with many other sheet metal shielding materials, but it has relatively lower elastic resilience than that of CuBe. However, some CuNiSn parts with certain profiles exhibit similar spring performance or compression set to those of CuBe in a certain deflection range, as shown in Figure 4.5 (Fenical et al., 2006; Laird Technologies, 2006a). This is because spring performance is not only dependent on material's elastic resilience, but also influenced in part by geometry, profile design, and load application manners. In other wards, the spring performance of CuNiSn can be improved by optimizing part geometry, profile design, and application manners or using the right niche.

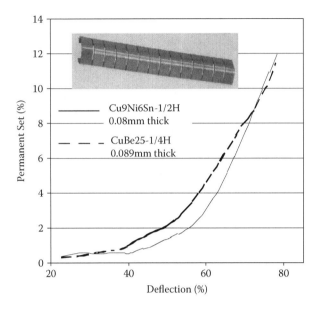

FIGURE 4.5 Comparison of permanent set vs. deflection of CuNiSn gasket and CuBe gasket: same profile, different strip thickness.

4.4.8 Shielding Effectiveness

A CuNiSn gasket generally has a slightly lower shielding effectiveness than that of CuBe in a wide range of frequencies, but the values at high frequencies are very close to each other, as shown in Figure 4.6 (Fenical et al., 2006; Laird Technologies, 2006a). This is mainly because the conductivity of CuNiSn is lower than that of CuBe. However, the shielding effectiveness of CuNiSn is pretty high compared with stainless steel and many other alloys.

4.4.9 Fatigue Strength

Figure 4.7 shows that the fatigue strength of the CuNiSn strip is close to that of the CuBe strip (Mitsubishi Electric, 2006). In the EMI gasketing application, an outstanding characteristic of CuNiSn is its ability to withstand cyclic stress. Cyclic condition here is produced by cantilever bending with the spring CuNiSn testing strip flexed in reverse bending $R = -1$. R is termed the stress ratio, which is the ratio of minimum to maximum stress. Fatigue strength is defined as the maximum stress that can be endured for a specified number of cycles without failure. Low cycle fatigue strength approaches the static strength. When the number of cycles exceeds one million to ten million, the fatigue strength fails to a fraction of the static strength. Similar to CuBe, CuNiSn alloy resists fatigue failure with high static strength, toughness, and an ability to diffuse strain by work hardening when the local stress concentration is over the alloy's yield strength.

Metal-Formed EMI Gaskets and Connectors

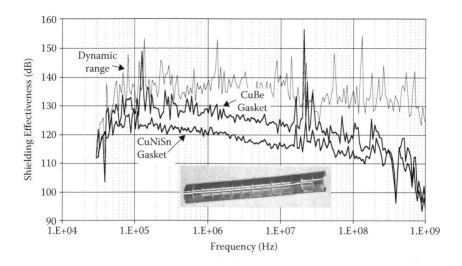

FIGURE 4.6 Shielding effectiveness of CuNiSn and CuBe gaskets measured by transfer impedance method.

4.4.10 Some Considerations to Improve Performance of CuNiSn Gasketing

Compared with CuBe, major drawbacks of CuNiSn gasketing may include: (a) a relatively low yield strength makes its elastic resilience not good enough for some applications where a large deflection range and minute permanent set are needed; (b) the allowed upper limitation of the deflection range is lower (around 40%–50%) than that of CuBe (up to 70%–80%); (c) a relatively low electrical conductivity makes its

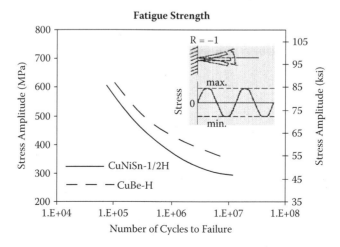

FIGURE 4.7 Fatigue strength of CuNiSn and CuBe alloys.

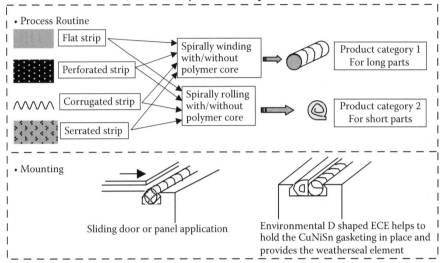

FIGURE 4.8 Spirally wound hybrid EMI gasketing.

shielding effectiveness is not as good as CuBe; and (d) CuNiSn gasketing is most suitable for small parts or some certain profiles to maintain the similar spring performance to CuBe.

To overcome these drawbacks, the following solutions can be considered: (a) modify the geometry design, and design new profiles more suitable for CuNiSn gasketing; (b) use multigauge, inlay or laminate materials to optimize mechanical and conductive performance of CuNiSn; and (c) combine with conductive polymer, flexible graphite, or other EMI shielding material to form hybrid gasketing profiles. Some examples are as shown in Figures 4.8 and 4.9. Figure 4.8

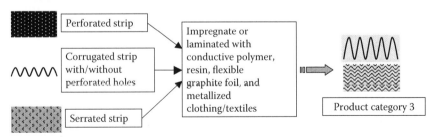

FIGURE 4.9 CuNiSn strip impregnated or laminated with conductive polymer, resin, flexible graphite foil, and metallized clothing/textiles.

shows possible process routines and mounting methods for spirally wound hybrid CuNiSn gasketing, which has some advantages, such as being beryllium free; improved shielding effectiveness due to enhanced conduction and low contact resistance; minimized compression set and stress relaxation; and can provide an EMI seal and an environmental seal. Figure 4.9 gives some examples for making hybrid CuNiSn gaskets by using impregnation or lamination with conductive polymer, resin, flexible graphite, or metallized clothing/textiles. The aim is to combine the advantages of each material to improve gasketing performance and extend application areas.

4.5 COPPER–TITANIUM ALLOY

The age hardening of copper–titanium (CuTi) alloys containing approximately 1–5 wt% Ti (around 1–6 at% Ti) has been recognized since the 1930s (Soffa and Laughlin, 2004). The electrical conductivity of the aged alloys falls somewhat below that of CuBe alloys. CuTi's mechanical and physical properties, however, are found to be comparable to CuBe alloys with better high temperature strength and superior stress relaxation behavior. Therefore, CuTi alloys have received a great deal of attention as ultra-high strength conductive materials for applications such as conductive springs, EMI shielding gaskets, and interconnections.

The age-hardened CuTi alloys are typically quenched from 800°C–900°C to room temperature prior to aging. The decomposition of supersaturated CuTi alloys containing 1–6 wt% titanium (Ti) during aging in the range of 350°C–550°C embodies a complex interplay of clustering and ordering effects in solid solutions as well as a subtle synergy between short-range order and long-range order, and the nanoscale prolate spheroids of the tetragonal Cu4Ti (D1a) phase from the onset of decomposition. The particle arrays appear to evolve through concomitant growth and coarsening under the influence of elastic interaction from the earliest stages of the decomposition, revealing CuTi alloys at high supersaturations representing spinodal alloys. Most precipitation systems including the CuTi system involve the precipitation of ordered phases within a disordered parent solution. As a result, there will be a complex interplay of the ordering and clustering. However, the emergence of the cellular microconstituent heterogeneous nucleation of the equilibrium phase at the grain boundaries and the interaction of the grain boundaries with particles must be suppressed. One approach is to systematically add ternary and quaternary alloying elements, which can influence the nucleation and growth processes and boundary migration involved in the genesis of the cellular colonies. This grain boundary engineering was successful in the classic CuBe alloy series using cobalt additions and has been applied to the age hardening of CuTi alloys with some promising results (Soffa and Laughlin, 2004). For example, Fe has been successfully used in CuTi alloys.

One typical commercial CuTi alloy is copper alloy C1990: Cu–2.9~3.4 wt% Ti–0.17~0.23 wt% Fe. It can provide high strength and excellent formability comparable to CuBe. Its stress relaxation resistance is better than CuBe. It can be used for many applications including battery terminal, antenna spring, and EMI shields.

4.6 STAINLESS STEEL

Stainless steels are iron-base alloys containing 10.5 wt%, or more chromium, and can be used for EMI gaskets where high elastic resilience and electrical conductivity are not required. Stainless steels have a high temperature performance superior to that of copper alloys. They are highly corrosion resistant, and perform well in harsh environments. Austenitic (300 series) stainless steels are most common for EMI gasketing, which offer 60 dB shielding effectiveness under 1 GHz, high strength, and fair formability at moderate cost. They are best for static load or low-cycle applications. Because stainless steel provides a corrosion-resistant surface, plating is usually unnecessary, but sometimes needed to improve solderability and galvanic compatibility to mating parts. In addition to EMI shielding, stainless steel is also commonly used for mechanical spring, switchblades, perforated panels, connector shields, reed valves, battery contacts, and underhood automotive components.

Typical stainless steels that can be used for EMI gasketing include austenitic, martensitic, and precipitation hardening stainless steels. Austenitic stainless steels containing chromium and nickel are identified as 300 series types. Alloys containing chromium, nickel, and manganese are identified as 200 series types. The stainless steels in the austenitic group have different compositions and properties, but many common characteristics (Specialty Steel Industry of North America [SSINA], 2002). They can be hardened by cold working, but not by heat treatment. In the annealed condition, all of them are essentially nonmagnetic, although some may become slightly magnetic by cold working. They have excellent corrosion resistance, unusually good formability, and increase in strength as a result of cold work. The mechanical and electrical properties of some typical austenitic stainless steels are shown in Table 4.7 (ASTM, 2000; Sandvik, 2002). Austenitic stainless steels can be cold worked to high tensile and yield strengths, while retaining good ductility and toughness, and this good ductility and toughness is retained at cryogenic temperatures. The commonly used Type 301 for EMI gasketing has good toughness even after cold rolling to high tensile strengths. Fatigue or endurance limits (in bending) of austenitic stainless steels in the annealed condition are about one-half the tensile strength (SSINA, 2002).

Martensitic stainless steels are straight chromium 400 series types that are hardenable by heat treatment. The martensitic grades are so named because when heated above their critical temperature (around 870°C) and cooled rapidly, a metallurgical structure known as martinite is obtained. They are magnetic, and therefore may not suitable for some applications where the magnetism will influence the function of electronic circuits. Table 4.8 gives the mechanical and electrical properties of some martensitic stainless steels (ASTM, 2006). Type 410 is the general-purpose alloy of the martensitic group. In the hardened condition, the steel has very high strength and hardness, but to obtain optimum corrosion resistance, ductility, and impact strength, the steel is given a stress-relieving or tempering treatment (usually in the range 149°C to 371°C). The martensitic stainless steels are generally selected for moderate resistance to corrosion, relatively high strength, and good fatigue properties after suitable heat treatment (SSINA, 2002).

Precipitation-hardening stainless steels are chromium–nickel types, with some containing other alloying elements such as copper or aluminum. They can be hardened

TABLE 4.7
Mechanical and Electrical Properties of Typical Austenitic Stainless Steels

Stainless Steel Type	Temper	Tensile Strength, min		Yield Strength 0.2% Offset, min		Elongation in 50 mm (min, %)	Electrical Conductivity (% IACS)
		(MPa)	(ksi)	(MPa)	(ksi)		
201 (S20100)							
	Annealed	655	95	310	45	40	2.5
	1/16 Hard	655	95	310	45	40	2.5
	1/8 Hard	690	100	380	55	45	2.5
	1/4 Hard (H01)	860	125	515	75	25	2.5
	1/2 Hard (H02)	1035	150	760	110	15	2.5
	3/4 Hard (H03)	1205	175	930	135	10	2.5
	Hard (H04)	1275	185	965	140	8	2.5
202 (S20200)							
	Annealed	620	90	260	38	40	2.5
	1/4Hard (H01)	860	125	515	75	12	2.5
205 (S20500)							
	Annealed	790	115	450	65	40	2.5
	1/16 Hard	790	115	450	65	40	2.5
	1/8 Hard	790	115	450	65	40	2.5
	1/4 Hard (H01)	860	125	515	75	45	2.5
	1/2 Hard (H02)	1035	150	760	110	15	2.5
	3/4 Hard (H03)	1205	175	930	135	15	2.5
	Hard (H04)	1275	185	965	140	10	2.5
301 (S30100)							
	Annealed	515	75	205	30	40	2.39
	1/16 Hard	620	90	310	45	40	2.39
	1/8 Hard	690	100	380	55	40	2.39
	1/4 Hard (H01)	860	125	515	75	25	2.39
	1/2 Hard (H02)	1035	150	760	110	15	2.39
	3/4 Hard (H03)	1205	175	930	135	10	2.39
	Hard (H04)	1275	185	965	140	8	2.39
	Extra hard (H06)	1525	221	1230	179	—	2.39
	Spring hard (H08)	1715	249	1425	207	—	2.39
302 (S30200)							
	Annealed	515	75	205	30	40	2.39
	1/16 Hard	585	85	310	45	40	2.39
	1/8 Hard	690	100	380	55	35	2.39
	1/4 Hard (H01)	860	125	515	75	10	2.39
	1/2 Hard (H02)	1035	150	760	110	9	2.39
	3/4 Hard (H03)	1205	175	930	135	5	2.39
	Hard (H04)	1275	185	965	140	3	2.39

(Continued)

TABLE 4.7 (CONTINUED)
Mechanical and Electrical Properties of Typical Austenitic Stainless Steels

Stainless Steel Type	Temper	Tensile Strength, min		Yield Strength 0.2% Offset, min		Elongation in 50 mm (min, %)	Electrical Conductivity (% IACS)
		(MPa)	(ksi)	(MPa)	(ksi)		
304 (S30400)							
	Annealed	515	75	205	30	40	2.39
	1/16 Hard	550	80	310	45	35	2.39
	1/8 Hard	690	100	380	55	35	2.39
	1/4 Hard (H01)	860	125	515	75	10	2.39
	1/2 Hard (H02)	1035	150	760	110	6	2.39
	Hard (H04)	1130	164	880	128	-	2.39
316 (S31600)							
	Annealed	515	75	205	30	40	2.39
	1/16 Hard	585	85	310	45	35	2.39
	1/8 Hard	690	100	380	55	30	2.39
	1/4 Hard (H01)	860	125	515	75	10	2.39
	1/2 Hard (H02)	1035	150	760	110	6	2.39
13RM19 (EN 1.4369)							
	Annealed	850	123	470	68	45	2.5
	1/2 Hard (H02)	1100	160	975	142	12	2.39
	3/4 Hard (H03)	1300	189	1150	167	10	2.39
	Hard (H04)	1500	218	1350	196	3	2.39
	Extra hard (H06)	1600	232	1440	209	2	2.39

by solution treating and aging to high strength. The mechanical and electrical properties of some typical precipitation hardening stainless steels are shown in Table 4.9 (ASTM A693-06, 2006). They have high strength, relatively good ductility, and good corrosion resistance at moderate temperatures. Their principle advantage is that gaskets can be fabricated in the annealed condition and then strengthened by a relatively low temperature (482°C to 620°C) treatment, minimizing the problems associated with high-temperature treatments.

4.7 OTHER MATERIALS OPTIONS

4.7.1 Phosphor Bronze

Phosphor bronzes are copper–tin–phosphorus alloys, usually available in a full range of mill-hard tempers. They are stronger than brass and weaker than 174 CuBe, and are less conductive than either. Phosphor bronzes have been widely used in connectors and relay contacts due to their low elastic modulus and high spring characteristics. However, phosphor bronzes generally degrade in bending workability when heavily worked for work hardening (harder tempers). Some new versions of phosphor

TABLE 4.8
Mechanical and Electrical Properties of Typical Martensitic Stainless Steels

Stainless Steel Type	Tempering Temperature (°C)	Tensile Strength, min (MPa)	(ksi)	Yield Strength 0.2% Offset, min (MPa)	(ksi)	Elongation in 50 mm (min, %)	Electrical Conductivity (% IACS)
403 (S40300)							
	Annealed	483	70	207	30	20	3.0
	Hardened/tempered	689	100	483	70	10	3.0
410 (S41000)							
	Annealed	480	70	275	40	15	3.0
	204	1310	190	1000	145	15	3.0
	316	1240	180	960	139	12	3.0
	427	1405	204	950	138	12	3.0
	538	985	143	730	106	12	3.0
	593	870	126	675	98	18	3.0
	650	755	109	575	83	20	3.0
420 (S42000)							
	Annealed	655	95	345	50	15	3.1
	204	1600	232	1360	197	6	3.1
	316	1580	229	1365	198	8	3.1
	427	1620	235	1420	206	4	3.1

bronzes with smaller grain size have improved strength and bending workability. However, the stress relaxation resistance is slightly degraded irrespective of reduced grain size (Mihara et al., 2004). Some high-strength phosphor bronzes can be used as alternative gasketing materials, which offer better performance than stainless steels although they are inferior to CuBe and CuNiSn spinodal alloys.

4.7.2 Brass

Brass (copper–zinc) is commonly used for electric connector parts where stress is very low. Brass has either good formability or moderate strength, but not both. Of the brass alloys, cartridge brass is the usual choice because it is available in thin strip form. Brass can be resistance welded to copper beryllium and is sometimes used as a rigid component in a spring or gasketing assembly.

4.7.3 Nickel Silver

Nickel silver is a kind of copper–zinc–nickel alloy that is used to make miniature parts where appearance is important. The material has strength comparable to phosphor bronze, but is less formable in spring tempers and has lower conductivity. It is commonly used for optical parts, mechanical springs and zippers, and it is especially suitable for board-level shielding cans (see Chapter 7 for details).

TABLE 4.9
Mechanical and Electrical Properties of Typical Precipitation Stainless Steels

Stainless Steel Type	Hardening/ Precipitation °F (°C)	Tensile Strength, min (MPa)	(ksi)	Yield Strength 0.2% Offset, min (MPa)	(ksi)	Elongation in 50 mm (min, %)	Electrical Conductivity (% IACS)
630 (S17400)							
	Solution (1050°C)	—	—	—	—	—	1.8
	900 (482)	1310	190	1170	170	5	2.2
	935 (496)	1170	170	1070	155	5	2.2
	1025 (552)	1070	155	1000	145	5	2.2
	1075 (579)	1000	145	860	125	5	2.2
	1100 (593)	965	140	790	115	5	2.2
	1150(621)	930	135	725	105	8	2.2
	1400 + 1150 (760 + 621)	790	115	515	75	9	2.2
631 (S17700)							
	Solution (1065°C)	1035	150	450	65	20	2.0
	1400(760) + plus 55(15) + 1050(566)	1240	180	1035	150	3	2.0
	1750(954) + minus 100(73) + 950(510)	1450	210	1310	190	2	2.0
XM12 (S15500)							
	Solution (1038°C)	—	—	—	—	—	2.0
	900 (482)	1310	190	1170	170	5	2.0
	935 (496)	1170	170	1070	155	5	2.0
	1025 (552)	1070	155	1000	145	5	2.0
	1075 (579)	1000	145	860	125	5	2.0
	1100 (593)	965	140	790	115	5	2.0
	1150(621)	930	135	725	105	8	2.0
	1400 +1150 (760 + 621)	790	115	515	75	9	2.0
XM13 (S13800)							
	Solution (927°C)	—	—	—	—	—	1.7
	950 (510)	1515	220	1410	204	6	1.7
	1000 (538)	1380	200	1310	190	6	1.7

4.7.4 OTHER HIGH STRENGTH AND HIGH CONDUCTIVE COPPER ALLOYS

These high strength and high electrical conductivity alloys can be classified into four categories: CuCr (copper–chromium), CuNiSi (copper–nickel–silicon), CuFe (copper–iron), and Cu(Fe,Co,Ni)P. Generally, materials with high strength tend to have low electrical conductivity, and vice versa. An effective way to improve both properties is to use a base alloy composition that causes precipitation hardening, in which the elements do not form a solid solution with the bulk metal (matrix), but instead increase

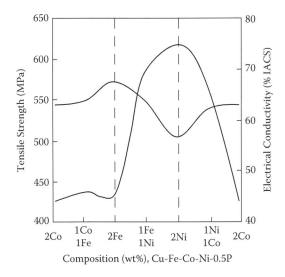

FIGURE 4.10 Effect of the combination of Fe, Co, and Ni on the tensile strength and electrical conductivity of the CuFeNiP alloy, aged at 440°C for 1 hour. (Modified from Yamamoto et al., 2000.)

the strength when they precipitate in the matrix. This purifies the matrix; therefore, the material tends to have high electrical conductivity (Yamamoto et al., 2000). In addition, formability and platability should be considered for application in EMI shielding.

Among these alloys, CuCr alloys have the highest electrical conductivity while the Cr precipitation increases the strength; however, they have inferior stampability or formability, and their castability is also limited because Cr is very reactive. CuNiSi alloys exhibit very high strength due to precipitation of NiSi compounds, and provide good electrical conductivity. However, their platability is not very good due to large precipitate deposits remaining on the surface that do not dissolve during the pre-cheaning or pickling process before plating. CuFe alloys have improved strength and electrical conductivity due to Fe precipitates, but the improvement in strength is insufficient.

Cu (Fe,Co,Ni) P alloys provide well-balanced properties of both good electricity and high strength as a result of precipitation of Fe_2P, Ni_5P_2, Ni_2P, Co_2P, and so forth. To obtain high strength and electrical conductivity in a precipitation-hardened alloy, however, it is important to precipitate the compounds efficiently, optimizing the weight ratio of the Fe, Ni, and P (phosphorus) to minimize any surplus alloy components not precipitating, as shown in Figure 4.10 (Yamamoto et al., 2000). Besides, the strength and electrical conductivity are greatly affected by proper precipitation hardening through optimal heat treatment control.

4.8 DESIGN GUIDELINE OF METAL GASKETS AND CONNECTORS

As mentioned earlier, electronic enclosures and many other electronic devices often require shielding gaskets and connectors to provide effective conduction and to prevent the transmission of undesired EMI through cracks around access doors.

A wide variety of metal shielding gaskets and connectors have been designed and commercialized for a variety of shielding and interconnection applications, spanning the aerospace, automotive, computer, electronics, medical, and telecommunications industries. The most common metal gaskets and connectors include shielding gaskets, grounding products, contacts, and connector shields. EMI shielding gaskets offer the best combination of high shielding effectiveness and low compression force available. When used properly, 90 to 110 dB or more of shielding effectiveness can be achieved with a year of service, assuming proper design of the gasketed joint.

4.8.1 Primary Approaches

In general, the success of a shielding gasket or a connector design is to specify the contact material providing optimum performance, ease of manufacturing, and cost benefits. Contact materials possess varying combinations of strength, conductivity, magnetic permeability, stress-relaxation resistance, formability, corrosion resistance, and galvanic compatibility. Choosing an alloy that best meets the complex needs of shielding and contact applications requires a greater awareness of the characteristics and interrelationships governing material performance. Therefore, the design process can be identified as two parts (Brush Wellman, 1988): first, determine the general class of alloy necessary to fulfill the requirements of a gasket or connector application; and, second, identify the temper that provides the optimum solution to the application's total operating and manufacturing parameters. In most shielding gasket and connector applications, the design factors that are dictated by the service requirements will involve the spring or normal force, resistance to permanent set on compression, insertion, temperature rise, elevated temperature stress relaxation, and fatigue. Once the critical design factors have been identified and a general class of alloy has been selected, it is necessary to consider the available alloy forms, their formability characteristics, and the economic implications of specifying one alloy over another. In each instance, material properties that dictate the performance of the various alloys must be considered, and the specific factors must be assessed in selecting the optimum temper for a given application.

As in shielding gaskets and electronic connectors, the critical performance required of the application focuses on ensuring the strength and conductivity characteristics of the finished part. The gaskets or connectors must provide a reliable contact over wide deflection ranges while withstanding multiple mechanical flexures. As a result, the shielding contact alloy must provide (Brush Wellman, 1988): (a) high elastic resilience (high yield strength and high modulus of elasticity) to fulfill the application's strength and spring force requirements; (b) excellent electrical and EMI conductivity; (c) high stress-relaxation resistance; (d) a fatigue limit ($R = 0$) that is a minimum of 50% of the yield strength; (e) excellent corrosion resistance; and (f) good manufacturability, including a 90-degree bend formability ratio of $R/t \leq 3$ in the transverse direction, a maximum strength by heat treating after forming, and plateability and soldarability.

Finite element analysis (FEA) matched with prototyping has become an effective tool for designing shielding gaskets and electronic connectors with many benefits: (a) Virtual prototyping of complex systems, saving costs, and speeding up the design

Metal-Formed EMI Gaskets and Connectors

cycle and time to market; (b) 3D analysis can be used to predict contact forces on all the mating surface, and compression and shearing moments on the gasket; (c) maximize the spring force while miniaturizing the part; (e) predict compression and insertion forces and fatigue cycle life; (f) evaluate failure modes; (g) predict material thinning and flat blanket behavior for complex tooling to shorten tool building time; (h) predict compression set and allowed load-deflection range; and (i) predict buckling between bolts on a cover plate to verify if a gasket is effectively closing the gap.

4.8.2 Some Special Design Options

There are other design options that can be considered to meet application requirements (Instrument Specialties, 1998):

- Zone annealed materials are hard temper alloys that have been softened in a localized area. This allows severe, small radius forms to be made in the annealed zone while preserving spring properties in stressed areas.
- Multigauge materials have two or more different thicknesses across the width of the strip. This allows one end of a part to be thick and rigid, while the other end is thin and flexible.
- Dual materials are made by joining two different metals using an electron beam or laser welding. The electron beam or laser-welded materials are often multigauge as well. For example, a dual material part might have a thin spring material at one end and a ductile solder terminal at the other end of a gasketing shield, with a thick section to provide stiffness for the side wall (Figure 4.11).
- Inlay materials have a strip of precious metal imbedded in the base material and rolled flush with the surface. The inlay area is used as an

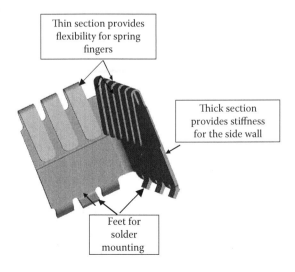

FIGURE 4.11 A design with dual materials for a gasketing shield.

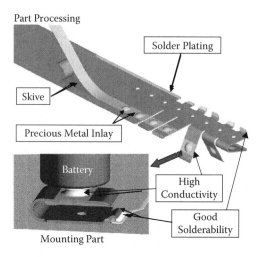

FIGURE 4.12 Inlay and selective plating used to make battery contacts.

electrical contact. Available inlay materials include alloys of gold, silver, silver cadmium oxide, platinum, and palladium. Cladding materials can be either precious metals or other solderable alloys to enhance solderability. They can be pressure bonded or plated onto the strip surface. Figure 4.12 shows an inlay and selective plating example used to make a battery contact.

4.8.3 Gasketing Design Guideline

During initial enclosure gasket interface design, it is important that the impedance between the mating surfaces be as nearly equal to the enclosure material as possible. Any significant difference in surface impedance through the gasket materials can produce fluctuations in current flow, resulting in the generation of EMI voltages. These fluctuating voltages can create the leakage of radiated energy into or from the enclosure. Therefore, the following features should be considered in the gasket interface design: (a) gasketed surfaces should be protected with a conductive coating such as tin, nickel, or zinc; (b) mating surfaces should be as flat as possible considering the manufacturing method, that is, bending or machining; and (c) mating surfaces should be cleaned of oxides prior to assembly of the gasketing material.

The mechanical aspects of EMI gasket interface design are important factors in ensuring a reliable EMI shield. Joint unevenness, or the degree of mismatch of mating surfaces, is one of the most important design considerations. An effective EMI gasket should make continuous and uniform contact with the mating surfaces. The shielding performance of a gasketed enclosure can be adversely affected by improper mating of the gasketed interface due to joint unevenness. The maximum joint unevenness is the dimension of the maximum separation between the flanges of the seam when the two surfaces are touching. The gasket, under these extreme conditions, undergoes its severest electrical tests at the minimum deflection. With joint unevenness such a

Metal-Formed EMI Gaskets and Connectors

critical factor, it is important to consider gasket deflection, compression set, shielding effectiveness, and environmental sealing needs when designing an EMI gasket or specific interface.

The minimum force should be applied on the gasket to establish good shielding effectiveness. The maximum recommended closure force is based on two criteria: maximum compression set of 10% and avoidance of possible damage to the gasket material when pressure exceeds the recommended maximum. Compression stops built into an enclosure or supplied with the EMI gasket aid in controlling the amount of deflection, thus minimizing the amount of compression set. Selection of a gasketing material for a joint that must be opened and closed is, to a large extent, determined by the compression set characteristics of the gasket material. Most resilient gasket materials will recover most of their original height after being deflected no greater than 25% to 30%. The difference between the original height and the height after the compression force is removed is the so-called compression set. Compression set increases as the deflection pressure increases.

It is necessary to consider the following rules during gasket design to avoid common problems (Instrument Specialties, 1998):

1. Control gasket compression force and deflection range. Enclosure designers need to know the compression force required to close gasketed joints. EMI shielding gaskets can achieve excellent shielding effectiveness when compressed approximately 20% deflection from free height. Additional compression makes little or no improvement in shielding performance. Variation of shielding effectiveness is only around 10 dB for most gaskets. It would seem logical, therefore, to design for 20% deflection, since this produces minimum compression or closure force. Unfortunately, real gasketed joints are not rigid and tend to bulge open between fasteners. This distortion can break contact and create a serious EMI leak. Since typical manufacturing tolerances are often as much as 20% of the gasket's working range, a routine variation in dimension could result in an EMI leak. Therefore, the gasket compression force and deflection range should be carefully controlled to avoid these problems. For instance, designing for as high as approximately 50% deflection from free height is a reasonable rule for most CuBe gaskets. However, excessive compression and deflection will fail or damage gasket fingers. The best way to eliminate excess compression and deflection is to use a positive stop. A problem often overlooked is crushing damage due to shock and vibration during shipping. It is advisable to remove shielded doors for shipment or block them open.
2. Use the largest size gasket possible for certain contact gap tolerance. Small-sized gaskets have a small deflection range. This means that close gap tolerances must be maintained in the shielded enclosure. The cost of holding close tolerances will greatly exceed any cost penalty for using larger shielding gaskets. In addition, small finger gaskets are disproportionately stiffer than large gaskets. This means that the enclosure must be considerably more rigid in order to avoid excessive distortion. Adding rigidity will usually raise the cost of an enclosure considerably.

3. Wipe toward the free end of the gasket finger. In most finger gasket installations, some wiping action occurs. It may be deliberate, or it may occur on the hinge side of a door when the gasket is used in compression. In either case, engaging the gasket from the wrong direction will destroy fingers and impair the shielding effectiveness. When a gasket is compressed, wiping in either direction will work.
4. Protect the gasket from snagging. When a shielded door is used for personnel access, the most common form of damage is from snagging. Fingers that catch on clothing, cables, or other objects may be bent or broken, and shielding effectiveness will be reduced. In general, gaskets should be mounted where they are least likely to be snagged. It is usually better to mount the gasket on a doorframe where some protection is offered. However, finger gaskets should never be mounted on the floor in a doorjamb.

These rules cover only the most common mechanical problems encountered in shielded enclosure and gasket design. There are many other considerations, including selection of an appropriate gasket type, enclosure, material surface finish, and closure mechanism. For those inexperienced at enclosure design, construction and testing of a prototype is highly recommended before finalizing the design for production.

4.9 FABRICATION PROCESS AND TYPES OF METAL GASKETS AND CONNECTORS

Metal gaskets and electronic connectors have been fabricated with a variety of processes, including progressive die forming, multiple-slide forming, notching and roll forming, photoetching, and with or without subsequent heat treating and plating (Laird Technologies, 2006b).

4.9.1 PROGRESSIVE DIE FORMING

Progressive die forming performs a series of operations at two or more die stations during each stroke of the press. The metal strip progresses through successive stations until a complete gasket is produced, as shown in Figure 4.13. One or more operations are done on the workpiece at each die station. As the outline of the workpiece is developed in the trimming or forming stations, connecting tabs link the workpiece to the transmitting strip until the workpiece reaches the last station where it is cut off and ejected from the die. Pilot holes that are engaged by pilot pins in the die to keep the workpieces aligned and properly spaced as they progress through the die. Progressive die is high-speed stamping process with high tool costs, low piece price, and close tolerance. It is good for high-volume production of loose pieces with scored or coiled strips. Progressive dies can be used to perform an almost endless variety of operations on one piece. These operations that can be combined in a progressive die including notching, piercing, coining, embossing, lancing, forming, cupping, drawing, and trimming.

Metal-Formed EMI Gaskets and Connectors

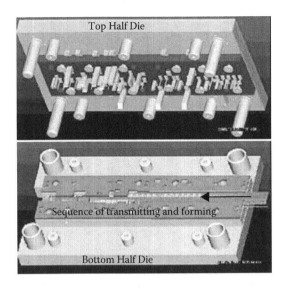

FIGURE 4.13 Progressive dies.

4.9.2 ROLL FORMING

Roll forming is similar in speed to progressive die stamping. Tooling used in roll forming includes the forming rolls and the dies for punching and cutting off the material. Tooling cost is comparable or less than progress die forming. It is preferred for certain part geometries. As shown in Figure 4.14, the roll forming of the metal strip into a desired final shape is a progressive operation in which small amounts of forming are performed at each pass or pair of rolls. The amount of change or contour in each pass must be restricted so that the required bends can be formed without elongating the metal strip. Generally, the number of passes depends on the properties of the material and the complexity of the gasket shape, in addition to the gasket part width, the horizontal center distance between the individual stations, and the part tolerances. The number of passes must be increased as the tolerance of the shape becomes tighter.

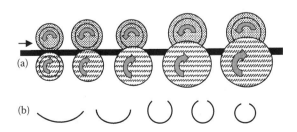

FIGURE 4.14 Roll-forming process. (a) Roll forming line; (b) example of cross-section of the formed gasket.

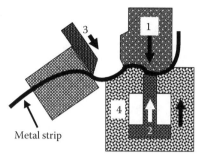

1-Center Post; 2-Pressure Finger
3-Cutoff Blade; 4-Front Tool

FIGURE 4.15 Cutoff blade and dies for multiple-slide machine.

4.9.3 Multiple-Slide Stamping

Multiple-slide forming is a process in which the workpiece is progressively formed in a combination of units that can be used in various ways for the automated fabrication of a large variety of simple and intricately shaped gasket parts from coil stock. Operations such as straightening, feeding, trimming, blanking, embossing, coining, lettering, forming to shape, and ejecting can all be done in one cycle of a multislide machine. Forming is generally limited to bending operations, but the four slides and center post permit the fabrication of very complex gasket parts. However, deep drawing is generally not done in the forming or press stations of a multiple-slide machine. Figure 4.15 illustrates the cutoff blade and dies for a four-slide forming machine. The cost of four-slide stamping is usually less than progressive dies. It is a fast process and preferred for small-sized parts and certain geometries with high volume production. Usually the setup times are long and general tolerances are relatively loose.

4.9.4 Photoetching

Photoetching is used to make flat blanks in copper alloys or stainless steels. The process is similar to printed circuit technology using inexpensive artwork instead of cutting tools. Its advantages include fast turnaround on design changes, works best for thin materials, and intricate shapes cost the same as simple shapes. Figure 4.16 shows the typical process of the photoetching.

4.9.5 Typical Part Profiles

There have been hundreds of metal gaskets and electronic connectors designed and commercialized. Figure 4.17 shows some typical product profiles, including finger stock gaskets and metal grounding products, metal connector shields, electronic contacts, and engineered assemblies (Instrument Specialties, 1998; Laird Technologies, 2006a, 2006b).

Metal-Formed EMI Gaskets and Connectors

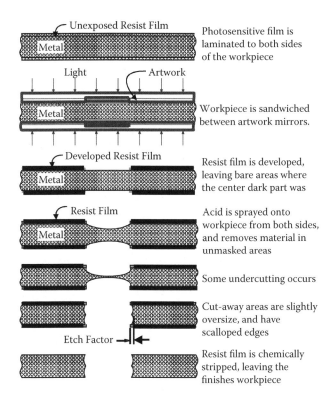

FIGURE 4.16 Schematic of photoetching process.

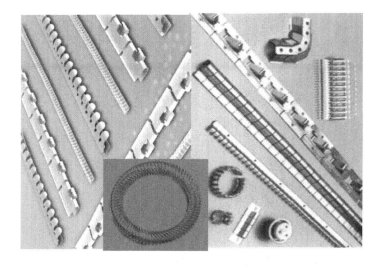

FIGURE 4.17 Typical part profiles of metal gaskets and electronic connectors.

4.10 MOUNTING METHODS AND SURFACE MATING ASSURANCE

The optimum condition for an EMI gasket mounting is a rigid machined flat surface with bolts located outside the gasket contact area. This condition, however, is not practical or possible in most shielding applications. Therefore, an EMI gasket that meets the required shielding effectiveness should be chosen to be adapted to interface constraints of the enclosure package. This is a major consideration in designing an effective EMI gasket. Figure 4.18 shows some common mounting methods for metal gaskets (Instrument Specialties, 1998; Laird Technologies, 2006b; Molyneux-Child, 1997).

- Slot mounted gaskets are easily installed using slots where bidirectional movement is required. The finger strip has retention tabs that mate with prepunched holes and is retained by spring force.
- Clip-on gaskets hold firmly in place due to their own spring characteristics. The strips are pushed onto the edge or flange of the door/lid or enclosure, and retained by spring force and optional lance features or locking perforations.
- Adhesive mounting provides an instant, pressure-sensitive bonding system, ideal for all-purpose contact strips for metal cabinets and electronic enclosures, particularly where space saving is crucial. The finger strips are double sided. The pressure-sensitive adhesive is attached and shipped with

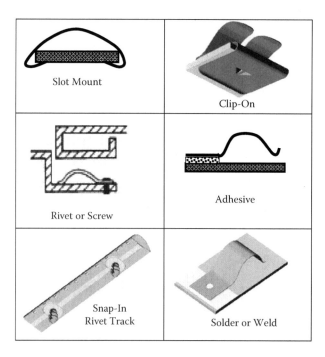

FIGURE 4.18 Typical mounting methods of metal gaskets.

release paper. The adhesive properties strengthen as it cures with age and remain unaffected by temperatures from −55°C to +121°C.
- Riveting produces tight, long-lasting installation. Mounting track can be attached by pushing rivets into prepunched holes. Either plastic or metal rivets may be used. Similarly, gasket fingers can also be mounted with screws.
- Welded mounting requires spot welding or other traditional welding techniques.
- Solder mounting requires normal low-temperature soldering techniques, including cleaning and fluxing of parts with common copper flux materials. It provides strong installation with the best joint conductivity.

The most common method of grounding between two mating surfaces is with a flat gasket mount. This method is particularly effective in those joints where there is an occasional need to open the enclosure for maintenance purposes. It is recommended that a positive stop be used to prevent overcompression of the gasket, which could increase compression set and reduce shielding effectiveness when the lid is secured after opening. This is because all gasket materials have an elastic limit and are subject to taking a compression set if overcompressed. The stops should be designed to the maximum suggested compression limits of the gasket material used.

For applications where space is at a minimum and wide flanges are not feasible (such as cast housings), mounting of the EMI gasket in a groove is an option. Where equipment would be subjected to submersion, or needed for an environmental protection, groove mounting of a metal gasket combined with conductive elastomer insert or gasket in cast housing is a viable solution. In many severe environment applications, it might be necessary to provide a double groove where the inside groove is designed for loading the metal gasket, while the outside groove provides a barrier against the environment, be it salt spray or chemical. In this application, the outside barrier is usually a conductive elastomer. In all groove-mounted gasket applications, care should be taken to ensure that the groove is sized to secure against overcompression of the gasket (25% maximum for most CuBe gaskets) and the width of the groove should be sufficient to allow gasket material to flow under compression. Use of a knife-edge contact helps reduce the required compression force of a closure. For groove applications using a knife-edge, it is recommended that the gasket material be a hollow O or D profile. The gasket under compression will deflect into itself minimizing the flow effect.

Consequently, the minimum contact resistance should be required when mounting a gasket in order to maintain conductivity with the housing flanges. For consistency of electrical contact, EMI shielding gaskets need to be compressed to a certain deflection range. Otherwise, poor conductivity between opposing flanges through the gasket will diminish shielding effectiveness. Increasing compression of the gasket will reduce contact resistance. However, overcompression can result in hard-to-operate doors, case bowing, or a premature compression set or even break the gaskets. The optimal compression load and deflection range varies between gaskets used in a similar application in accordance with their material composition and profile configuration (Hudak, 2000).

4.11 SUMMARY

The appeal of metal gaskets and connectors is a result of their robust construction, various configuration, and high level of shielding effectiveness in a variety of applications. The materials used for making the gaskets and connectors require various properties, including responsiveness to formation of complexes peculiar to various assemblies; initial strength to reach contact pressures above certain levels in contact areas; ability to maintain these pressures in the long term; and fatigue resistance for high-cycling applications.

CuBe has long set the benchmark for EMI shielding and elastic performance. Other alloys come with compromises in terms of shielding effectiveness and elastic performance. However, beryllium-free alloys are options with the advent of severe environmental requirements.

The success of a shielding gasket or a connector design depends on specifying the contact material providing optimum performance, ease of manufacturing, and cost benefits. Choosing an alloy that best meets the complex needs of new shielding and contact applications requires a greater awareness of the characteristics and interrelationships governing material performance.

Metal gaskets and electronic connectors have been fabricated with a variety of processes, such as progressive die stamping, multislide stamping, notching and roll forming, photoetching, heat treating, and plating. Hundreds of metal gaskets and electronic connectors have been designed and commercialized. The product profiles mainly include finger stock gaskets and metal grounding products, metal connector shields, electronic contacts, and engineered assemblies.

Because EMI gaskets that meet the required shielding effectiveness should be chosen to be adapted to the interface constraints of the contact components, quite a few mounting methods have been developed and utilized.

REFERENCES

ASTM A666-00. 2000. *Standard specification for annealed or cold-worked austenitic stainless steel sheet, strip, plate, and flat bar*. West Conshohocken, PA: ASTM international.

ASTM A693-06. 2006. *Standard specification for precipitation-hardening stainless and heat-resisting steel plate, sheet, and strip*. West Conshohocken, PA: ASTM international.

Brush Wellman. 1988. *Connector Design Guide: Material Specification in the Design of Connectors and Interconnections*. Cleveland, OH: Brush Wellman.

Brush Wellman. 2002. *Guide to Copper Beryllium*. Cleveland, OH: Brush Wellman.

Cribb, W. R. 2006. Copper spinodal alloys for aerospace. *Advanced Materials & Processes*. http://www.asminternational.org/pdf/amp_articles/AMP200606_Spinodal.pdf

Fenical, G., C. Tong, and P. Graustein. 2006. *EMI gasketting: The switch to recyclable clean copper*. Interference Technology. http://Shielding, pp 1–3 Accessed March 9, 2007.

Hudak, S. 2000. A guide to EMI shielding gasket technology. *Compliance Engineering*. http://www.ce-mag.com/ARG/Hudak.html

Instrument Specialties. 1998. *Engineering Design and Shielding Product Selection Guide* (acquired by Laird Technologies). Delaware Water Gap, PA: Instrument Specialties.

Laird Technologies. 2006a. *Recyclable clean copper (RCC)*. http://www.lairdtech.com/downloads/RCCcatalogwithoutbleeds-Oct2006.pdf

Laird Technologies. 2006b. *Fingerstock, gaskets and metal grounding products.* http://www.lairdtech.com/downloads/FingstockGasketsandMetalGroundingProductsCatalog.pdf

Mason. 2005. *High strength copper alloys: AEM 400.* Birmingham, UK: B Mason & Sons Ltd. http://www.bmason.co.uk

Mihara, K., T. Eguchi, T. Yamamoto, and A. Kanamori. 2004. Effect of metallographic structures on properties of high-performance phosphor bronze. *Furukawa Review* 26: 44–48.

Mitsubishi Electric. 2006. *High-strength electrical conducting spring materials with excellent cost performance.* Kanagawa: Mitsubishi Electric Metecs Co., Ltd. http://www3.metecs.co.jp/seihin/12_mx96r.pdf

Molyneux-Child, J. W. 1997. *EMC Shielding Materials.* Oxford: Newnes.

Sandvik. 2002. *Stainless steel for spring and other demanding applications.* S-3411-ENG. Sweden: Sandvik AB.

Soffa, W. A., and D. E. Laughlin. 2004. High-strength age hardening copper-titanium alloys: Redivivus. *Progress in Materials Science* 49: 347–366.

Specialty Steel Industry of North America (SSINA). 2002. *Design guidelines for the selection and use of stainless steel.* Washington, D.C.: Specialty Steel Industry of North America.

Takano, H., Y. Yamamoto, and C. Tong. 2006. Development of high-strength HCL 307 copper-alloy strip for electrical components. *Hitachi Cable Review* 25: 22–25.

Yamamoto, Y., G. Sasaki, and K. Yamakawa. 2000. High-strength and high-electrical-conductivity copper alloy for high-pin-count leadframes. *Hitachi Cable Review* 19: 65–70.

5 Conductive Elastomer and Flexible Graphite Gaskets

5.1 INTRODUCTION

Conductive elastomers have been used successfully in electromagnetic interference (EMI) shielding applications mainly focusing on EMI gasketing. The primary function of elastomeric gaskets is to provide sufficient conductivity across the enclosure-gasket-lid junction or other gasketing joints, and to act as a seal or barrier to prevent fluid or moisture intrusion into the shielded components. Partly because of their high costs, conductive elastomers were initially limited to military and aero space applications. With the advancements in digital electronics and miniaturization of electronic components, however, the increasing electromagnetic compatibility (EMC) requirements have promoted demands for cost-effective elastomeric materials in commercial applications.

Conductive elastomers are typically made from silicone, fluorosilicone, ethylene propylene diene monomer (EPDM), and fluorocarbons, loaded with conductive fillers that are either inherently conductive or made electrically conductive by plating. As mentioned earlier, conductive elastomers not only provide EMI shielding but also can function as a pressure and moisture seal.

Conductive elastomers have been preferred and regulated by the U.S. military through MIL-G-83528 (1992). The elastomer gaskets are usually more expensive than other gaskets, such as wire mesh, stainless-steel fingers, and fabric-over-foam. Most conductive elastomer manufacturers supplement their military-grade offerings with commercial-grade materials that feature the same elastomeric binders with similar conductive fillers, such as glass plated with silver, or copper plated with silver. Gaskets can be produced with forming in place, extruding, or molding into sheets, tubes, or complex geometries. A variation of the fully loaded conductive elastomer gaskets are commonly fabricated with the co-extruded version, which features a conductively loaded outer liner over a nonconductive hollow inner core (Hudak, 2000).

The shielding effectiveness of conductive polymers range from 120 dB for military-grade products down to about 60 dB for the commercial grades. Variation in their shielding effectiveness is related to the compression force applied and the amount of conductive filler loading. Conductive elastomer gaskets require a minimum deflection of 10% to 30% from free height in order to function properly, and their maximum deflection is limited to 50%, depending on the geometry of the part. Compression load values tend to be among the highest of all gasketing products, ranging from about 35 kg/m to 150 kg/m. Compression set values at 100°C range from 5% for hollow-core extrusions up to 30% or more for some solid-core constructions (Hudak, 2000; Instrument Specialties, 1998).

Another similar class of EMI gasketing material, flexible graphite, is comparable to or better than conductive elastomer materials. The shielding effectiveness of flexible graphite gaskets can reach up to 125 to 130 dB due to their low electrical resistivity and high specific surface area (Luo and Chung, 1996). Its high thermal conductivity, low CTE, high temperature resistance, and excellent chemical resistance of flexible graphite make it appealing for use in EMI shielding.

This chapter will give a brief review about conductive elastomer gaskets and flexible graphite gaskets, including raw material selection and the elastomer fabrication process; conduction mechanism and processing optimization; a gasket design guideline; and gasket fabrication, installation, and application.

5.2 RAW MATERIAL SELECTION AND CONDUCTIVE ELASTOMER FABRICATION

Silicones and fluorosilicones have historically been used as the polymer base for conductive elastomers. The rationale behind their use has been their ability to provide reliable service at temperature extremes, good resistance to compression set and ozone. Fluorosilicones combine most of the attributes of silicone with resistance to petroleum and hydrocarbon fuels. The threat of nuclear and chemical warfare and the need for seals for applications such as the drilling and automotive industries have created a demand for fluorocarbons and EPDM-based conductive elastomers (Instrument Specialties, 1998). The filters used in conductive elastomers range from carbon powder to pure silver. The need for reliable elastomeric gaskets to meet the stringent military EMI requirements, and grounding and corrosive environments, have expanded the envelope for the filters to plated particles, such as silver-plated copper, aluminum, nickel, glass, and inert aluminum, nickel, and nickel-coated graphite.

5.2.1 BASE BINDER MATERIALS

Conductive elastomers are mostly based on premium grade polymers, which can provide low volatile contents and temperature extremes within the polymer families (Instrument Specialties, 1998).

5.2.1.1 Silicone

Silicone is a semiorganic elastomer with excellent resistance to extreme temperatures, and can operate continuously at temperatures ranging from −62°C to 260°C for solid and −75°C to 205°C for closed-cell sponge elastomers. Silicones also have excellent resistance to the effects of aging and ozone, and have outstanding compression set properties. However, because of their relative low physical strength and abrasion resistance, silicones are poor candidates for dynamic seal applications. They are used primarily for dry heat static seals. Although the silicone melts considerably in petroleum lubricants, this is not detrimental in most static sealing applications.

5.2.1.2 Fluorosilicone

Fluorosilicones combine most of the attributes of silicone with resistance to petroleum oils and hydrocarbon fuels. Similar to silicones, fluorosilicones are most commonly used for static applications and are seldom recommended for dynamic seals due to their low tensile strength and poor abrasion resistance. They are used primarily in aircraft fuel systems over a temperature range of –65°C to 175°C.

5.2.1.3 Ethylene Propylene Diene Monomer (EPDM)

EPDM is used frequently to seal phosphate esters and has excellent resistance to hydraulic fluid, alkalis, silicone oils and greases, ketones, and alcohols. However, it is not recommended for petroleum oils or diester lubricants. EPDM has a temperature range of –55°C to 150 °C. It is compatible with polar fluids that adversely affect other elastomers.

5.2.1.4 Fluorocarbon

Fluorocarbon has the same characteristics as silicone with improved resistance to petroleum oils, fuels, and silicone oils. Most fluorocarbons become quite hard at temperatures below –20°C, but they do not easily fracture, therefore they are serviceable at much lower temperatures. Fluorocarbons can be used at temperatures up to 205°C for long periods and higher temperatures for shorter periods, and have resistance to aging and ozone.

5.2.1.5 Natural Rubber and Butadiene-Acrylonitrile

The treated natural rubber has good resistance to acids and alkalies and can be used to 106°C. It is resilient and impervious to water. However, rubber tends to crack in a highly oxidizing or ozone atmosphere, and swell in the presence of oils. Comparably, butadiene-acrylonitrile resists swelling in the presence of most oils. It has moderate strength and heat resistance but is not generally suited for low-temperature applications.

5.2.2 CONDUCTIVE FILLERS

Traditionally, pure metallic powders such as silver, nickel, or copper were used extensively in conductive elastomers; however, these have been superseded by silver-plated metal or composite powders on the grounds of cost. The most common fillers used include passivated aluminum, silver-plated aluminum, copper, nickel, and ballotini. Ballotini is glass or ceramic microspheres that have been coated with metallic silver. The ballotini approach should be avoided in designs involving vibration conditions or EMP (electromagnetic pulse), but ballotinis are good for weight and cost saving (Molyneux-Child, 1997). Pure nickel flake is a good all-purpose filler if only reasonable attenuation is needed. Carbon or graphite could be employed, but due to their the poor shielding effectiveness they are usually limited to loading to electrostatic discharge and grounding, or others where good conductivity is not required. In addition, composite powders, such as nickel-plated graphite, have become commonplace, which affords a useful cost advantage over pure nickel with little reduction in performance. Other fillers have been used and developed, including conductive wire, mesh, cloth, and carbon nanotube (CNT).

As the electrical and mechanical properties of conductive elastomers are highly dependent on loading levels of the conductive fillers, the base binder must be loaded with conductive fillers to a sufficient level to create gasket materials with the required EMI shielding performance. Therefore, the selection of suitable conductive fillers and elastomer binders is crucial to obtain the high-performance conductive elastomer gasket materials that meet specific EMI application requirements.

5.2.3 Fabrication of Conductive Elastomer Materials

In addition to gaskets, electrical conductive elastomer materials have been fabricated for various applications including EMI shielding, adhesives, devices, and sensors. A variety of processes have been employed including doping of the elastomer, vacuum metallization and plating on the surface of the elastomer, implantation of metallic ions, and compounding with conductive fillers. The mixing of an elastomer with conductive filler particles is one widely used method because compounds of elastomers with conductive fillers can be easily shaped into various complicated shapes. Moreover, the electrical and mechanical properties of conductive elastomers can be beneficially adjusted through adequate control of the dynamics of the mixing process and the resulting load levels of the conductive fillers in the elastomer matrix.

TABLE 5.1
Some Conductive Elastomer Materials Classified by MIL-G-83828

Materials Type	Polymer	Filler	Hardness Score A	Applied Temperature Range (°C)
A	Silicone	Silver/copper	65	−55 to + 125
B	Silicone	Silver/aluminum	65	−55 to + 160
C	Fluorosilicone	Silver/copper	75	−55 to + 125
D	Fluorosilicone	Silver/aluminum	70	−55 to + 160
E	Silicone	Silver	65	−55 to + 160
F	Fluorosilicone	Silver	75	−55 to + 160
G	Silicone	Silver/copper	80	−55 to + 125
H	Silicone	Silver	80	−55 to + 160
J	Silicone	Silver	45	−55 to + 160
K	Silicone	Silver/copper	85	−55 to + 125
L	Silicone	Silver/nickel	75	−55 to + 125
M	Silicone	Silver/glass	65	−55 to + 160

Based on the application requirements, the conductive elastomer material can be formulated and mixed using the proper ingredients. These ingredients usually include (a) resin—a base elastomer, either silicone, fluorosilicone, or EPDM; (b) catalyst, for cross-linking the resin and provides the required elasticity; and (c) fillers, which provide the physical properties of the elastomer, including its conductivity.

The mixed conductive elastomer materials should meet the specified application requirement, such as MIL-DTL-83528C standard, with balanced properties, and are suitable for the subsequent process of gasket products. Table 5.1 shows some typical conductive elastomer materials classified by the MIL-G- 83828.

5.3 CONDUCTION MECHANISM AND PROCESSING OPTIMIZATION OF CONDUCTIVE ELASTOMER MATERIALS

5.3.1 Conduction Mechanism and Process Parameters

In formulating a conductive elastomer, a resilient polymer is combined with highly conductive rigid particles. Electrical conductivity in the conductive elastomer is accomplished through a certain level of conductive particle-to-particle contact or close proximity. In the uncured material, the conductive particles are considered to be touching or near touching with little electrical conductivity. During the curing process, the polymer shrinkage due to the chemical action of curing–cross-linking and the physical action of thermal contraction as the elastomer is cooled to room temperature causes the particles to be pressed and percolated together, forming the conductive path.

In general, the resistivity of the conductive elastomer is closely related to the resistivity of the filler. For example, carbon black, which in pure powder form has a volume resistance of approximately 0.1 Ω-cm, can produce conductive polymer composites with resistivities down to 1 Ω-cm (Sichel, 1982). However, silver powder with a resistance of only 1.5×10^{-6} Ω-cm can be used to make polymer composites with volume resistances as low as 10^{-2} Ω-cm (Bhattacharya, 1986; Rosner, 2000). This difference indicates metal coatings and fillers may be suitable for EMI shielding applications, whereas carbon-black loaded elastomers might only be used for electrostatic discharge (Mother and Thomas, 1997).

Numerous factors can affect the conductive properties of the filler loaded elastomer composite, including filler type, loading level, particle distribution, particle size, intrinsic conductivity of filler, process conditions, plating, and compounding. However, the most basic factors are particle conductivity, loading level, and particle shape. Therefore, the design and process for a conductive elastomer should involve the selection of an appropriate type of conductive filler, tailoring of the particle shape and size distribution, the optimization of the filler concentration, and the selection of a suitable binder. The filler concentration or volume loading level, which are relevant to the percolation regime of the composite at which the composite changes from an insulator to a conductor, is especially important.

The effect of filler loading on the conductivity or resistivity of the conductive elastomer composite follows a nearly universal pattern, regardless of which fillers

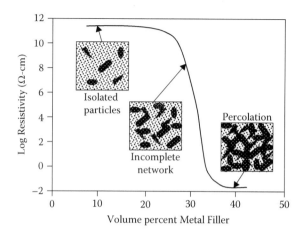

FIGURE 5.1 Variation of resistivity of conductive elastomers as a function of metal filler volume, approaching the percolation threshold.

are chosen. At low filler loadings, where the filler particles are isolated with no contact between them, the composite remains an insulator although its dielectric properties may change significantly (Kalyon et al., 2002). Upon the increase of the filler loading and the resulting close proximity of the conduct filler particles, electrons can travel the polymeric gap between the conductive particles, allowing current to flow. This is called the hopping and tunneling effect (Bigg, 1984; Kalyon et al., 2002). The ability of an electron to jump a gap under a given voltage field increases exponentially with decreasing separation between particles. As the filler loading is increased further and reaches a critical point, the conductive particles contact one another, and a continuous network is established. The sudden and precipitous drop in resistivity is achieved through the generation of a continuous network, which is defined as percolation threshold (Weber and Kamal, 1997). This phenomenon is illustrated in Figure 5.1, which shows the resistance of a polymer-silver composite as a function of the volume of silver spheres. As seen in this graph, network formation occurs with a slight increase in filler volume and yet results in a tremendous change in the electrical resistance (Rosner, 2000). A simplified view of this percolation transition is also shown in the Figure 5.1. At low filler loadings, the filler particles are isolated in an electrically insulating resin. Little or no change in resistance occurs because electrons moving through the composite still encounter the insulating polymer. As filler volume increases, the conductive particles become more crowded and are more likely to come into contact with one another. Finally, at the percolation threshold, a majority of filler particles are in contact with at least two of their nearest neighbors, thereby forming a continuous chain or network. An electrical charge can now pass through the composite via this network without encountering the high-resistance polymer resin. Additional filler loading beyond the percolation threshold does not greatly reduce the resistivity of the composite (Bhattacharya, 1986; Rosner, 2000).

The percolation behavior is primarily affected by the surface chemistry of the filler and its wettability by the polymer melt, the rheology of the polymer melt, the parameters of the solidification process, the crystallinity of semicrystalline polymers, and the parameters of the mixing process (Kalyon, 2002; Weber and Kamal, 1997). For a certain filler volume, the shape of the particle plays a critical role in where percolation occurs. The more structured or elaborately shaped the particle, the more likely it is to contact a nearest neighbor and form a continuous network. Perfectly spherical fillers, which arguably have the least elaborate and least structured shape, can require as much as 40% loading to reach the percolation threshold. Carbon-black particles are more irregularly shaped and often have long branches reaching out from the main body of the particle. These moderately structured fillers can require anywhere from 5% to 35% loading to reach the percolation threshold (Sichel 1982). In other words, the greater the aspect ratio of the particles, the smaller the loading level needed to reach the percolation threshold. On the other hand, the broader the particle size distribution of the solid particles, the greater the loading level needed (Fiske et al., 1994). Therefore, the smallest loading level necessary to achieve percolation occurs with high aspect ratio particles. Conversely, the highest filler loading level at which percolation occurs is achieved with a broad size distribution of spherical particles. In addition, the mixing process plays an important role in the percolation and conductivity of the conductive elastomers. For instance, with an increase in mixing time and hence the specific energy input incorporated during the mixing process, the homogeneity of the spatial distribution of the conductive particles in the binder is improved and the conductive particles are better coated with the binder. The improved homogeneity of the spatial distribution of the conductive particles, and uniform coating thickness, leads to the insulation of the conductive particles from one another. Thus the formation of a conductive network, that is, the percolation, is hindered, and the resistivity of the composite increases (Kalyon, 2002).

Moreover, fabrication process of the conductive elastomers should be optimized to balance their properties and meet the product performance requirements.

5.3.2 Appearance Properties and Performance Evaluation

In order to design a conductive elastomer gasket, optimal material selection usually needs to balance various property requirements under all possible operating conditions to achieve the most benefits with fewest disadvantages. Most of the properties of gasketing materials are interrelated, and improving one property through formulation or process adjustment will affect one or all others to some extent. During the formulation and fabrication of conductive elastomer gaskets, the conductive elastomer is usually characterized and evaluated on the basis of its priority properties, such as physical and mechanical properties, electrical properties, and sometimes thermal properties.

5.3.2.1 Physical and Mechanical Properties

The physical and mechanical properties typically include specific gravity, tensile strength, hardness, elongation, modulus, tear strength, compression/deflection behavior, and compression set.

The specific gravity or density of conductive elastomer materials generally ranges from 1.2 to 4.8 g/cm^3, depending on the base elastomer and filler combination. It is

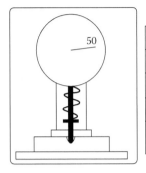

10	Softer	Sponge
30	Soft	Silicone
50	Medium	Neoprene
80	Harder	Tire
100	Hard	Not Compressible

FIGURE 5.2 Type A durometer and typical hardness values of elastomers.

a valuable quality control tool for monitoring the variation of filler loading or other additions in the conductive elastomer composite. The specific gravity is tested in accordance with ASTM D792. For some allocations where weight restriction is sensitive, light materials are preferred if they can meet the requirements of EMI, EMP, and vibration.

The hardness of conductive elastomers is important when considering the desired formability and compressibility of the EMI shielding gasket. Its value relates directly to the gasket sealing performance in compression modulus; resistance to extrusion and shape forming; and conformance to mating surface and compression deflection. Soft materials are highly compressible, conforming well to mating surfaces and requiring low forces to effectively seal a joint (Callen and Mah, 2002). Otherwise, if the gasket is too hard, it cannot deflect enough to fill the gap well. Therefore, an appropriate hardness will make the gasket tend to bite into the mating surfaces to create a lower interface resistance. The hardness is measured in accordance with the ASTM D2240 test method by a device called the type A durometer. Figure 5.2 illustrates this device and gives hardness values of some elastomers (Instrument Specialties, 1998). The hardness of a given elastomer is determined by the amount of force required by an indenter to penetrate a given sample of a flat piece of the sample. The higher the reading, the harder the elastomer.

The conductive elastomers with different fillers and polymer systems would have different durometer hardness values; moreover, the hardness usually increases rapidly with increasing filler loading levels.

Tensile strength and ultimate elongation are common measures for the strength of conductive elastomers. The tensile strength reflects the amount of force required to rupture a sample by application of tension. The elongation is the amount of extension or stretch of a sample. It is measured simultaneously with tensile strength and is reported as the amount of extension at time of rupture. This property primarily determines the stretch that can be tolerated during the installation of a gasket. It is a good measure of the ability of a material to recover after it has been stretched and also a useful indicator of the cure state. When a compressive force is applied to a gasket, it is compressed in one direction and stretched in the other direction perpendicular to the compressive force. A material with low tensile strength and ultimate

Conductive Elastomer and Flexible Graphite Gaskets

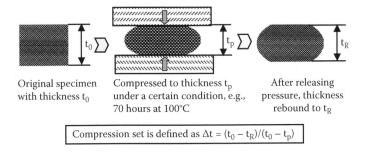

Original specimen with thickness t_0 | Compressed to thickness t_p under a certain condition, e.g., 70 hours at 100°C | After releasing pressure, thickness rebound to t_R

Compression set is defined as $\Delta t = (t_0 - t_R)/(t_0 - t_p)$

FIGURE 5.3 Illustrated definition of compression set for conductive elastic gaskets.

elongation is more susceptible to damage by compression and does not recover as well from strain. Similarly, tear strength is usually used to characterize and the initial tearing behavior of the conductive elastomer material in the direction normal to the direction of the stress.

In addition, modulus is used to characterize the stress required to stretch an elastomer sample to a predetermined elongation, most frequently 100%. The modulus generally increases with an increase in hardness. Modulus is usually taken as a major design factor for high pressure applications, as it characterizes the resistance of elastomer materials to deformation and shape forming such as extrusion. Moreover, modulus values are also used for process control, being a good indicator of the cure state of the compounds.

As mentioned in Chapter 4, optimal compression force and deflection range applied on a metallic or conductive elastomer gasket indicate how well the gasket can compensate for the mating surface unevenness. Another related property, compression set, is measured in accordance with ASTM D395 Method B and is expressed as the percentage of the original deflection that the sample fails to recover, as shown in Figure 5.3 (Instrument Specialties, 1998). For conductive elastomer material, compression set is a function of base polymer, filler loading, and shape of the filler particles. Compression set has its effect, but another property, compression relaxation, is more dominant for conductive elastomer design. When the gasket is compressed, the elastomer material pushes back against the compressive force. This push-back is caused by the internal rearrangement of the molecules during the stress. Excessive stress relaxation or compression set can compromise the environmental sealing capabilities of the material (Instrument Specialties, 1998).

5.3.2.2 Electrical Properties

The electrical conductivity of conductive elastomers is usually characterized by volume resistivity, which is the most important requirement for conductive materials, and it has a direct influence on the shielding properties of a gasket. Moreover, the elastomer must be capable of conforming to the unevenness of the mating surfaces to provide the least resistance across the interface. In addition, vibration usually increases the volume resistivity. This is because vibration can affect the filler particle distribution, bonding between the polymer and filler particle, and clamping

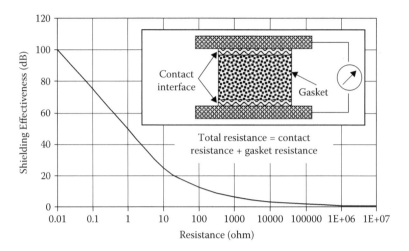

FIGURE 5.4 Schematic relationship of resistance vs. shielding effectiveness for a typical conductive elastomer gasket.

force. Basically, mechanical and electrical properties are interdependent variables of conductive filler loading that work against each other. For example, a stronger and softer gasket would provide lower conductive properties as the loading of the filler is decreased. Conversely, if the elastomer is loaded to produce very low volume resistivity to optimize shielding effectiveness, the material will become comparably hard and exhibit low tensile strength and elongation. Therefore, a compromise should be made between the desired physical and electrical properties of the conductive elastomer gasket with an appropriate filler loading level (Callen and Mah, 2002).

The electrical conductivity of the conductive elastomer gasket and contact resistance between the mating surfaces dominate the shielding effectiveness of the gasketing joint. Figure 5.4 shows the relationship between total resistance and shielding effectiveness of a typical conductive elastomer gasket. The total resistance is the sum of the bulk resistance (reciprocal of the electrical conductivity) and the contact resistance. The higher the conductivity and the lower the junction resistance, the better the shielding performance. These properties are crucial for conductive elastomer gaskets and other parts to provide grounding and bonding as well as protection against EMP. Protection against EMP is basically no different than normal EMI shielding except for the higher level of current encountered and the limited frequency spectrum involved. In order for any EMP shield/seal to be effective, it must have the basic ability to conduct high currents without any significant change in volume resistivity. As the current flows through the conductive material, heat is generated proportional to the basic internal resistance of the elastomer. If the shield material is unstable, resistivity will change to such a degree that the seal may cease to conduct currents and thus become ineffective (Instrument Specialties, 1998). In general, the resistivity of conductive elastomers is also affected by the oxidation of the metallic particles with a long shelf life or during heat aging. However, the rate of change in volume resistivity is different for different materials. In the case of plated metallic particles loaded materials, the rate of change depends on many factors, such as the plating process, plating thickness, the degree to which particles have been coated, and the oxidation-resistant treatment.

5.3.2.3 Thermal Properties

The maximum operating temperature should be limited to the point where the conductive elastomer will keep its properties within reasonable tolerances. This limitation depends on both the elastomer base material and the conductive filler used. If the temperature is too low, the conductive elastomers become brittle, and this change is reversible. Conversely, a very high temperature results in loss of electrical conductivity and degradation of physical properties, and the effect is also irreversible.

Among the thermal conductivities of conductive elastomer materials, thermal conductivity is essential information when designing any electronics, since excess heat can damage electronic components. In many applications, it would be a significant benefit if EMI shielding could also assist in the dissipation of heat from electronic components. The thermal conductivity of the electrically conductive elastomers typically ranges from 0.85 W/m-K to 2.6 W/m-K, whereas the elastomer matrix materials, with no filler particles, have thermal conductivities ranging from 0.8 W/m-K to 1.2 W/m-K. Therefore, the elastomer matrix usually contributes more to the thermal conductivity of electrically conductive elastomers than the filler particle. Although the filler particles have a small effect on the thermal conductivities, the affecting trend of filler materials on the thermal conductivity has been found in the order of silver > copper > aluminum > nickel > carbon > glass. The elastomers with nonmetal core filler particles with metal plating or coating have lower conductivities of about 1 W/m-K, whereas the ones with all metal filler particles are about 2 W/m-K. This is because the plating or coating is very thin, and the core material dictates more of the particle thermal properties than the coating. Therefore, electrical properties usually cannot be used to infer thermal properties of elastomers with different filler particles.

5.4 CONDUCTIVE ELASTOMER GASKET DESIGN GUIDELINE

An optimal conductive elastomer gasket, usually made from the conductive, resilient, and conformable vulcanized elastomer composite, can function as an EMI shield and an environmental seal. When applied between the mating surfaces, the gasket can follow small deflections, flow around small flaws, and conform to flange surface irregularities providing low interface resistance between the mating enclosure flanges, while simultaneously providing moisture, pressure, and/or environmental sealing. When designing an optimal conductive elastomer gasket, several interrelated factors should be considered, including flange joint geometry and unevenness, applied compressive force, optimal material selection, gasket profile, and application environment. Each of these factors is related to the others in terms of initial flange design parameters, available closure force, flange geometry, and sealing gasket material. A change in one will affect one or more of the others. The relationship should be considered when designing a new gasket or evaluating possible changes in an existing flange, shielding material, or bolt loading requirements.

5.4.1 Flange Joint Geometry and Unevenness

EMI gasketing flanges are joint parts of the enclosure to retain and provide strength and rigidity for the gasket. The flange surface geometry could be flat, recessed, grooved or differently shaped to completely enclose the gasket. Mating metal surfaces in contact with EMI gasket should be as conductive as possible, and galvanically compatible to avoid any corrosion or reaction between the gasket and the flange. If the chemical reaction with the environment is unavoidable, the reaction products should themselves be electrically conductive, easily penetrative, or removed by mechanical abrasion or other cleaning methods (Chomerics, 2000; Instrument Specialties, 1988).

By definition, a gasket is necessary only where an imperfect mating surface exists. When rigid flange surfaces are brought together, slight surface irregularities or unevenness on each surface always exist and prevent them from meeting completely at all points. The degree of joint unevenness or misfit is defined as the maximum separation that exists anywhere between the two flange joint surfaces when they are just touching. The joint unevenness is equal to this maximum and minimum difference as illustrated in the Figure 5.5. When the joint gap is filled with an elastomer gasket, the maximum compressed gasket height is at the point of maximum joint unevenness. In practice, the design of the gasket joint is complicated because the mating surfaces are not perfect dimensionally or electrically. Because gasketing materials are deformable, the contact points of two surfaces become enlarged as the clamping force is applied and more contacts begin to appear. The resistance to total current flow depends on the effective area of contact. This area can vary with the number of contact points and the applied force. Additional resistance, called constriction resistance, is encountered when current lines of flow are bent together and flow is forced through narrow contact areas. Moreover, surface irregularities can introduce another effect on the contact resistance due to possible radiation through

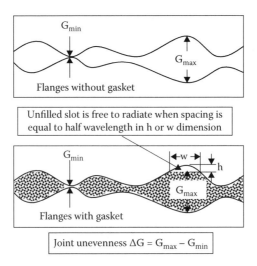

FIGURE 5.5 Flange joint unevenness with and without gasketing.

Conductive Elastomer and Flexible Graphite Gaskets

the gaps between points in contact, as shown in Figure 5.5. As discussed in Chapter 1, the cutoff frequency for radiation parallel to the long dimension of the slot will depend on the gap height (h). The cutoff frequency for polarization perpendicular to the slot will be determined by the width of the gap (w). This radiation will be small if the spacing with respect to wavelength is small or the gasket material in the slot is lossy. Conversely, the radiation can be large if the opposite conditions are true, or if the mechanical tolerances are large enough to create an air gap between the EMI gasket and the mating flange (Instrument Specialties, 1998). In order to minimize the gap effects, an appropriate gasket thickness should be determined, which is mainly influenced by applied clamping/compressive force, electrical performance, flange characteristics, fastener spacing, and the properties of the gasket materials.

5.4.2 Applied Clamping/Compressive Force and Deformation Range

In order to provide low contact resistance and high electrical conductivity across the gasketed flange joint, as shown in Figure 5.5, an EMI gasket must make full contact at the point of maximum separation between the mating surfaces to be sealed. In most shielding applications, a minimum compressive force or squeeze load is required at the point of maximum separation to assure an effective EMI shielding and a pressure or moisture seal. Moreover, the compressive force applied to the gasket must be great enough to compress it so that the difference between the maximum and minimum gasket height is equal to the joint unevenness. Therefore, an optimal clamping/compressive force must be chosen when designing a gasketed flange joint. Under this compressive force, the gasket operates within a specified height or deflection range. Staying within this deflection range assures adequate contact force for electric conductance while preventing damage to the gasket (Instrument Specialties, 1998). In addition to the compression force, the deformation of a gasketed flange joint system is mainly influenced by gasket stiffness, flange stiffness, fastener spacing, and fastener assembly or span type.

Gasket stiffness, E_g, is the amount of force required to compress a unit length of gasket at unit distance laterally, in the unit of kg/mm/m. It can be obtained from a linear approximation slope of load versus deflection test curve. For better accuracy, nonlinear finite element analysis can be used calculating it. Flange stiffness, E*I, is a measure of a flange's resistance to bending per unit of applied moment. E is elastic modulus (MPa), and I is the moment of inertia, which measures the stiffness of a geometrical cross-section independent of material. Gasket compression, Δh, is the amount of compression applied to a gasket during installation. With elastomer gaskets, deflection is often expressed in percent of free height, or in distance (mm). Design compression Δh_D, is usually halfway between minimum and maximum compression, as shown in Figure 5.6a. Compressing at least Δh_{min}, assures good contact; while compressing less than Δh_{max} assures no damage. Fastener span type refers to the end conditions of a gasketed region between a pair of fasteners. In an edge type span, as shown in Figure 5.6b, the end fasteners are located adjacent to the flange edge. The flange deflection can be expressed as (Instrument Specialties, 1998):

$$\Delta d = \Delta h_D [2e^{w/2}(1+e^w)\cos(w/2)]/[1+e^{2w}+2e^w\cos(w)] \quad (5.1)$$

where $w = L_f [E_g/(4EI)]^{1/4}$.

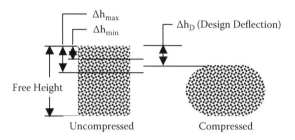

(a) Gasket design deflection

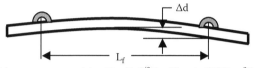

(b) Edge type span: $\Delta d = \Delta h_D [2e^{w/2}(1+e^w)\cos(w/2)]/[1+e^{2w}+2e^w\cos(w)]$,
where $w = L_f [E_g/(4EI)]^{1/4}$

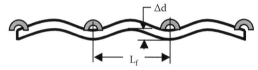

(c) Middle type span:
$\Delta d = \Delta h_D \{2e^{w/2}[(e^w-1)\cos(w/2)+(e^w+1)\sin(w/2)]\}/[e^{2w}+2e^w\sin(w)-1]$,
where $w = L_f [E_g/(4EI)]^{1/4}$

FIGURE 5.6 Schematic illustration of gasket installation and fastener span types.

In a middle type span (Figure 5.6c), the end fasteners are shared by another gasketed region. Middle type spans have smaller deflection than edge type spans of equal size, because the flange is locked in rotation at the fasteners. For this situation, the flange deflection can be expressed as (Instrument Specialties, 1998):

$$\Delta d = \Delta h_D \{2e^{w/2}[(e^w - 1)\cos(w/2) + (e^w + 1)\sin(w/2)]\}/[e^{2w} + 2e^w \sin(w) - 1] \quad (5.2)$$

where $w = L_f [E_g/(4EI)]^{1/4}$.

Fastener spacing, L_f, is the distance between bolts, clamps, or hinges, which can be calculated by (Chomerics, 2000):

$$L_f = 2\,[4EI/E_g]^{1/4} \quad (5.3)$$

For the flange with rectangular sections, the moment of inertia is $I = bh^3/12$, where b is width of the enclosure flange in contact with the gasket and h is thickness of the flange. E_g of the seals is typically 70–100 MPa (Chomerics, 2000). Use of proper fastener spacing will keep flange deflection (Δd) small enough to prevent loss of contact at span center. Spacing is very critical because Δd varies approximately with the third power of L_f. Increasing the number of fasteners is a major improvement in a flange joint design (Chomerics, 2000; Tecknit, 1998).

5.4.3 Gasket Profile and Materials Selection

Selection of the proper conductive elastomer gasket profile largely depends on application, manufacturability, installation, and material properties. When using flat gaskets, assembling holes should not be located closer to the edge than the thickness of the gasket, or a web cutter wider than the gasket thickness. Moreover, flat gaskets should not be deflected more than about 10%, compared with 15% to 30% for molded and solid extruded gaskets and 50% for hollow gaskets. Groove designs for shaped configurations are effective because gasket deflection can be controlled and larger deflections can be accommodated. Groove designs also provide metal-to-metal flange contact and require fewer fasteners, thereby minimizing the number of paths where direct leakage can occur. Hollow gasket configurations are very useful when large gaps are encountered or where low closure forces are required. Hollow gaskets will compensate for a large lack of uniformity between mating surfaces because they can be compressed to the point of eliminating the hollow area. The gasket configurations are usually limited to some typical shapes, which can be made by current manufacturing technologies, such as die cut, molding, extrusion, and form-in-place. In addition, as mentioned earlier, gasket geometry is largely determined by the largest gap allowed to exist in the junction. The ultimate choice in allowable gap tolerance is a compromise between cost, performance, and the reliability required during the life of the device (Chomerics, 2000).

Selection of the proper material depends on which design parameters are open to adjustment. Minimum compressive force, deflection characteristics, groove dimensions, and the gap between the mating surfaces are all interrelated, and each one is fixed by the assigned or restricted values of the other. There is a general guideline for gasket design (Instrument Specialties, 1998): (a) to increase compressibility, a thicker or softer gasket, or combination of both, should be used; (b) a thinner or harder gasket, or combination of both, will increase pressure applied on the gasket and need high clamping force; and (c) to accommodate great flange joint unevenness, a thicker or softer gasket should be used, or the clamping force is increased. The limitation for increasing the clamping force is the ability of the flange material itself to transmit additional pressure without being deformed.

Table 5.2 shows some typical materials that can be selected for conductive elastomer gaskets (Instrument Specialties, 1998). However, the design requirements of the installation will usually narrow the choice considerably, particularly if the basic geometry of the enclosure is already established, or if military EMI shielding specifications are involved. In addition to size and shape choices dictated by the enclosing structure and the joint geometry, the suitability of EMI gasket materials is also greatly influenced by shielding effectiveness, clamping force, service life, and environment compatibility requirements. Shielding performance of conductive elastomer gasket materials vary greatly, which is closely related to intensity and frequency of the EMI; the predominance of electrical (E), magnetic (H), or plane (P) fields; and system power and signal attenuation requirements.

As discussed earlier, the clamping force required for a given deflection depends on the hardness of elastomers. The harder the material the greater the required force. In sponge-type elastomers, the cross-sectional area shrinks as the height decreases

TABLE 5.2
Materials Selection for Typical Conductive Elastomer Gaskets

Conductive Elastomer Materials	Elevated Temperature	Salt Fog/ Shipboard	Jet Fuels/ Hydraulic Fluids	Oil Fields/ Automotives	Nuclear Biological Chemical (NBC)	Electromagnetic Pulse (EMP)	Electrostatic Discharge (ESD)	Light Weight/ Airborne	Low Closure Force	Commercial	Cellular Phones	Connectors	Waveguides	Military
F.SIL + Ag/G	X		X											
EPDM+C					X		X							
SIL + Al		X												
F.SIL + Al		X												
SIL + Ag/G									X	X				
SIL + Ag/Al	X	X				X		X	X	X		X		X
FC + Ag/Ni				X										
SIL + Ni/C		X	X					X	X	X	X	X		X
F.SIL + Ag/Al	X	X	X	X		X								X
SIL + Ag/Cu						X						X		X
SIL + Ag	X					X		X	X			X	X	X
SIL + Ag/Ni						X						X		X
EPDM + Ni/C		X			X					X				
SIL+C							X							
F.SIL + Ag/Cu			X	X		X						X		X

Environmental Application

Material									
F.SIL + Ag/Ni	X	X	X	X		X		X	X
F.SIL + Ag/Ni			X	X					
EPDM + Ag/Ni					X				
F.SIL + Ni/C		X	X			X	X		
SIL + Ag/Cu						X		X	X
EPDM + Ag/Al		X			X	X			X
F.SIL + Ag	X				X			X	X

Note: C, carbon; Al, passivated aluminum; Ag/Al, silver-plated aluminum; Ag/Cu, silver-plated copper; Ag/G, silver-plated glass; Ag/Ni, silver-plated nickel; Ni/C, nickel-coated carbon; Ag, silver.

under the clamping force. In solid conductive elastomers, the cross-sectional area remains virtually the same, except it performs flowing laterally in the unconfined direction under the increased clamping force. Deflection characteristics of conductive elastomers are profile specific and a selection of a proper shape should be made to stay within the constraints of flanges and applied clamping force. Solid conductive elastomer materials usually stand up better to high clamping forces, environmental pressures, and repeated opening and closing of the joint. Unlike sponge elastomers, solid conductive elastomers do not actually compress. They accommodate pressures by changing shape rather than volume. This is an important difference in flange joint design requirements between the two materials types, since additional gland volume must be allowed for the potential expansion of the elastomer under heat and/or pressure. Greater flange strength must often be provided to allow for increased clamping force requirements. If low clamping force is needed, however, the use of hollow gasket profiles in conjunction with softer elastomers will significantly reduce the clamping force required (Instrument Specialties, 1998).

Service life of conductive elastomer gaskets will vary depending on the environment, which is basically influenced by (a) resistance to corrosion or structural deterioration due to environmental factors, such as detrimental chemicals and fluids, ozone aging, and temperature extremes; (b) the ability to maintain resiliency and integrity against repeated compression/deflection cycles; and (c) inherent resistance to impact and abrasion.

Compatibility between the gasket and the mating flanges is another area must be given proper attention when designing a conductive elastomer gasket. This involves the appropriate selection of a gasket material compatible with the joint material that employs a compatible or corrosion-resistant metal in its composition, or a seal design that reduces or excludes environmental moisture from the metallic regions of the gasket and flanges. Proper material selection for effective EMI shielding depends on the total environmental envelope within which the gasketing seal will be expected to function. The major environmental conditions include temperature; aging; pressure/vacuum; fluid compatibility; corrosion potential; and nuclear, biological, and chemical environments.

When the temperature decreases below allowable limits, the elastomeric properties are lost and the material becomes very hard and brittle. High temperatures also affect the properties of elastomers in the same way as low temperatures. Deterioration with time or aging relates to the type of polymer and storage conditions. Exposure to ultraviolet light, radiation, heat, oxygen, and moisture may cause deterioration of elastomers whether installed or in storage. Resistance to deterioration in storage varies greatly among elastomers. Conductive elastomer seals are rarely used for high-pressure systems, with the exception of waveguide seals. Pressure has a bearing on the choice of material and hardness. Softer materials are used for low-pressure applications, whereas high pressure may require a combination of material hardness and proper design. Outgassing (vaporization) and/or sublimation in a high vacuum system can cause seal shrinkage (loss of volume) resulting in possible loss of sealing ability. When properly designed and confined, an O-ring, molded shape, or a form-in-place seal can provide adequate environmental sealing, as well as EMI shielding for vacuum (to 1×10^{-6} Torr) applications. Consideration must be given to the basic compatibility between the elastomer gasket and any fluids with which it may come in prolonged contact. Common forms of corrosion include galvanic, pitting, and crevice corrosion. However, galvanic corrosion is the major concern for the use of conductive elastomers. In addition, exposure to nuclear, biological, or chemical environments is a complex phenomenon. In order to survive in the these environments, the elastomer should have low permeability, resistance to nuclear hardening, or loss of memory due to breakage of cross-linking chains (Chomerics, 2000; Instrument Specialties, 1998).

5.4.4 Flange Materials and Joint Surface Treatment

As mentioned earlier, conductive elastomer gasket materials are usually selected to produce low volume and low surface resistivity in a wide variety of environments. It is equally important to make both mating surfaces highly conductive and to minimize surface contact resistivity over an extended time period during operation in a variety of external environments. This is generally accomplished by selecting highly conductive flange materials and providing surface finishes that resist the formation of hard oxides or other insulating or semiconducting films on the flange surfaces.

Flanges are normally made of the same material as the basic enclosure for reasons of economy, weldability, strength, and resistance to corrosion. Wherever possible, flanges should be made of or plated with materials with the highest possible conductivity (Chomerics, 2000). In those cases where the oxide buildup is both rapid and hard, such as with copper and aluminum flange, it is best to protect these surfaces with some form of plating or other surface protection that will produce a relatively soft oxide film that can easily be abraded by minimal mechanical contact. Coatings such as tin, cadmium, and zinc form such soft oxides. Gold plating offers absolute resistance to any oxide formation and is an outstanding conductor but is very expensive and rather soft. Silver is also an excellent conductor and its oxides are themselves fairly conductive, however, silver is an expensive metal and while harder than gold it is still subject to wear and abrasion. Nickel, which is much less expensive, is a good conductor and is hard and extremely resistant to corrosion making it an excellent all-around plating material for EMI flanges. Which surface coating is most suitable for a given application depends on the base flange material and the compatibility of the intended coating with it. Nickel plating is generally recommended for aluminum parts, although cadmium and tin are widely accepted. Cadmium, however, is only used in some special applications because it is against RoHS regulations. Zinc plating is primarily used with steel (Instrument Specialties, 1998).

Chemical coatings and films that exhibit relatively good electrical conductivity include Oakite36, Alodine 1000, Iridite 14, and Iridite 18P. These films may be useful to protect conductive surfaces on an interim basis, but, in general, they are not as durable as electrodeposited plated finishes. Never, under any circumstances, should anodized mating surfaces be used for EMI shielding service. Anodized surface coatings are very hard and nonconductive. All protective coatings having conductivity less than the metal surfaces being bonded must be removed from the contact area of the two mating surfaces. When interim protective coatings are necessary, they should be easily removable from the mating surfaces when required. Certain protective metal plating such as cadmium, tin, or silver need not be removed. The noble metals (gold and its alloys) provide improved conductivity over cadmium. In all cases, given the corrosive atmosphere present in many applications, some surface treatment is better than no protection at all. Most other protective surface coatings, such as paint and anodizing, are nonconductive and must be completely removed for an effective EMI flange surface chains (Chomerics, 2000; Instrument Specialties, 1998).

5.5 CONDUCTIVE ELASTOMER GASKET FABRICATION

Conductive elastomer materials provide an extensive range of EMI gasketing and environmental sealing solutions. A variety of conductive elastomer gasket configurations have been fabricated, including various extruded profiles, molded preform and die cuts, molded and shaped components, form-in-place, screen-printed gaskets, vulcanized metal covers or flanges, and reinforced shielding seals with oriented wires and metal meshes. Some typical fabrication processes will be discussed next.

5.5.1 Extrusion

As shown in Figure 5.7, the extrusion process forces the conductive elastomer material under pressure through a shaped die to produce the desired profiles. Depending on the type of extruder, the profile is either cured in a heated die or cured as it passes through a high temperature oven, forming solid or hollow strips, and from flat sheets to complex cross-section components produced in cost-effective continuous lengths.

A variety of differently dimensioned sections of each of the profiles can be produced. Compared with the equivalent solid cross-section, hollow sections are relatively

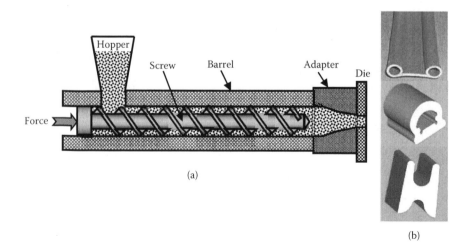

FIGURE 5.7 (a) Extrusion process and (b) extruded profiles.

favorable because of their extra compressibility and the profile uses less or saves the metal-filled elastomer material. U channel and H sections are commonly fabricated for use with sheet metalwork enclosures; both sections can be mounted along the edge of the metal sheet or strip. They are particularly made for cabinet doorframes. In addition, extruded profiles can readily be cut and spliced to form a continuous seal; this can offer cost advantages over a molded seal, unless the quantities are large. EMI or environmental performance is not sacrificed by this splicing or die-cut method to of making the gaskets.

When laying out the channel section to take the extruded profile, the grooves must provide a radius equal to or greater than half the strip width in the corners. Alternatively, a silver-loaded conductive silicone adhesive can be cured at room temperature. Mitered joints at the corners are another possibility with each length of the profile laid in the grove or channel provided, and glued to the next length to ensure continuity (Molyneux-Child, 1997).

5.5.2 Molding

There are several kinds of molding processes, including compression or transfer molding, comolding into intricate shapes, rotary or injection molding, and mold-in-place on metal or plastic substrates.

Transfer molding requires an extra pot in which the conductive elastomer material will be placed. The mold is then closed and the elastomer material is expelled from the barrel to the cavities through runners and gates. This approach enables O and D cross-section rings to be easily formed. Virtually any cross-section can be joined to form a ring, although if it is a complex section or has a volume requirement, molding should be seriously examined. Flat washers can be constructed out of rectangular profiles, where transfer molding would exhibit more advantages than die stamping from sheet stock. Comparably, injection molding is the most elaborate and automatic way of manufacturing complex gasket profiles. It is particularly suited for high quantities.

Die-cut gaskets from conductive sheeting preform have high dimensional accuracy, however, considerable loss of expensive material may be caused for many configurations (sometimes up to 80% of the sheet material is wasted using this technique; Molyneux-Child, 1997). The die-cut tools are not particularly expensive and the technique lends itself to relatively small quantities of gaskets for waveguides or flange mounting applications. Die cutting gaskets into sheets is one of the most common finishing jobs. The sheet can receive pressure-sensitive tape (PST) to manufacture self-adhesive gaskets.

Whereas die cutting thin sheets of conductive elastomer is adequate for small batch runs, gaskets required in thicker sections or in a more complex shape would be better molded in the volume-conductive elastomer. Molding might be a better approach for higher volume requirements, once all design changes have taken place and the fabrication process finalized for production.

The mold-in-place process is used to create high aspect ratio gasket profiles (height is bigger than width) on metal or plastic substrates with a defined profile or shape. It is an automated process with efficient material utilization.

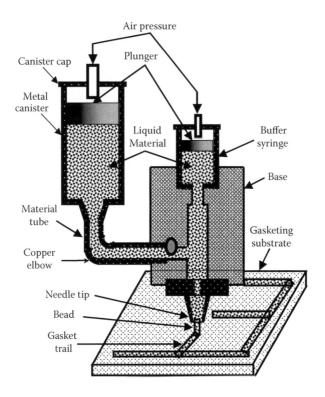

FIGURE 5.8 Process of form-in-place gaskets.

5.5.3 FORM-IN-PLACE AND SCREEN PRINTING

Form-in-place gasketing is the liquid application process of gasketing material onto a substrate via programmable automation. As shown in Figure 5.8, the conductive elastomer can be directly dispensed over the part to form the gasket directly in place. This process is highly accurate, fast, and repeatable. The liquid material is usually supplied in one or more parts, and can be precisely metered and mixed. Once applied, the form-in-place gasket is cured either by heating or at room temperature, depending on the chosen material and other design factors. Form-in-place gaskets are typically comprised of a foamed, gelled, or unfoamed elastomer resin, such as silicone urethane or other similar polymers, and are used as carriers for conductive fillers. The filled resin is lined onto one or more mating surfaces of an enclosure or other similar substrate to provide an EMI shielding gasket. Alternatively, an unfilled elastomer resin can also be lined onto the enclosure and then coated with a conductive outer layer, such as silver or another similar alloy. Form-in-place gaskets have been commonly applied with the proper machinery to most contours and mating surface patterns. However, they have some disadvantages like many other conductive elastomer gaskets. Because they are only partially filled with conductive materials, the form-in-place gaskets typically require high compressive forces between the mating enclosure surfaces to ensure that adequate grounding contact is made

Conductive Elastomer and Flexible Graphite Gaskets 149

with the conductive particles contained within the elastomer resin. This may limit their application when the electronic enclosures are both smaller and designed with increasingly thinner wall thickness, and achieving the necessary compressive forces without flexing or damaging the enclosure becomes more difficult.

Similarly, screen printing can also be used to form flat gaskets directly on back panels for input/output. This method requires perfectly flat receiving flanges. The manufacturing process can be used on any flat substrate able to withstand the curing temperature. It is possible to print secondary environmental sealing as well as compression stops in addition to the conductive gasket. Flatness of mating components is a critical requirement for this printed technique because of the inherent limitation on thickness caused by the printing process. Printed gaskets allow large contact areas to the flange with minimal thickness, minimizing and maintaining impedance of the joint and therefore maintaining even current flow over a wide frequency range. The flexibility of this screening technique provides various opportunities to meet selective customer requirements. A single print will produce a 0.30 mm thick gasket, whereas an additional print results in 0.50 mm thickness. This would normally be used in producing a raised sealing bead on the gasket profile. Besides, the very low tooling costs involved in screen printing, in contrast to conventional molding processes, results in a highly economic gasket system tailor-made to application requirements. Gaskets can be manufactured as convenient sets on the same temporary release carrier, resulting in improved inventory control. Special configurations can be printed directly onto a printed circuit board (PCB), flexible circuit, cover plate, or flange. Sealing bead widths down to 1.5 mm are possible in special circumstances giving weight-saving advantages.

5.5.4 REINFORCED SHIELDING GASKETS WITH ORIENTED WIRES AND METAL MESHES

Figure 5.9 illustrates the winding process of conductive elastomer gaskets reinforced with oriented wires. Imbedding oriented wires in an elastomer material is usually processed in silicone matrix impregnated with conductive aluminum or Monel wire, or metal meshes, which are chemically bonded to the silicone and convoluted to minimize compression set. The reinforced gaskets can be then manufactured by a rolling and cutting process. Metal wires are first sandwiched between silicone layers. Then raw slabs are cut and cured. The final strips are then obtained by slicing the slabs. In general, only rectangular cross-sections can be fabricated using this process.

5.6 CONDUCTIVE ELASTOMER GASKET INSTALLATION AND APPLICATION

Common conductive gasketing installation methods include positioning in a groove, interface fit, bonding with adhesives, ribbed profiles, bolt-through holes, vulcanized mounting, and covering with a compression stop, as shown in Figure 5.10. The flange with a groove is ideal for positioning the gasket if a suitable groove can be provided at a relatively low cost. This method provides several advantages (Instrument Specialties, 1998): (a) metal-to-metal contact of mating flange surfaces provides a

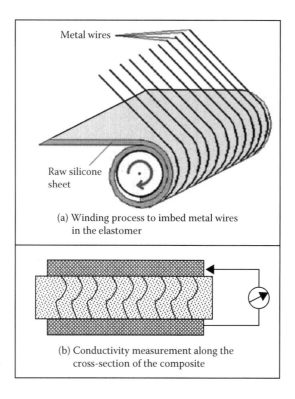

FIGURE 5.9 Conductive elastomer gaskets reinforced with oriented wires. (a) Winding process to imbed metal wires in the elastomer; (b) conductivity measurement along the cross-section of the composite.

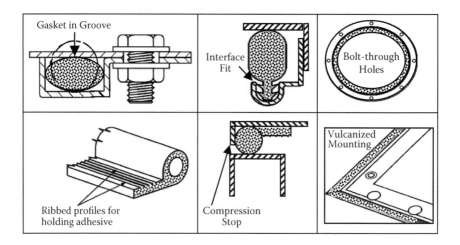

FIGURE 5.10 Typical mounting techniques of conductive elastomer gaskets.

compression stop and prevents overcompression and deflection of the gasket material; (b) it is a cost-effective method due to reducing assembly time; and (c) it is the best overall seal for EMI, EMP, salt fog, NBC, and fluid with metal-to-metal flange contact and reduced exposure of the seal element to attack by outside elements. For conductive elastomers that are incompressible, however, proper allowance for the material flow must be provided in the groove design. The underfill and overfill of the groove will affect the overall performance of the gasket. A 95% volume/void ratio is generally recommended. Overstressing of the elastomer may degrade the electrical and physical properties of the conductive elastomer material. In addition, complex fasteners are often required to make junctions more compact. Screw or bolt fasteners should be mounted outboard of the gasket to eliminate the need for separate shielding gaskets under each individual fastener (Chomerics, 2000).

For applications such as face seals, or where the gasket must be retained in the groove during assembly interface, the interface fit method is an excellent and inexpensive choice. The gasket is simply held in the groove or against a shoulder by mechanical friction.

The bonding and adhesive method can be used to attach the gasket to one of the mating flanges by the application of pressure sensitive or permanent adhesives. A suitable conductive adhesive is always preferable for mounting EMI gaskets to provide adequate electrical contact between the gasket and the mounting surface. However, some specially designed rib profiles with longitudinal grooves on the bonding surface allow the use of nonconductive adhesives.

Bolt-through holes are a common and inexpensive way to hold an EMI gasket in position. Locator bolt holes can be accommodated in the tab in flat gaskets.

The vulcanized-to-the-metal technique offers a homogeneous one-piece gasket with superior conductivity between the gasket and the metal. In this case, the seal element is vulcanized directly to the metal flange or cover under heat and pressure.

Access-door gaskets are suitable for positioning in an environment where friction, abrasion, and impact need to be considered. The gaskets in such an environment should be positioned so that they receive little or no sliding or side-to-side motion when being compressed. In addition, the objective of any well-designed mounting method is to achieve a junction that will transmit all signals across the interface with little or no loss of signal strength as well as having no effect on the characteristics and pulse shape of the signal being transmitted. Many factors such as surface oxides, dirt, grease, and other contaminants increase junction resistance. Other factors such as nonparallelism of flanges, surface roughness, and flange misalignment cause reactive losses, which can have a major effect on system performance (Instrument Specialties, 1998).

A waveguide flange is the simplest of the waveguide components. However, a poorly designed waveguide flange can eventually cause more system degradation than malfunctioning electronic parts. For flat-cover flanges, a die-cut gasket incorporating expanded copper foil to control gasket creep into the critical waveguide opening provides an excellent EMI seal. Raised lips around the gasket cutout improve the power and pressure sealing capability of this compression-type gasket design. Choke flanges may be sealed with molded circular D-section gaskets and contact-type flanges with molded rectangular D-section gaskets in suitable retaining grooves.

Die-cast flange joints with cast or machined grooves provide higher shielding than overlap sheet-metal flange joints. The precession of cast flange cannot be achieved in sheet-metal flanges. The sheet-metal flanges can be reinforced with metal strips, bars, or angles; however, sheet-metal flanges cannot offer the same rigidity and reliance as cast flanges. A machined groove is always preferable over a cast groove. The machined surface should not be too smooth to permit the rolling of the seal or too rough to permit joint leakage.

The thickness of a flange is more or less determined by the amount of stiffness necessary to prevent excessive bowing of the flange between fastener points. Fewer and wider spaced but larger diameter fasteners will require a thicker flange to prevent excessive flange deflection and potential leakage of fluid or electromagnetic energy. Another consideration is the "electrical" spacing of fasteners to prevent possible signal leakage from minute spaces between fasteners centers. To prevent this, bolt or fastener spacing should be a minimum of ¼ wavelength of the primary operating frequency (Instrument Specialties, 1998).

Flat gaskets are ordinarily used with sheet-metal or machined flanges. Location of the fastener holes in the flanges should be designed at least 1.5 times the bolt diameter from the edge of the flange to avoid tearing when the metal is punched. In the case of drilled holes, the positions of holes should not be less than the thickness of the material from the edge of the flange. In those cases where the holes must be placed closer to the edge, ears or slots should be considered.

An O-ring gland or a gland for any type of seal consists of a groove in the flat surface plus the mating surface that covers it when assembled. Groove configuration may vary according to the requirements of the assembly. However, rectangular-type groove shapes have proven to be advantageous over other configurations. It is desirable that the O-ring not be excessively stretched or compressed when installed in the gland. As the gasket stretches, its cross-section diameter diminishes accordingly. Also, some elastomeric parts age more rapidly while under stress, which is another reason to minimize stretch. For a short-term application or where the O-ring material is not prone to stress cracking, the stretch should be limited to less than 5% (Molyneux-Child, 1997).

Conductive elastomers are, by their nature and composition, pressure-sensitive materials, that is, the conductivity across the joint increases with the increase in applied load and the resulting squeeze. The EMI seal/shield should always be designed in such a way as to ensure that it is subjected to the minimum squeeze sufficient to provide the required mechanical and electrical performance.

Providing EMI shielding and environmental sealing, conductive elastomers have been widely used in many areas including the electronic, electrical, telecom, wireless communication, housing, medical, and automotives industries.

5.7 COMPARABLE FLEXIBLE GRAPHITE GASKETS

As discussed thus far, EMI shielding gaskets can be made from metallic and conductive elastomer materials. The conductive elastomers usually require the use of an expensive conductive filler, such as silver particles, in order to attain a high shielding effectiveness while maintaining resilience. With increasing filler content, the resilience of the conductive elastomer decreases rapidly. Moreover, the shielding effectiveness of these elastomer materials degrades in the presence of moisture or solvents. In addition, the elastomer matrix limits the material's temperature resistance, and the thermal expansion mismatch between filler and matrix reduces the thermal-cycling resistance. Comparably, flexible graphite gasketing material has advantages over conductive elastomer in many facets, such as conductive, resilient, chemically inert, and thermally resistant properties. Although it is not as ductile as silicone, it does not suffer from stress relaxation. As the electrical conductivity (especially that in the plane of the sheet) and specific surface area are both quite high in flexible graphite, the effectiveness of this material for shielding is exceptionally high (Luo et al., 2002).

Therefore, the high thermal conductivity, low coefficient of thermal expansion (CTE), high-temperature resistance, and excellent chemical resistance make flexible graphite an attractive candidate for EMI shielding applications.

5.7.1 Flexible Graphite and Its Properties

Flexible graphite is a flexible sheet made by compressing a collection of exfoliated graphite flakes (called worms) without a binder. During exfoliation, an intercalated graphite flake expands typically by over 100 times along the c axis of honeycomb graphite crystalline structures. Compression of the resulting worms (like accordions) causes the worms to be mechanically interlocked to one another so that a sheet is formed without a binder.

Figure 5.11 illustrates the typical fabrication process of flexible graphite (Garlock Sealing Technologies, 2000). Flexible graphite is prepared from flakes of purified

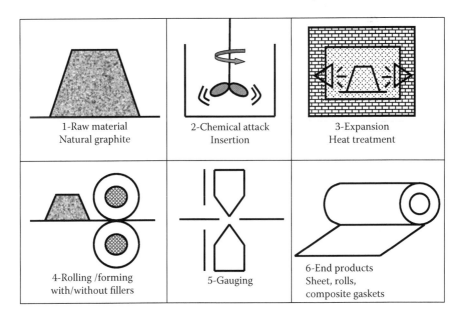

FIGURE 5.11 Fabrication process of flexible graphite.

natural graphite. This allotropic variety of carbon can form into lamellar compounds by the insertion of certain atoms or molecules in the crystalline structure of the graphite. This is used in the preparation of flexible graphite by adding an acid. Then the insertion compound thus formed undergoes a thermal shock at a very high temperature that regenerates a highly expanded form of graphite by the sudden departure of the inserted element. This expanded graphite, which has a very low specific gravity (a few grams per liter), can self-agglomerate by a simple mechanical action without any binder. Therefore, a flexible or semirigid material in roll or sheet form, with or without conductive fillers, can be produced through rolling or pressing. In order to adapt flexible graphite to the needs of specific applications of EMI gasketing, certain conductive additives or fillers can be inserted into the flexible graphite during processing through careful selection of raw materials and processes.

Due to the c axis of the honeycomb microstructure of exfoliated graphite, being somewhat connected perpendicular to the sheet, flexible graphite is electrically and thermally conductive in the direction perpendicular to the sheet although not as conductive as the plane of the sheet. These in-plane and out-of-plane microstructures result in resilience and impermeability to fluids perpendicular to the sheet. The combination of resilience, impermeability, and chemical and thermal resistance makes flexible graphite more appealing than conductive elastomers for use as a gasket material for high-temperature or chemically harsh environments.

In summary, flexible graphite provides the following performance (Garlock Sealing Technologies, 2000; Luo et al., 2002):

- High electrical conductivity and high specific surface area (skin effect)—Make the shielding effectiveness of flexible graphite exceptionally high, up to 125–130 dB.
- High thermal conductivity and low CTE—Make flexible graphite a good thermal dissipation material.
- High creep and stress relaxation resistance—Flexible graphite has little stress relaxation, a very low creep rate in its service temperature range, and holds up to high seating pressures.
- Excellent chemical inertness—Flexible graphite, which contains no binders, withstands most chemicals, mineral acids, and solvents.
- High temperature resistance—In a reducing or inert atmosphere, flexible graphite retains its characteristics over a broad range of temperatures, from −196°C to +2500°C (sublimation temperature 3500°C). In the presence of oxygen, graphite forms CO or CO_2 from 450°C (550°C for the grade with passive inhibitor). The rate of oxidation depends on the area in contact with the oxidizing medium.
- Good compressibility—Flexible graphite is compressible up to a density of 2.25 g/cc.
- Good resilience and elastic recovery—Flexible graphite is highly resilient and has a capacity for elastic recovery of about 10% in volume, making it possible to maintain compressibility and tightness over a broad temperature range.

- Good lubrication—Flexible graphite is self-lubricating and has low coefficient of friction (less than 0.10 with steel).
- Easy processing—Flexible graphite is easy to cut using conventional manufacturing equipment.
- Long service life and least aging—Flexible graphite is one of the nonmetallic materials least affected by aging. This property is not influenced by temperatures.
- Apparent anisotropy—Flexible graphite is anisotropic. Its mechanical properties and conductivity are very different between the directions parallel and perpendicular to the flakes. Due to its microstructure involving graphite layers that are preferentially parallel to the surface of the sheet, flexible graphite is high in electrical and thermal conductivities in the plane of the sheet.
- Excellent radiation performance—Flexible graphite resists aging from high levels of radiation: no visible effect at 10^{22} neutron/cm^2.
- Excellent removability—Flexible graphite adheres very little to the surfaces in contact with it.

With these remarkable properties, flexible graphite is effective for many electronic applications, including EMI gasketing, resistive heating, thermoelectric energy generation, and heat dissipation. It is comparable to or better than conductive elastomer materials for EMI gasketing. As a heating element, flexible graphite provides temperatures up to 980°C, response half-time down to 4 sec, and heat output at 60 sec up to 5600 J. Its absolute thermoelectric power through-thickness is –2.6 µV/°C. Flexible graphite is effective as a thermal-interface material when the thickness is low (0.13 mm), the density is low (1.1 g/cm^3), and the contact pressure is high (11.1 Mpa; Luo et al., 2002). These applications result from the flexibility and compliance of flexible graphite, in addition to its electronic and thermal behavior. Compliance is particularly important for the use of flexible graphite as shielding gaskets or interface materials, whether the interface is electromagnetic, thermoelectric, or thermal.

5.7.2 Fabrication and Application of Flexible Graphite Gaskets

Flexible graphite gaskets have been developed and processed with a variety of profiles, as shown in Figure 5.12.

- Flexible graphite sheets and laminates—Plain or reinforced sheets and laminates are used for various gasketing applications. The reinforcements include nickel, stainless steel or other metal foils, wires, meshes, perforated sheets, Mylar, and other inserts.
- Flexible graphite foil in rolls—Homogeneous sheet in continuous roll form, slit ribbon, and adhesive backed sheets have been produced for applications of cutting gaskets, metal gaskets and spiral-wound gasket filler, or other applications such as thermal liner, heat conductor, solid lubricant, and metal fabrication. The graphite sheet or rolls can be processed with active protection (such as zinc coating) or passive protection.

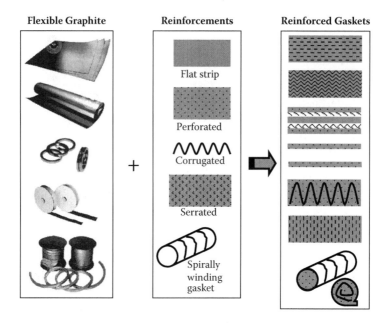

FIGURE 5.12 Processing of typical flexible graphite gaskets.

- Graphite tape and die-formed rings—Corrugated or crinkled flexible graphite tape, die-formed high-density rings can be used as valve stem packing, gaskets, or seals.
- Flexible graphite compression packing—Braided flexible graphite compression packing can be used for applications in pump and valve packing.

Flexible graphite-cut gaskets can fit standardized flanges with flat or raised faces, and narrow or wide tongue and groove faces. Some advantages of flexible graphite gaskets over conductive elastomer gaskets include (Garlork Sealing Technologies, 2000) they fit in place of standardized gaskets made of conventional materials; they have very good resistance to extreme temperature (from cryogenic levels to 450°C); they have good chemical inertness; they remain tight at fire test conditions; and the seating pressure applied during installation is maintained over time because of the very low level of relaxation, in the order of 6% to 8% at most.

For instance, flexible graphite gaskets can replace elastomer O-rings for static applications in high and low temperatures, highly corrosive fluids, and radioactive environments. For nuclear applications, flexible graphite gaskets can be used in steam generators, heat exchangers, all types of valves, and ships and submarines. In the chemical and petrochemical industry, flexible graphite gaskets can be used for valves, autoclaves, pipes, heat exchangers, corrosive circuits, cryogenics, heat-transfer fluids, and high- and low-pressure steams. Flexible graphite gaskets can be installed on enameled sealing contact areas and on technical polymers. Flexible graphite can also be used for blocks and plates for handling pieces of molten glass.

Flexible graphite is feasible and promising for EMI gasketing or other electronic applications in the future. Its shielding performance, for instance, is comparable to or better than conductive elastomer materials, such as silicone filled with AgCu particles (Luo et al. 2002).

5.8 SUMMARY

Conductive elastomer gaskets, made from conductive, resilient, and conformable vulcanized elastomer composites function as EMI shields and environmental seals. When applied between the mating surfaces, they can follow small deflections, flow around small flaws, and conform to flange surface irregularities providing low interface resistance between the mating electronic enclosure flanges, while simultaneously providing moisture, pressure, and environmental sealing.

The shielding effectiveness of conductive polymers ranges from 60 dB to 120dB. Variations in their shielding effectiveness is related to the compression force applied and the amount of conductive filler loading.

Conductive elastomers are, by their nature and composition, pressure-sensitive materials. The conductivity across the joint increases with the increase in applied load and the resulting squeeze. The EMI shield/seal should always be designed in such a way as to ensure that it is subjected to the minimum squeeze sufficient to provide the required mechanical and electrical performance.

Flexible graphite gasketing material, however, is comparable to or potentially better than conductive elastomer materials. The shielding effectiveness reaches up to 125 to 130 dB due to its low electrical resistivity and high specific surface area. In addition, its high thermal conductivity, low CTE, high temperature resistance, and excellent chemical resistance make it very attractive for increasing use in EMI shielding.

REFERENCES

Bhattacharya, S. 1986. *Metal-filled Polymers*. New York: Marcel Dekker.
Bigg, D. M. 1984. The effect of compounding on the conductive properties of EMI shielding compounds. *Advances in Polymer Technology* 4: 255–266.
Callen, B. W., and J. Mah. 2002. *Practical considerations for loading conductive fillers into shielding elastomers.* http://www.conductivefillers.com/papers/callen_article_13.pdf
Chomerics. 2000. *EMI Shielding Engineering Handbook*. http://www.chomerics.com/products/documents/catalog.pdf
Fiske, T., S. Railkar, and D. Kalyon. 1994. Effects of segregation on the packing of spherical and non-spherical particles. *Powder Technology* 81: 57–64.
Garlock Sealing Technologies. 2000. Flexible graphite. http://www.garlock.eu.com/pdfs/Products-OK/Gasketting-OK/Flexible%20Graphite%20-%20CEFIGRF.pdf. Accessed on June 3, 2008.
Hudak, S. 2000. A guide to EMI shielding gasket technology. *Compliance Engineering*. http://www.ce-mag.com/ARG/Hudak.html
Instrument Specialties. 1998. *Engineering Design and Shielding Product Selection Guide* (acquired by Laird Technologies). Delaware Water Gap, PA: Instrument Specialties.
Kalyon, D. M., E. Birinci, B. Karuv, and S. Walsh. 2002. Electrical properties of composites as affected by the degree of mixedness of the conductive filler in the polymer matrix. *Polymer Engineering and Science* 42: 1609–1617.

Luo, X., and D. D. L. Chung. 1996. Electromagnetic interference shielding reaching 130 dB using flexible graphite. *Carbon* 34: 1293–1294.

Luo, X., R. Chugh, B. C. Biller, Y. M. Hoi, and D. D. L. Chung. 2002. Electronic applications of flexible graphite. *Journal of Electronic Materials* 31: 535–544.

Mather, P. J., and K. M. Thomas. 1997. Carbon black/high density polyethylene conducting composite materials. Part I. Structural modification of a carbon black by gasification in carbon dioxide and the effect on the electrical and mechanical properties of the composite. *Journal of Materials Science* 32: 401–407.

MIL-G-83528. 1992. Gasketing material, conductive, shielding gasket, electronic, elastomer, EMI/RFI, general specification for MIL-G-83528.

Molyneux-Child, J. W. 1997. *EMC Shielding Materials*. Oxford: Newnes.

Rosner, R. B. 2000. Conductive materials for ESD applications: An overview. *Electrical Overstress/Electrostatic Discharge Symposium Proceedings*, 121–131.

Sichel, E. 1982. *Carbon Black-polymer Composites*. New York: Marcel Dekker.

Tecknit. 1998. *Electromagnetic Compatibility Design Guide*. http://www.tecknit.com/REFA_I/EMIShieldingDesign.pdf

Weber, M., and M. Kamal. 1997. Estimation of the volume resistivity of electrically conductive composites. *Polymer Composites* 18(6): 711–725.

6 Conductive Foam and Ventilation Structure

6.1 INTRODUCTION

The electromagnetic interference (EMI) shielding of large-area openings or apertures for access to electronics, ventilation, and displays is one of most important requirements to ensure the shielding integrity of electronic equipment enclosures. In addition to panel meters, digital displays, oscilloscopes, status monitors, mechanical indicators, or other readouts, the most critical display shields against EMI are large-area, high-resolution monitors (cathode ray tubes; Carter 2004; Grant 1984). EMI shielding of these large apertures of the integrated enclosures are generally more challengeable than those encountered for cover plates, doors, and small apertures, such as connectors, switches, and other controls in which the majority of the opening is covered by a continuous homogeneous conductive plate. Therefore, a variety of window panels made from conductive foam, waveguide honeycomb, metallized fabrics, and other ventilation structures have been developed and utilized to meet the EMI shielding requirements of the large-area openings and ventilation apertures.

Among the conductive foams, some of the most common materials are the low cost and highly conductive fabric-over-foam EMI shielding panels and gaskets, which are generally made from the resilient, open-cell polyurethane foam clad with metallized fabric. This unique foam structure provides high shielding performance and allows the shields to fit snugly into uneven gaps and joints by low enclosure forces in operation.

This chapter will give a brief review regarding the waveguide aperture and ventilation panel; conductive foam and integral window structure; and metallized fabrics and fabric-over-foam shielding materials.

6.2 WAVEGUIDE APERTURE AND VENTILATION PANEL

The major purpose of constructing a waveguide aperture and ventilation panel in an EMI shielding system is to improve the thermal environment for the electronic components, such as processor core logic (PCL) in chassis enclosures, while significantly enhancing EMI shielding performance. The PCL components usually consist of the processor, chipset, memory, and graphic components. The chassis enclosure built around the PCL must provide increased airflow and lower internal air temperatures to meet increasing thermal dissipation demands. Although typical chassis design usually provides general EMI containment and cooling to the PCL components, increasing thermal and EMI requirements call for an additional waveguide aperture or ventilation panel to facilitate this performance and maintain the balance between thermal and EMI performance (Intel, 2001).

6.2.1 Waveguide Aperture and Its Design

A waveguide is essentially a hollow conducting tube, such as apertures of a honeycomb, that acts as a filter for EMI. As discussed in Chapter 1, a waveguide aperture is generally designed such that all frequencies of interest are greatly attenuated by the waveguide. The shielding performance of a waveguide structure is governed by the surface geometry of the apertures (maximum cross-section dimension, d, and depth, h), the shape, and the total number of apertures. These factors are related to a combination of two parameters: cutoff frequency (f_c), which determines the maximum possible frequency of effectiveness; and shielding effectiveness (SE), which determines the magnitude of the EMI attenuation and is a function of frequency. The cutoff frequency is the frequency beyond which the waveguide no longer effectively contains EMI. The shielding effectiveness of a waveguide represents the amount of EMI attenuation that the waveguide offers at a given frequency. The shielding waveguide pertains to honeycomb-type materials commonly used to make air vent panels for electronic enclosures. For these panels, the absorption loss of EMI for a single waveguide is reduced by a factor related to the total number of apertures in the panel. The total shielding effectiveness including the reflection factor can be estimated by Equation (1.39): $SE\ (dB) = 20 \log\ (f_c/f) + 27.3\ h/d - 10 \log n$. Based on this equation, Figure 6.1 illustrates the high levels of shielding effectiveness for waveguide-below-cutoff samples with different single-aperture size and moderate values for h/d. The effect of the number of apertures (n) is apparent. Figure 6.2 and Figure 6.3 illustrate the shielding effectiveness of waveguide-below-cutoff examples with different numbers of apertures when $h/d = 1$ and $h/d = 3$, separately. A change

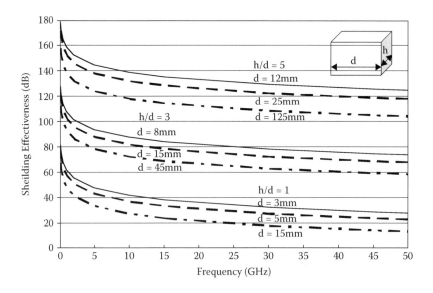

FIGURE 6.1 Estimated shielding effectiveness of waveguide-below-cutoff samples with single aperture ($n = 1$).

Conductive Foam and Ventilation Structure

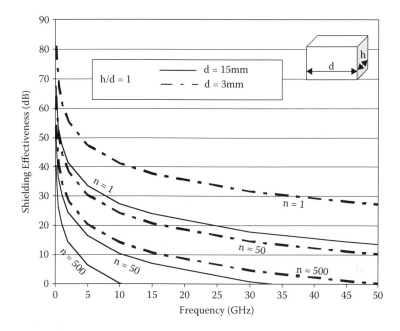

FIGURE 6.2 Estimated shielding effectiveness of waveguide-below-cutoff examples with different numbers of apertures when $h/d = 1$.

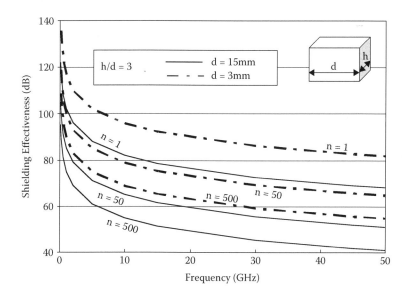

FIGURE 6.3 Estimated shielding effectiveness of waveguide-below-cutoff examples with different numbers of apertures when $h/d = 3$.

from n to $2n$ in a waveguide panel decreases the shielding effectiveness by approximately 3 dB for frequencies below the cutoff frequency. As discussed in Chapter 1, the shape of the aperture is another factor. For instance, a change from round holes to square holes reduces shielding effectiveness by approximately 8 dB. And changing the aperture geometry from round to square also reduces the cutoff frequency. However, the benefit of square holes or honeycomb structures is that they can generally be used to improve the open-area percentage considerably and reduce material and tooling costs. By adjusting the aperture size of the square hole or honeycomb, the cutoff frequency can be adjusted and the depth further adjusted to achieve desired EMI performance (Intel, 2001). Moreover, the cutoff frequency (f_c) should be optimized. An f_c that is much higher than necessary for a particular design may burden the design with unnecessary additional costs in excess material. An f_c that is too low may limit the performance of the waveguide below the desired level.

For some applications, implementation of a waveguide panel may be difficult or expensive and the enclosure design may not immediately need the extra shielding effectiveness that a waveguide provides. For such cases, the use of perforated metal sheet or wire mesh may be flexible and cost-effective options. Figure 6.4 illustrates the estimated shielding effectiveness of waveguide-below-cutoff perforated metal sheet or wire screens with different numbers of holes when $h/d = 0.01$. For wire screens, it is assumed that the wires make good electrical contact at each crossover or intersection. When good electrical contact is not made at each intersection, the mesh will not provide the expected shielding results. In general, the shielding effectiveness of mesh screens

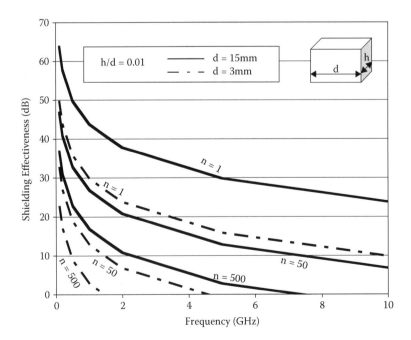

FIGURE 6.4 Estimated shielding effectiveness of waveguide-below-cutoff wire screens with different numbers of holes when $h/d = 0.01$.

depends upon (Canter, 1999) the diameter or width of the wire mesh; the wire material properties and associated surface treatment; the contact resistance of the wires at the intersections, which is in part a function of the wire tension produced by the process equipment; the maximum spacing or separation distance of the wire, which yields the aperture dimension; the total number of openings; and the size of the overall aperture covered by the mesh. The specific estimation of the shielding effectiveness for the wire screen can be obtained from Eqations 1.30–133. (See Page 22, Chapter 1.)

On the other hand, the primary function of a waveguide panel is to facilitate improved cooling of the electronic components by waves of reduced pressure drop at a given airflow rate. Thermal performance of a waveguide aperture panel is governed by the waveguide insert geometry (percent open area, hole shape, depth, insert size, number of holes in the insert) and by the geometry of the approach section of the duct. Each of these features has an impact on the overall flow resistance of the insert. Airflow resistance and EMI containment must be balanced in determining the proper geometry for the waveguide. Generally, increasing the open-area percentage reduces the pressure drop of the waveguide. Square and honeycomb holes generally have better thermal performance than round holes due to the higher density with which square and honeycomb holes can be manufactured. There is a trade-off with EMI shielding performance utilizing square and honeycomb holes; however, increasing the depth can compensate for much of this degraded EMI shielding performance. There is also a pressure drop trade-off with hole depth, but the pressure drop is less sensitive to increased hole depth than reducing the open area. In practice, an open area of greater than 70% is both desirable and achievable while maintaining satisfactory EMI performance (Intel, 2001).

In addition, there are many choices available for the trade-off between cost and EMI shielding performance when designing a waveguide panel. When larger holes are used, depth must be added to achieve the same EMI shielding performance as a design with smaller, shallower holes. Fewer deep holes may prove to be more expensive due to the more material required than several shallow holes for a fixed area; however, larger holes provide a greater percentage of open area resulting in a higher performance thermal solution.

Besides, use of waveguides may lead to increased acoustic noise because waveguide apertures can generally be larger with a larger percentage of open area. Moreover, flow generating devices, such as a fan, should not be placed in the direct proximity of the waveguide as this may introduce additional noise. Care should be taken in the system design to minimize the acoustic noise originating from or transmitting through the waveguide apertures.

A waveguide panel insert can be mounted to a shielding enclosure using screws, rivets, or snap-attach features. Where utilized, EMI, structural, safety, and security considerations should be taken into account.

6.2.2 Materials Selection and Fabrication Options

Waveguide and ventilation materials must be made of electrically conductive materials, and the contact interface in the assembled enclosure should be galvanically compatible to eliminate occurrence of galvanic corrosion. Knitted wire mesh, metallized woven screen and conductive foam, perforated metal, and honeycomb have been used for ventilation materials to protect large openings. Different mechanical and electrical properties, as well as some subtle and crucial mechanical differences in the fabrication and construction of these materials result in rather dramatic differences in their shielding effectiveness.

As discussed previously, materials homogeneity is crucial and closely linked to the surface transfer impedance of the shielding materials. In relatively thin materials, the size and surface geometry of the holes in the material panel determines the shielding effectiveness. In general, if the hole size is well defined and the surrounding material characteristics are homogeneous, the material will exhibit a well-behaved surface transfer impedance and uniform shielding characteristics.

From this point of view, perforated materials provide better shielding than woven materials, and woven materials provide better shielding than knitted materials. Unfortunately, perforated materials are typically about 65% open and do not have airflow characteristics as good as screening materials, which range up to about 85% open (Brewer, 2001). However, the drawback of screening materials used for ventilation openings is uncontrolled contact impedance between the wire strands. With lower strand contact pressure, vibration, or deformation, and/or material corrosion at the strand contact points, the surface transfer impedance increases and shielding performance can be severely degraded.

As mentioned earlier, considerable improvement in aperture shielding effectiveness can be obtained, for instance, by increasing the diameter of the wire or the thickness of the material, especially when $h/d > 1$. At this point, the opening begins to act as a waveguide being operated below its cutoff frequency. However, this will increase material and fabrication costs to some extent.

Hence, for lower frequency applications where only moderate shielding effectiveness is required, the least costly approach is to use perforated materials, or woven or knitted screening. But for critical applications requiring highly effective shielded ventilation panels and outstanding airflow characteristics in a restricted area, honeycomb materials combine the best airflow with the best EMI shielding performance.

As shown in Figure 6.5, honeycomb can be fabricated and constructed of small, adjacent, electrically connected tubes/waveguides made mainly from thin metal foil strips assembled in parallel or other processes (Pflug et al., 2002). The metals used include aluminum, steel, and magnesium alloys. Other materials have also been developed, such as zinc, and other metal-coated plastics and conductive foams.

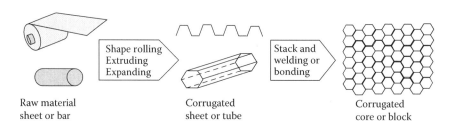

FIGURE 6.5 Fabrication process of honeycomb panels.

The honeycomb has up to 97% open area with greater directional characteristics compared with holes in thin materials. However, the airflow through the honeycomb is not as turbulent, and the directional characteristics allow honeycomb vent panels to be constructed in drip-proof or visually secure configurations.

Although high performance honeycomb can provide excellent EMI attenuation, the overall shielding effectiveness of the enclosure depends on the proper installation of the ventilation panels in the enclosure. Any vent panel that is not adequately bonded into the enclosure behaves as a lossy antenna structure sitting in a hole. Grounding the panel at one point will reduce the antenna efficiently, and may even solve radiation problems at low frequencies, but it will not eliminate leakage from the rest of the seam. To be effective in shielding applications, honeycomb metal panels must have a metal frame with good surface conductivity, with sufficient stiffness and enough fixing holes to correctly compress a conductive gasket around its entire perimeter to avoid creating additional apertures that would degrade shielding effectiveness. Alternatively, they can be glued in place using conductive adhesives (Armstrong, 2007).

The best honeycomb installation methods include welding, brazing, soldering, or riveting, but prevent easy access for maintenance and repair. Installing closely spaced threaded fasteners or clamps permits but does not facilitate field removal. Consequently, ventilation panels generally use EMI gaskets to maintain contact across the seams.

Honeycomb metal panels can also be used to shield apertures where optical energy is emitted, such as lamps and lasers. They have even been used to shield display screens, which can then only be seen from directly in front. This can benefit security-conscious applications. Besides, fiber-optic cables can also be passed through the honeycombs, as long as the cables are totally metal free (Armstrong, 2007).

6.3 CONDUCTIVE FOAM AND INTEGRAL WINDOW STRUCTURE

Metallized conductive plastic foams have been developed for fabricating high performance vent panels and other EMI shields. Combining advantages of metal honeycomb and plastic foam materials, integral window structure provides high application performance and multifunctions with EMI shielding, thermal ventilation, and dust filtering. These vent panels can be used for many applications, such as communications enclosures or cabinets, aeronautics and general aerospace cooling applications, medical equipment, manufacturing systems, and military applications.

6.3.1 Conductive Plastic Foam and Vent Panels

Compared to metal honeycomb, conductive plastic honeycomb panels are constructed of nickel/copper-plated polyurethane foam, which is light weight, dent resistant, and provides increased airflow along with EMI shielding.

Figure 6.6 illustrates the construction of a conductive plastic honeycomb panel with (a) processing polycarbonate honeycomb before metallization; and (b) after plating with Cu/Ni and edging constructed with conductive foam or conductive elastomer gaskets. Typically, round polycarbonate straws are glued together to form cells

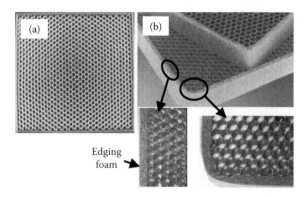

FIGURE 6.6 Conductive plastic honeycomb: (a) Polycarbonate honeycomb before metallization; (b) plated with Cu/Ni and edging constructed with conductive foam. (Modified from Laird Technologies, 2003.)

from 3.2 mm to 6.4 mm and panel thickness from 6.4 mm to 12.7 mm, and then plated with a conductive base layer of ~0.8 μm thick electroless Cu; top layer of ~0.3 μm thick electroless Ni to improve corrosion resistance; and flame protection plating when needed, V0 coating or intumescent coating (Fenical 2007). The conductive plastic honeycomb does not have the rigidity and strength of metal honeycomb, but it weighs less and should also cost less while having similar attenuation characteristics to metal honeycomb.

Special features can be machined with conductive foam and attached to the honeycomb, such as recesses and rabbet cuts to the honeycomb panel. The compressible conductive foam bonding around the perimeter of the panel can typically accommodate large variations and shelf width or vent panel opening tolerances, while also providing improved electrical contact to the enclosure. Vent panels with the frameless design also allow for greater airflow by as much as 10% to 20% per unit surface area of vent panel than framed vent panels. This increased percentage can be critical for cooling many of the densely packed enclosures currently being designed.

Figure 6.7 shows the ventilation performance-pressure drop of the nickel/copper-plated conductive plastic honeycomb, which was the same as the aluminum honeycomb. Shielding effectiveness testing also showed similar EMI performance between the conductive honeycomb and the Al honeycomb (Fenical, 2007; Laird Technologies, 2003).

Moreover, the nickel/copper-plated plastic honeycomb material with a conductive foam band around the perimeter can eliminate the need for costly frames, as well as the associated attachment hardware and much of the installation labor.

In addition, the conductive foam with smaller cell size can be sculpted into a variety of profile shapes for edging the conductive plastic honeycomb panel or EMI shielding gaskets. Superior compression set values can be obtained; in many cases the sculpted edging foam is 50% better. Compression force and surface resistivity can be improved or adjusted. Figure 6.8 shows variation of electrical resistance and application load of a typical conductive foam with controlled compression deflection percentage (Fenical, 2007; Laird Technologies, 2003).

Conductive Foam and Ventilation Structure

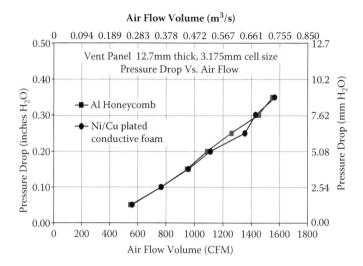

FIGURE 6.7 Comparison of ventilation performance of Ni/Cu-plated conductive foam versus Al honeycomb. (Modified from Laird Technologies, 2003.)

Other forms of conductive foams, as shown in Figure 6.9, can be made by bonding the polyester mesh to top and bottom surfaces of urethane foam, and then metallizing with copper/nickel or silver/copper/nickel. The metallized foam is laminated to full width, and the release liner with random coat adhesive. Moreover, reticulated urethane foam allows for uniform plating throughout the product while providing a more direct conductive path and strong compression set resistance (Fenical, 2007; Laird

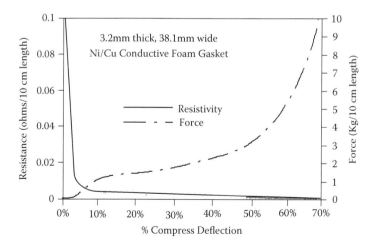

FIGURE 6.8 Variation of electrical resistance and application load of typical conductive foam with controlled compression deflection. (Modified from Laird Technologies, 2003.)

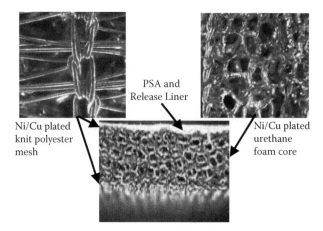

FIGURE 6.9 Reticulated urethane foam clad with polyester mesh allows for uniform plating throughout the product while providing a more direct conductive path and strong compression set resistance. (Modified from Fenical, 2007.)

Technologies, 2003). These conductive plastic foams are cost effective. In addition to vent panels, they can be constructed with different profiles, including sheets, shaped gaskets, and also can be hot wired, die cut, convoluted, and laminated to other substrates, such as pressure-sensitive adhesive. Conductive foam has x-, y-, and z-axis conductivity for maximum EMI shielding attenuation. The shielding effectiveness can be over 100 dB up to 10 GHz. The material is ideal for computer, router, and telecom input/output shielding. Conductive foam is RoHS compliant, and due to its low compression set (5% to 20%), it has a long life (over 1 million cycles).

6.3.2 Integral EMI Window Structure

As shown in Figure 6.10, a typical integral EMI window structure is designed to provide EMI shielding with positive grounding; airflow straightening for maximum cooling efficiency; and high dust arrestance with low pressure drop. The honeycomb air filter enhances EMI shielding performance by using an EMI shielding stainless steel mesh that is bonded to the aluminum frame with a conductive silver/copper adhesive caulk. In addition, an EMI shielding gasket is installed around the frame perimeter. The gasket ensures full conductivity of the frame to the equipment. Moreover, the honeycomb provides waveguide apertures that are designed to reflect and absorb EMI noise. The honeycomb pattern maintains minimal airflow impedance, and provides a straightening effect for even air distribution. Meanwhile, the flame-retardant, open-cell polyurethane foam filter media is good for high dust arrestance. It is removable, cleanable, and reusable (Universal Air Filter, 2007).

In this case, high levels of EMI shielding can be achieved through optimal fabrication of components in the window structure: (a) a stainless steel mesh that captures EMI/RFI noise and is bonded to the filter frame with a conductive adhesive to assure grounding; (b) a honeycomb pattern of apertures that are designed to reflect and

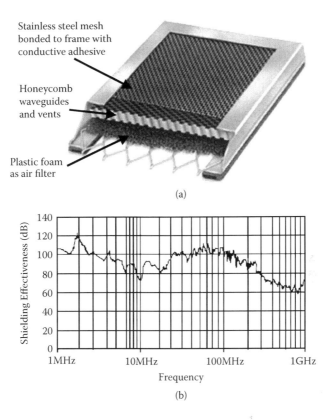

FIGURE 6.10 (a) Integral EMI window structure and (b) its shielding effectiveness. (With permission of Universal Air Filter, www.uaf.com.)

absorb EMI noise without restricting airflow; (c) a metallized fabric or Monel metal on neoprene foam gasket that can be installed around the frame perimeter; and (d) an open cell polyurethane foam that can be used as the filtering media in the structure. This media can be specially coated to provide improved fire retardancy and fungi resistance. It has deep loading, a high dust holding capacity, and low air resistance. Special metal finishes can be made to obtain galvanic compatibility.

The integral EMI window structure can be designed and used for specific applications including electronics, telecom, and datacom equipment.

6.4 METALLIZED FABRICS AND FABRIC-OVER-FOAM

Because of their high surface conductivity, anticorrosive performance, and excellent shielding effect, metallized fabrics have been widely used for EMI shielding applications. Metallized fabrics can be used for garments or wraps of fabric-over-foam products. Due to its lower cost and ease of installation, fabric-over-foam has become a popular alternative to traditional metal finger stock gaskets, especially in high volume, commercial electronics.

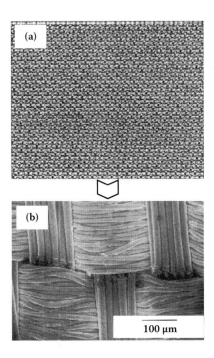

FIGURE 6.11 Metallized fabrics: (a) Ni/Ag-coated nylon fabric; (b) enlarged SEM image.

6.4.1 METALLIZED FABRICS

Metallized fabrics are made from metallization on a core of a woven or knitted fabric material fabricated from polymer fiber, such as polypropylene, nylon, polyester or aramid fibers, glass fiber, carbon fiber, silica, or alumina fibers. Typical core fiber diameters range from 10 to 250 μm, covered with a thin metallic layer ranging in thickness from 0.1 to 5 μm. Metallizing processes used to produce such materials usually include vacuum vaporizing, solution emerging, melt spraying, and plating. The most commonly used metallized fabrics are highly conductive and corrosion-proof materials, such as woven, non woven, ripstop, and taffeta coated with nickel/copper or nickel/silver woven, nonwoven, ripstop and taffeta, as illustrated in Figure 6.11 for instance, which provide a variety of mechanical and electrical properties for EMI shielding.

In fact, metallized fabrics can offer a unique combination of properties that is proving of great interest in a wide variety of EMI shielding solutions, such as useful shielding performance, ease of fabrication, transparency to air, partial transparency to visible light, and low specific weight. Metallizing does not change a fabric's characteristics apparently; drape, texture, tear strength, shrinkage, elongation, and other properties stay remain similar (Molyneux-Child, 1997). Basically, the shielding performance of metallized fabrics depends on and is attributed to a combination of several factors: low surface electrical resistivity (0.01 to 0.35 Ω per square depending on the fabric and its surface coating); complete surface coverage of the fabric by metal,

Conductive Foam and Ventilation Structure

eliminating the contact resistance and the transmission by induction that occurs at the crossing points of conventional woven metal fabrics; and the finite thickness of the fabrics and fine gaps between resulting in a certain amount of waveguide behavior at high frequency. In addition, several microscopic fibers spun together compose a multithread. These threads are woven to constitute a material that is coated with a continuous layer of metal. The different microscopic fibers, as well as the crossing points of the threads, are bonded together with a metallic corrosion-resistant sheath. This assures high conductivity and good shielding. The threads of multithread fabrics are thus substantially opaque, but not airtight.

Metallized fabrics have been mainly used to produce fabric-over-foam products, gaskets, and shielding tapes. They are also used for making shielding garments, ESD (electric static discharge) garments and packing materials, shielding tents and shielding rooms. Due to their pliable and tough property, stable and good transparency, and excellent capabilities of electric conduction, they are suitable for many other electronic applications, such as ion PDP (plasma display panel) TV frames, shielding materials for front panels, shielding display windows, optical filters, and shielding nets.

6.4.2 Fabric-over-Foam

Since the appearance of fabric-over-foam in the end of 1980s, there has been a proliferation in the use of conductive fabric-over-foam gaskets for EMI shielding. Advances in plating adhesion, availability of UL 94V0 (vertical burn, specification 0) flame-retarded material, and profile design have made fabric-over-foam the gasket of choice for grounding and EMI shielding solutions in a majority of indoor electronics housing applications. Compared to metal-based spring finger gaskets, such as copper, beryllium (CuBe), stainless steel, and other high strength and high conductive alloys, fiber-over-foam has the advantages of low cost and ease of installation; therefore, its popularity has increased, especially for high-volume, commercial electronic shielding applications.

Fabric-over-foam can be made from urethane, EPDM (ethylene propylene diene monomer), polypropylene, neoprene, or 100% conductive foam in open- or closed-cell structure, wrapped with Ni-, Cu-, Ag-, Au-, Sn-, or C-coated woven, nonwoven, knit, and random fiber fabric, and attached with pressure-sensitive adhesive (PSA) and clips. The most popular forms of fabric-over-foam include nickel/copper- or nickel/silver-plated nylon woven or nonwoven fabric wrapped around a urethane foam core. Fabrication of fabric-over-foam is easy and the tooling cost is low. Figure 6.12 illustrates (a) the process of fabric-over-foam and (b) typical profiles of fabric-over-foam products (Laird Technologies, 2003). Table 6.1 shows the typical performance of fabric-over-foam. Generally, the urethane foam performs well enough in applications with few closure cycles or low shear activity. However, fabric-over-foam is not well suited for applications where long-term performance at elevated temperatures is crucial and compression set unacceptable. The performance of most urethane foams starts to degrade above 70°C. When most fabric-over-foam gaskets with urethane foam cores are tested for compression set according to ASTM D-3574, the results are in the 20% to 30% range (Chomerics, 2000; Laird Technologies, 2003). When long-term gasket performance is critical or extreme situations are present, the fabric-over-foam gasket is usually replaced by CuBe or other metal fingerstock gaskets.

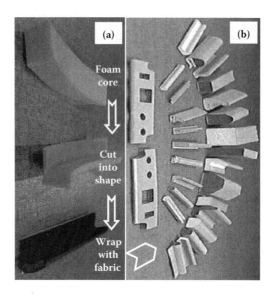

FIGURE 6.12 (a) Fabrication process of fabric-over-foam and (b) typical profiles of the products. (Modified from Laird Technologies, 2003.)

However, some fabric-over-foam gaskets have been developed that are particularly effective in extreme situations. For instance, a fabric-over-foam gasket made from soft silicone foam has a typical compression set of 1% to 2% when tested to 70°C and less than 10% when tested to 125°C. Compared with metal gaskets, fabric-over-foam exhibits the following characteristics (Chomerics, 2000; Laird Technologies, 2003):

TABLE 6.1
Typical Performance of Fabric-over-Foam

Shielding effectiveness	Typically between 80–100 dB
Compression set	5%–15% urethane and 10%–30% TPE
Loads/forces	0.9kg/m to 30 kg/m
Cycle life	High (compression); low to moderate (shear)
Attachment	PSA, clip-on, I-bar, Groove
Installation	Fast and easy
Galvanic compatibility	Great (matched with surfaces like zinc-plated steel or stainless steel)
Resistance to shear	Moderate
Temperature range	−40°C to 70°C
Cost	Low to moderate
Recyclability	Yes

- As a fabric-over-foam gasket is compressed, more surface area of the gasket contacts the mating flange, allowing for reduction in the ground path resistance. This is not always the case with metal gaskets, especially with stiffer versions. The actual flange contact point at the top of the metal gasket remains relatively narrow until highly compressed.
- In addition to excellent shielding effectiveness, fabric-over-foam gaskets provide air sealing and ease of installation.
- In most cases, the material and fabrication costs will be less for fabric-over-foam than for CuBe finger stock with similar profiles.
- When shear forces are a concern, options with fabric-over-foam can be used to get the same physical performance as metal gaskets. Grooves can be designed in the flanges to reduce insertion forces on gaskets. Uniquely shaped fabric-over-foam gaskets are also available for use on flat flange surfaces that will work very well in shear applications.

Moreover, the availability of new, high-performance foam has expanded and will enhance the applications of fabric-over-foam EMI gaskets. This allows more flexibility when designing gaskets for cost-critical computing equipment and telecommunications equipment.

When designing and selecting fabric-over-foam gaskets, the following factors should be considered (Chomerics, 2000; Dye and Niederkorn, 2000; Hug, 2001; Laird Technologies, 2003).

1. Electrical conductivity and surface resistivity. The electrical conductivity and surface resistivity of fabric-over-foam gaskets are largely determined by the conductivity and thickness of the metal coating on the fabric. Because the metal thickness around the fiber is considerably less than the skin depth necessary for absorption to be significant, the reflection rather than absorption is the dominant shielding mechanism in fabric-over-foam gaskets. The base coating metals chosen in fabric-over-foam gaskets are highly conductive metals such as copper and silver, and the surface protective metals chosen are those that are also conductive, like nickel or tin.
2. Compression force and deflection range. The compression force and deflection range are largely determined by the selection of the foam core, although the size and shape of the gasket also have effects on it.
3. Compression set. Compression set of fabric-over-foam depends on the type of foam, gasket size and shape, application temperature, and service time.
4. Galvanic corrosion. The galvanic corrosion performance of the fabric-over-foam EMI gasket is determined by the metal used for coating the fabric. If a problematic metal is used as the fabric coating, the metal may be somewhat protected by additional insulating coatings. Such coatings reduce the galvanic corrosion rate, but they do not change the inherent corrosion potential, because it is determined solely by the metal itself.
5. Environmental corrosion. The effect of environmental corrosion is the oxidation or reduction of the metal on the gasket's surface, which can reduce

the electrical conductivity by an amount sufficient to lessen the gasket's shielding effectiveness. The environmental corrosion properties of the fabric-over-foam EMI gasket are determined by the type and thickness of the metal coating, and by the extent of protection of the metal coating resulting from additional coatings. Because such protective coatings typically increase the surface resistivity of the gasket, having an inherently inert metal is preferable to having a highly protected sensitive metal.

6. Flame rating. EMI gaskets have ratings of HB (horizontal burn) and UL 94V0 (vertical burn, specification 0). The flame rating of the gasket is determined by the selection of all the components in the gasket and how they interact in the flame. In general, to achieve a high flame rating, such as UL 94V0, flame-retardant additives must be added to the gasket. These additives usually result in increases of the gasket's compression force and compression set. Some advanced flame-retardant EMI gaskets are significantly improved in this regard.

7. Attachment method. Bonding with PSA is the most common attachment method in use for fabric-over-foam, although other methods such as rivets, press-on clips, and friction-fit in slots are also used. When PSA is used to attach the gasket, care must be taken to ensure that the initial tack and the final bond strength are sufficient for the application. The width and thickness of the PSA must also be such that acceptable electrical performance is achieved. Particular care must also be taken to ensure that the adhesive properties are acceptable for the operating temperatures of the final device. Some PSAs may flow or disbond at elevated temperatures, causing movement of the EMI gasket or a decrease in the electrical contact between the EMI gasket and the attachment piece.

8. Abrasion protective coatings are applied in very thin layers over the conductive coatings, making the metals able to withstand the flexing inherent in the gasket's function. Even under severe abrasion, nickel/copper- and tin/copper-coated gaskets show only a small increase in resistance, similar to that of silver-coated fabrics.

To summarize, optimal designing and selecting of fabric-over-foam require analyses of multiple electrical, mechanical, and physical material properties. Assuming that a product provides the proper form, fit, and function to satisfy mechanical requirements, the electrical, physical, and chemical properties of the final solution are the most critical features in the selection process.

Many shielding enclosures are required to meet twenty-year life cycle tests where shielding degradation over time, due to galvanic incompatibility and corrosive gases, must be an important design consideration. Analyses of environmental factors have confirmed that a tin-over-copper or nickel-over-copper construction for fabric-over-foam gaskets provides shielding effectiveness superior to silver-coated or overplated-silver solutions.

In addition, aesthetically colored fabric-over-foam offers an attractive design alternative to EMI shielding options by providing a gasket that is colored to match the outside of a chassis. The gasket is comprised of a polyester film material that is applied to one side of the foam core. The metallized fabric is then wrapped in a manner so that it bonds with two parallel edges of the polyester film material. Bonding the fabric to polyester film ensures the contact path from the chassis to the component. The polyester is colored to match application requirements (Laird Technologies, 2003).

Aesthetically colored fabric-over-foam shielding gaskets are ideal for applications such as consumer gaming electronics and home computers where shielding is needed in a location that may be visible on the outside of the chassis. These gaskets are typically used in applications where the input/output connectors, such as video, audio, and USB, are located on the front of the chassis.

6.5 SUMMARY

A variety of window structures made of conductive foam, waveguide honeycomb, metallized fabrics, and ventilation panels have been developed and utilized to meet the EMI shielding requirements of large-area openings, displays and ventilation apertures.

Conductive fabric-over-foam has been widely used for EMI shielding applications where no environmental seal, complex profile, or demanding mechanical durability is needed. It can be the lowest cost option. The cladding can consist of nylon thread coated with conductive metals woven into a fabric and wrapped over soft urethane foam. Alternatively, a woven fabric may be metallized with nickel-copper or other metal coating, then wrapped around the foam core.

Flexible, conformable fabric-over-foam maintains close contact with surfaces with minimal compression for low closure-force applications. It provides snug contact over irregular surfaces and bends around corners, making it good for electronic enclosures such as doors and access panels. Less demanding EMI applications include grounding contact pads in cell phones and laptop computers. Fabric-over-foam gaskets can also shield the input/output backplane of laptops and personal computers.

The manufacturing process limits fabric-over-foam to relatively simple cross-sectional profiles such as squares, rectangles, and D shapes. When a more complex profile is necessary, fabric-over-foam may not be the best solution.

REFERENCES

Armstrong, K. 2007, March. Design techniques for EMC. Part 4–Shielding (screening) first half. *The EMC Journal*, 31–46.
Brewer, R. 2001. *How to choose ventilation shielding materials.* http://www.electronic-products.com/ShowPage.asp?SECTION=3700&PRIMID=&FileName=FEBINS1.FEB2001
Carter, D. 1999. *A look at metal barriers with discontinuities for related Tecknit products.* Tecknit Europe Ltd. http://www.tecknit.co.uk/catalog/emcholes.pdf
Carter, D. 2004. *RFI shielded windows.* Tecknit Europe Ltd. http://www.interferencetechnology.com/ArchivedArticles/shielding_aids/02_ag_04.pdf?regid=
Chomerics. 2000. EMI Shielding Engineering Handbook. http://www.chomerics.com/products/documents/catalog.pdf
Dye J. M., and R. Niederkorn. 2000. Selecting the right fabric-over-foam EMI gasket. *Compliance Engineering.* http://www.ce-mag.com/ARG/Dye.html
Fenical, G. 2007. New development in shielding. *Conformity.* http://www.conformity.com/artman/publish/article_174.shtml

Grant, P. 1984. *Tecknit design guideline to EMI shielded windows.* http://tecknit.com/REFA_ I/E-Windows.pdf

Hug, K. 2001. Shielding: Fabric-over-foam EMI gaskets. *Compliance Engineering.* http://www.ce-mag.com/archive/01/11/Hug.html

Intel. 2001. *EMI waveguide apertures*, version 1.0. Intel Corporation. http://www.lairdtech.com/downloads/FoFCataloghleb.pdf

Laird Technologies. 2003. *Metallized conductive products.* http://www.lairdtech.com/downloads/FOFCataloghleb.pdf

Molyneux-Child, J. W. 1997. *EMC Shielding Materials.* Oxford: Newnes.

Pflug J., B. Vangrimde, and I. Verpoest. 2002. *Continuously produced honeycomb cores.* http://www.mtm.kuleuven.ac.be/Research/C2/poly/phds/jp/jp_sampe_us_1.pdf

Universal Air Filter. 2007. *Dual EMI honeycomb air filters.* http://www.uaf.com/Details.aSpx?NavID=174

7 Board-Level Shielding Materials and Components

7.1 INTRODUCTION

The continued miniaturization of surface-mounted devices (SMDs) and the increasing density of the printed circuit board (PCB) have made it necessary to shield different zones of a device from one another to achieve the desired levels of functional performance in the electronic system. Among different levels of shielding methods, board-level shielding is a low-cost way and can be used to reduce the electromagnetic interference (EMI) shielding requirements for the overall enclosure, even completely removing the need for enclosure-level shielding (Armstrong, 2004).

A variety of board-level-type shields have been designed and used to eliminate EMI radiation from entering or exiting sections of a PCB. In all these shielding methods, the ground plane of the PCB makes up one side of the six-sided can or cover required to create a Faraday cage or shield cover. The cavities on the inside of the cover are metallized on the surface and electrically sealed to the PCB via a conductive elastomeric gasket. The front and back housing are brought together by mechanical fasteners, which apply the necessary force to the gasket interface. The most common board-level shielding method is to attach the can or cover directly to the PCB, resulting in a finished PCB that has the shielding integrated into it. This method primarily uses solder-attached perforated metal cans with and without removable lids. A can is soldered to the ground trace on a PCB, directly over the electrical components that need to be shielded. Such cans offer extremely high levels of shielding effectiveness, are typically very reliable, and are widely used in the industry. They are often installed in a fully automated fashion via a surface-mount technology (SMT) process at the same time the components themselves are installed onto the PCB, using solder paste and a reflow process. Cans are extremely difficult to remove and replace if the shielded components need to be reworked, because of the complicated desoldering and resoldering process (Livington, 2004). The development of other board-level shielding methods, such as metallized thermoforming plastic shields, offer the promise of overcoming the problems associated with metal-can types of shielding, which incorporate high-performance plastic shield materials that have been metallized to provide the requisite EMI shielding effectiveness.

In addition, electrical components in a PCB that need to be shielded often generate heat while operating. Since the performance of these components often degrades at elevated temperatures, it is typically desirable to facilitate the transfer of heat away from them. This is can be done through thermally enhanced board-level shields.

Therefore, a great effort has been made to develop various board-level shields that are removable and compatible with single or multicompartment designs, sometimes thin in profile and lightweight, and allow for thermal dissipation of shielded components. This chapter will review the status and future trends of board-level shielding technology, including design method and materials selection, shield types and their manufacturing technology, methods of conductive coating on board-level shields, and thermally enhanced board-level shielding methods.

7.2 BOARD-LEVEL SHIELDING DESIGN AND MATERIALS SELECTION

As discussed earlier, EMI is a major problem in integrated electronic circuits including PCB. To eliminate EMI, the design should either remove the EMI source or protect the circuit from EMI by shielding. Radiated EMI emission can be minimized through proper circuit design and board layout. However, with increasingly high frequencies, the wavelengths are so short that many board circuit traces have lengths that make them act as efficient one-quarter or one-half wavelength antennas. Thus, board-level EMI emissions by circuit design alone cannot eliminate the need for board-level EMI shielding. Moreover, proper materials selection is an important portion for an optimal board-level shield design to meet whole system electromagnetic compatibility (EMC) requirements.

7.2.1 Basic Principles of Board Circuit EMC Design

Basically, it is less expensive to solve radiated EMI problems at the board level than at a later stage. More important, given the amount of electronic circuitry and the higher clock speeds in digital devices, it is rare that simple board layout can solve EMI shielding problems without some form of suppression. Adding filters and attenuator beads onto a PCB board, as well as using multilayer boards with separate ground planes, can offer some emission control.

In addition to component selection and circuit design, good PCB layout is an important factor in EMC performance. Since the PCB is an inherent part of the system, EMC enhancement by PCB layout does not add extra costs to the finished product. However, most PCB layouts are restricted by board size and the number of copper layers (Lun 2005). Some layout techniques may apply to one type of circuit but not another. Much of it will depend on proper PCB layout.

7.2.1.1 PCB Layout

A PCB is constructed using a series of laminates, tracking, and prepreg layers in a vertical stack. In multilayer PCBs, most designs require placement of the signaling tracking on a higher outer layer for easier debugging of the board. The following are some general guidelines for PCB layout (Intel, 1999; Lun, 2005): (a) increase the separation between tracks to minimize crosstalk by capacitive coupling; (b) maximize the PCB capacitance by placing the power and ground in parallel; (c) place sensitive and high frequency tracks far away from high noise power tracks; and (d) widen ground and power tracks to reduce the impedance of both power and ground lines. Segmentation is an effective method to use in physical separation for reducing the coupling between different types of circuit, particularly by the power and ground tracks.

7.2.1.2 Power Decoupling

Localized decoupling can reduce noise propagating along the supply rail. The use of large bypass capacitors connected at supply entries to the PCB will help as a low frequency ripple filters and potential reservoirs for sudden power demands. In addition, decoupling capacitors should be connected between the power and ground at the integrated circuit as close as possible to the pins. This helps to filter out switching noises from the integrated circuit. Whether with the reference ground plane on multilayer PCBs or ground traces on single-layer PCBs, a current path is present from the load back to the power supply source. The lower the impedance of the return path, the better the EMC performance of the PCB. Long return paths can create mutual coupling because of the radio frequency (RF) current from load to source (Lun, 2005). Therefore, the return path should be as short as possible and the loop area as small as possible.

7.2.1.3 Trace Separation

Trace separation is used to minimize the crosstalk and noise coupling by magnetic flux coupling between adjacent traces on the same PCB layer. As shown in Figure 7.1, for multilayer boards, the power and ground planes can be distributed between various signal layers to improve coupling between the signal and ground planes. For inner traces on multilayer boards greater than four, it is a good practice to limit the trace length on external layers. For traces that are not treated as transmission lines, that is, clocks and strobes, keep the length of the trace conductor to 1/20 of a wavelength of the highest harmonic of interest (Intel, 1999). In clock circuits, local decoupling

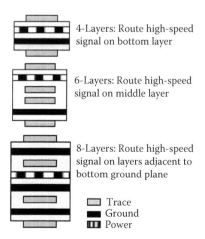

FIGURE 7.1 Trace, power, and ground layer stacking.

capacitors are very important in reducing noise propagating along the supply rail. But the clock lines also need protection from other EMI sources, otherwise jittering clock signals will cause problems elsewhere in the circuit.

7.2.1.4 Grounding Techniques

Grounding techniques apply to both multilayer and single-layer PCBs. The objective of grounding techniques is to minimize the ground impedance, thus reducing the potential of the ground loop from circuit back to the supply.

On a single-layer (single-sided) PCB, the width of the ground track should be as wide as possible, at least 1.5 mm. Since the star arrangement is impossible on single-layer PCBs, the use of jumpers and changes in ground track width should be kept to a minimum, as these cause changes in track impedance and inductance. On a double-layer (double-sided) PCB, the ground grid/matrix arrangement is preferred for digital circuits because this arrangement can reduce ground impedance, ground loops, and signal return loops. Another scheme is to have a ground plane on one side, and the signal and power line on other side. In this arrangement, the ground return path and impedance will be further reduced and decoupling capacitors can be placed as close as possible between the integrated circuit supply line and the ground plane. High-speed circuits should be placed closer to the ground plane, whereas the slower circuits can be placed close to the power plane. In some analog circuits, unused board areas are covered with a large ground plane to provide shielding and improve decoupling. But if the copper area is floating (i.e., not connected to ground), it may act as an antenna, and it will cause EMC problems. In multilayer PCBs, it is preferable to place the power and ground planes as close as possible on adjacent layers to create a large PCB capacitor over the board. The fastest and most critical signals should be placed on the side that is adjacent to the ground plane, while noncritical signals should placed near the power plane (Lun, 2005).

When the circuit requires more than one power supply, the idea is to keep each power separated by a ground plane. But on a single-layer PCB, multiground planes are not possible. One solution is to run power and ground tracks for one supply separate from the others. This helps to avoid noise coupling from one power source to the other. All metallized patterns should be connected to ground, otherwise these large metal areas can act as radiating aerials (Intel, 1999; Lun, 2005).

7.2.1.5 Chassis Construction and Cabling

The system case reduces EMI by containing EMI radiation. The case material and discontinuities in the case are main factors that impact the effectiveness of the case in reducing EMI. To mitigate case discrepancies, wherever possible, use round holes instead of slotted holes. Round holes provide the greatest airflow volume for the least amount of EMI leakage. Input/output (I/O) shielding should be grounded at as many points as possible (Lun, 2005).

Cables make excellent antennas. To reduce the impact of cables on EMI radiation, provide adequate grounding for all cables; ground both ends of cables to the chassis; keep all cables away from chassis seams to prevent coupling; and use ferrite beads to attenuate common mode noise on I/O cables (Intel, 1999).

Board-Level Shielding Materials and Components

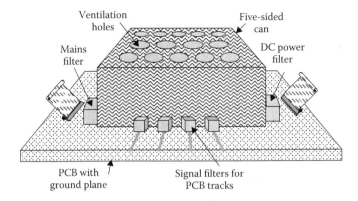

FIGURE 7.2 A traditional board-level shield: a totally enclosing metal can with shielded connectors and feed-through filters mounted in its walls.

7.2.2 BOARD-LEVEL SHIELDING DESIGN WITH PROPER MATERIALS SELECTION

Board-level shielding can be designed and accomplished in several different forms, depending on the application requirements and mounting methods. For example, a board-level shield can be a simple metal sheet or foil layer, a five-sided metal, or a metallized plastic enclosure, based on the Faraday cage principle with waveguide ventilation design.

Figure 7.2 illustrates the typical board-level shielding. A five-sided conductive shielding can is placed over a circuit zone on a PCB and electrically bonded via holes in the PCB at multiple points around its perimeter to a plane layer inside the PCB or on its bottom layer (Armstrong, 2004). The can is normally stamped with a hexagonal series of round holes or apertures for ventilation during assembling and application. Therefore, a six-sided Faraday cage is formed as a shielding enclosure, part of which is embedded in the PCB itself. The shielded connectors and feedthrough filters are mounted in its walls. Furthermore, for the lowest cost, all signals and powers should be brought into the shielded area of the PCB through tracks, avoiding wires and cables. Moreover, the PCB equivalents of bulkhead mounting shielded connectors and bulkhead mounting filters should be used. The PCB track equivalent of a shielded cable is usually called a stripline. Where striplines enter an area of circuitry enclosed by a shielded can or cover, it is sufficient that their upper and lower ground planes and any guard tracks are bonded to the soldered joints of the shielding can on both sides, close to the stripline. All unshielded tracks must be filtered as they enter a shielded PCB area. It is often possible to get valuable improvements using PCB shielding without such filtering, but this is difficult to predict, so filtering should always be included in the design. All PCB-mounted filters should ideally be lined up with their centers along the line of the wall of the shielding can, which will probably need a little cutout in it to accommodate the components. Using surface-mounted devices rather than leaded devices allows the can cutout size to be minimized, improve shielding effectiveness (Armstrong, 2004; Lun, 2005).

Generally, metal shielding cans can provide extremely high shielding effectiveness in a wide range of frequency systems. These metal cans have been made of tin- or zinc-plated steel (trivabent chromate or chromium free electrogalvanized steel is prefered for environmental compliance) stainless steel, tin-plated aluminum, brass, copper beryllium, nickel silver, or other copper alloys. Plating may be specified to improve the attenuation performance of a shield, while tin and solder plating are often used to enhance solderability preservation. And dielectric coatings can be designed for insulating purposes.

Tin-plated cold-rolled steel, stainless steel, and other high permeability materials are considered the best choices for shielding low frequency applications, such as transformers and coils, whereas nonferrous materials such as tin-plated copper, phosphor bronze, and copper beryllium are preferred above 200 MHz frequency applications, such as microprocessors, fast switching devices, and power suppliers. Tin-plated aluminum provides the lightest weight (3.3 g/cc density) in metal cans, with outstanding electrical conductivity (45% IACS), and excellent thermal conductivity (190 W/mK) in metal cans CuBe, CuNiSn, and CuTi can be specified for applications where top spring properties are required. Nickel silver is expected to be the preferred alloy for the majority of board-level shielding applications (Majr Products, 2007). The properties of some typical nickel silver alloys are shown in Table 7.1. The materials can take the place of traditional plated steel and other copper alloys

TABLE 7.1
Mechanical Property and Electrical Conductivity of Common Nickel Silver Alloys

Nickel Silver Alloy	Temper	Tensile Strength (MPa)	Yield Strength 0.2% Offset (MPa)	Elongation (%)	Electrical Conductivity (% IACS)
CuNi18Zn10 (C73500)					
	Annealed (TB00)	330–435	103–276	11 to 37	8
	1/4 Hard (H01)	386–476	193–400	10 to 25	8
	1/2 Hard (H02)	435–517	338–476	3 to 16	8
	3/4 Hard (H03)	476–545	407–503	2 to 5	8
	Hard (H04)	503–579	462–538	1 to 3	8
	Extra Hard (H06)	545–621	496–559	1 to 2	8
	Spring (H08)	586–641	510–579	≤1	8
	Extra Spring (H10)	Min 607	Min 538	≤1	8
CuNi10Zn25 (C74500)					
	Annealed (TB00)	330–450	Min 145	Min 34	9
	1/4 Hard (H01)	385–505	Min 310	Min 25	9
	1/2 Hard (H02)	460–565	Min 413	Min 12	9
	3/4 Hard (H03)	505–610	Min 465	Min 6	9
	Hard (H04)	550–650	Min 515	Min 4	9
	Extra Hard (H06)	615–700	Min 558	Min 3	9
	Spring (H08)	655–740	Min 586	Min 1	9

TABLE 7.1 (CONTINUED)
Mechanical Property and Electrical Conductivity of Common Nickel Silver Alloys

Nickel Silver Alloy	Temper	Tensile Strength (MPa)	Yield Strength 0.2% Offset (MPa)	Elongation (%)	Electrical Conductivity (% IACS)
CuNi18Zn17 (C75200)					
	Annealed (TB00)	355–450	124–221	29–42	6
	1/4 Hard (H01)	400–496	179–441	14–35	6
	1/2 Hard (H02)	455–552	331–538	6 to 22	6
	3/4 Hard (H03)	510–593	476–565	4 to 12	6
	Hard (H04)	538–627	517–621	3 to 7	6
	Extra Hard (H06)	593–676	586–669	3 to 4	6
	Spring (H08)	621–696	607–682	1 to 2	6
	Extra Spring (H10)	Min 662	Min 655	≤2	6
CuNi12Zn24 (C75700)					
	Annealed (TB00)	360–430	130–240	Min 32	8
	1/4 Hard (H01)	410–490	280–340	Min 23	8
	1/2 Hard (H02)	430–510	330–420	Min 8	8
	3/4 Hard (H03)	490–570	360–450	Min 5	8
	Hard (H04)	510–590	410–515	Min 3	8
	Extra Hard (H06)	540–630	465–565	Min 1	8
	Spring (H08)	560–650	480–580	—	8
	Extra Spring (H10)	Min 650	Min 580	—	8
CuNi12Zn29 (C76200)					
	Annealed (TB00)	393–517	138–276	32–49	9
	1/4 Hard (H01)	448–559	172–483	20–50	9
	1/2 Hard (H02)	517–627	414–607	6 to 30	9
	3/4 Hard (H03)	572–676	552–655	4 to 16	9
	Hard (H04)	621–724	607–710	3 to 6	9
	Extra Hard (H06)	696–786	669–779	1 to 3	9
	Spring (H08)	752–841	738–827	≤1	9
	Extra Spring (H10)	Min 786	Min 703	≤1	9
CuNi18Zn20 (C76400)					
	Annealed (TB00)	379–428	Max 250	Min 27	6
	1/4 Hard (H01)	476–600	Min 250	Min 9	6
	1/2 Hard (H02)	538–655	Min 410	Min 3	6
	3/4 Hard (H03)	610–680	Min 510	Min 3	6
	Hard (H04)	634–752	Min 586	Min 2	6
	Extra Hard (H06)	703–807	Min 600	Min 1	6
	Spring (H08)	744–848	Min 641	Min 1	6

(Continued)

TABLE 7.1 (CONTINUED)
Mechanical Property and Electrical Conductivity of Common Nickel Silver Alloys

Nickel Silver Alloy	Temper	Tensile Strength (MPa)	Yield Strength 0.2% Offset (MPa)	Elongation (%)	Electrical Conductivity (% IACS)
CuNi18Zn27 (C77000)					
	Annealed (TB00)	421–552	159–283	39–48	6
	1/4 Hard (H01)	476–600	303–572	11 to 41	6
	1/2 Hard (H02)	538–655	441–641	8 to 24	6
	3/4 Hard (H03)	621–696	579–690	6 to 13	6
	Hard (H04)	634–752	621–731	3 to 6	6
	Extra Hard (H06)	703–807	696–786	1 to 2	6
	Spring (H08)	745–848	738–814	≤1	6
	Extra Spring (H10)	Min 799	Min 792	≤1	6

across the board for those applications above 15 MHz where shielding effectiveness becomes more of a skin (conductivity) effect than a bulk effect. Nickel silver alloys are less conductive than copper or brass but far stronger; more corrosion resistant than steel, copper, or brass; and exhibit outstanding aesthetic qualities and excellent solderability with appropriate fluxing. Therefore, there is no need for tin or other platings to improve the solderability or corrosion resistance of nickel silver alloys. In addition, because tin plating is not needed, there are no tin whisker issues to worry about, which is crucial in some critical applications.

Metal cans made from tin-plated steel sheet are typically soldered onto the ground plane of PCB boards. Some are designed with snap-on lids so that adjustments may easily be made, test points accessed, or chips replaced with the lid off. Such removable lids are usually fitted with spring fingers around their circumference to achieve a good shielding when they are snapped in place. The drawback of metal can shielding is obviously (a) the apertures created by the gaps between the ground-plane soldered connections; (b) any apertures in the ground plane, for example, clearances around through leads and via holes; and (c) any other apertures in the five-sided can, for example, ventilation, access to adjustable components, displays, and so forth. Seam soldering the edges of a five-sided can to a component-side ground plane can remove one set of apertures at the cost of a time-consuming manual operation.

For some applications like wireless systems, metal cans may be less compatible than other shielding solutions because the solder is susceptible to cracking from tension or shock on the circuit board. From this point of view, molded conductive plastic covers provide greater flexibility and easy pin mounting. With a slim-profile metal or metallized plastic substrate, the cover has a conductive coating or plated surface and features integral conductive elastomer gaskets for shielding components or entire unity of PCBs. The gaskets mate with board trace and, together with a PCB's

TABLE 7.2
Typical Shielding System in Board-Level Shielding

Shielding Type	Application	Features	Shielding Effectiveness	Advantages
Metal cans	Soldered over PCB components	Pick and place assembly, difficult removal	40–100 dB	In-line application
Plastic covers	Pin or screw mount over PCB components	Conductively plated or painted, conductive elastomer molded to cover	30–100 dB	Removable; cover individual components or entire PCBs
Form-in-place gaskets	Crosstalk and perimeter shielding	Robotic dispensing on thin-walled enclosures parts	75–120 dB	Fast programming for new applications
Space gaskets	Component shielding and PCB grounding; pin or screw mount	Conductive elastomer molded over plastic frame	75–120 dB	Easy assembly

conductive back plane layer, provide shielding like the Faraday cage. In place of plating or paint, a thin, continuous layer of form-in-place conductive elastomer can provide a shielding surface. The elastomer can also be molded into detailed wall patterns for isolating components from one another under the cover.

The automated form-in-place EMI gaskets of conductive elastomer can be applied in very small cross-sections with limited spaces and high volume situations. In multiple boards, spacer-type EMI gaskets of plastic and conductive elastomer provide an EMI barrier between components to protect against crosstalk. The molded elastomer gasket mates with board traces or other conductive surfaces. Using design tools such as finite element analysis (FEA), the gasket profiles are designed with deflection and compression ranges that provide maximum performance. The typical performance of board-level shielding and gasketing is shown in Table 7.2 (Quesnel, 2008).

7.3 BOARD-LEVEL SHIELDING COMPONENTS AND THEIR MANUFACTURING TECHNOLOGY

Surface-mount technology (SMT) is the preferred manufacturing technique for assembling virtually all integrated electronic devices. This process involves screen printing solder paste down onto bare PCBs; picking and placing the components, including shielding cans or covers, onto the board; and then running the boards through a solder reflow oven to assemble all components together.

During the SMT assembling, the metal cans are placed over the other components as the last step before the board goes through the reflow oven, and then mechanically and electrically attached to the PCB using the same solder process as the other electronic components. The hexagonal series of round holes or apertures normally stamped on metal cans provide more even heat transfer during the solder reflow process.

As PCB designs become more compact with higher density and lighter weight, metal cans are becoming less attractive as a shielding option and being replaced by metallized thermoforming plastic shields. These plastic shields can be formed with multicavities, flexible geometries, and stylish contours, and therefore will be desirable shielding solutions for many device housings.

7.3.1 METAL CANS

Metal cans have been widely used for board-level shielding with a variety of types and features, such as single-piece, multipiece, and multicompartment cans with or without removable lids. Once the metal can without a removable lid is soldered in place, it greatly hinders the ability to inspect component solder joints after reflow and to rework the area if it is faulty. To solve this problem, some metal cans have incorporated removable lids into the can design. However, this feature often adds size and complexity to the shield and could compromise the shielding effectiveness of the can. Also, depending on how well the lid fits on the can's frame, and the snapping mechanism used to hold it in place, there may be issues associated with the lids popping off during mechanical, shock, or vibration tests (Quesnel, 2008). Therefore, alternative or improved solutions are always needed for board-level shielding.

7.3.1.1 Single-Piece Shielding Cans

Single-piece cans are the most effective option for board-level shielding. They are directly soldered to the PCB and offer appealing shielding effectiveness combined with excellent coplanarity for ease of assembly. They can be formed in a wide range of sizes and incorporated in a variety of design features to enhance performance and ease of use. The shields can be packaged in tape and reel for automatic assembly with standard pick-and-place equipment (Fenical and English, 2006). There are three typical types: single-piece etched and bent can with half-etched bend lines (Figure 7.3), drawn shields (Figure 7.3), and lid-removable shields (Figure 7.4).

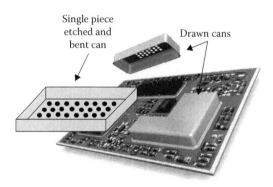

FIGURE 7.3 Illustration of single-piece cans including an etched and bent shielding can, and drawn cans.

Board-Level Shielding Materials and Components 187

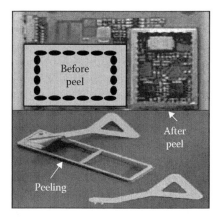

FIGURE 7.4 A single-piece shielding can with removable lid.

The single-piece etched and bent can is usually prepared with half-etched bend lines, and four edges are fold by bending process. To minimize corner seams, there are several options for corner design (Figure 7.5):

- The interlocking corner is usually used on shields with height over 2.5 mm, and two or more tabs are used depending on height and minimum gap requirements. Using the interlocking corner, the corner seam can be reduced to improve shielding effectiveness, and the reduced nesting and tangling will improve coplanarity during automatic packaging assembly.
- The tapered corner helps prevent nesting during automated packaging and is usually used on shields with height less than 2.5 mm.
- There are several ways to form a shield with an internal corner, including the wrap around corner or miter corner. For ease of tooling and to maintain mechanical strength, the miter corner is preferred. For shielding performance, the method that leaves the shortest linear gap is usually the best choice.

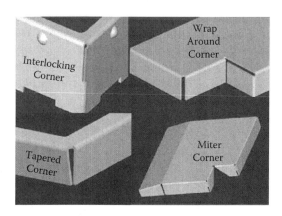

FIGURE 7.5 Corner designs of the single-piece shielding can.

The dimension control is critical for single-piece etched and bent cans. Maximum ventilation hole size is around 3 mm. For radio frequency, the hole size required is less than 1/4 the wavelength, typically 1/10 to 1/20 the wavelength to provide excellent shielding. Minimum hole size is 1 mm for general mechanical tool reliability. Holes are used to improve heat dissipation during soldering of the components, but the number of holes impacts tooling price. Dimension tolerances should be well controlled. Height is often the most critical dimension. The flatness determines if the shield can be used with SMT, the preferred electronic assembly method. The flatness must be better than the height of the solder paste for proper attachment. Typical solder paste height is 0.145 mm, therefore the can should be maintained at 0.13 mm flatness or better. The shield flatness is even more critical in double-sided board applications. For instance, a lifted corner can break the surface tension of the solder that holds the shield on the PCB while the PCB and the shield are inverted in the reflow oven. This can cause the shield to fall off the PCB in the reflow oven.

Drawn cans are designed to provide additional near-field and far-field EMI shielding effectiveness at high frequencies (>5 GHz) by eliminating the corner openings of the etched and bent shields. Pressed from a single piece for complete integrity and optimized attenuation in high-frequency applications, single-piece drawn shielding cans can be manufactured with hard press tooling in a wide variety of materials, usually including tin-plated cold-rolled steel, stainless steel, and nickel silver. Moreover, drawn cans can be manufactured with cooling perforations or vent holes, which will reduce or eliminate any improvement in shielding effectiveness. Therefore, they are more commonly seen as a contiguous conductive barrier. This type of can is usually specified in large volume applications. For drawn cans the cost of pieces can be less than that of an etched and bent can of similar dimensions but the initial setup and tooling cost can be prohibitive for small runs. Compared with the etched and bent cans, drawn cans are usually more costly, and automation of their packaging is more difficult because of draw lip, as the draw lip occupies valuable PCB real estate. In addition, residual stresses resulting from extreme forming can make it difficult to manufacture parts with good flatness. For these reasons, drawn cans are used in limited applications (Fenical and English, 2006).

If access to the components on the shielded PCB is needed during the early manufacturing and testing phase of a program, or infrequently thereafter, lid-removable cans can be a good option. These cans, produced in a variety of materials, have a solid lid with etched lines that allow peel off when access to the components enclosed within the can is necessary. This allows inspection without subjecting the board to further thermal excursion to remelt the solder, which could affect the enclosed components and joints. Peeling off the lid is achieved using a simple hook tool or tweezers, as shown in Figure 7.4 (Majr Products, 2007). After component repair, the original lid is replaced with a snap-on, solder-on, or stick-on replacement lid (Fenical and English, 2006). However, the major concern about lid-removable cans is that the flatness of the lip after peeling might not be well controlled.

The typical SMT process can be used to assemble all of these one-piece cans, as shown in Figure 7.6.

Board-Level Shielding Materials and Components

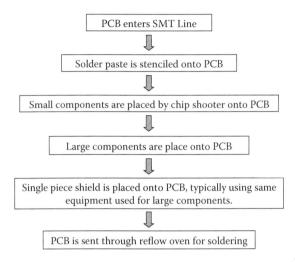

FIGURE 7.6 Assembly process flow of the single-piece shielding can.

7.3.1.2 Two-Piece Cans

If more frequent access to under-shield components is needed, a two-piece shield should be used. Two-piece cans are designed to protect sensitive components from noise while giving access for rework and adjustments. As shown in Figure 7.7, a two-piece can consists of a frame to form the sidewall of the enclosure and a removable fingered lid, which provides easy access to the enclosure contents and high attenuation in service (Majr Products, 2007). The shield lid is usually snapped into the frame with spring fingers and locking dimples. In addition, the two-piece design allows internal walls to be used to create multiple shielded compartments from one frame (Fenical and English, 2006).

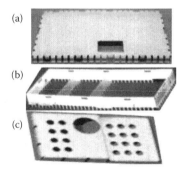

FIGURE 7.7 Two-piece, multicomponent, and multilevel cans. (a) Lid with spring fingers; (b) multicomponent frame; (c) multilevel shield.

7.3.1.3 Multilevel Shielding and Multicompartment Cans

A multicompartment can is an effective alternative to multiple cans. It is cost effective to produce and offers significant savings in terms of initial cost, assembly time, and real estate management. A matrix of walls, all folding easily from a single sheet and complemented with a separate lid, provides an efficient enclosure with enhanced accessibility for rework or the tuning of devices. Internal walls provide multiple shielded components. This is the best solution for circuit boards with multiple circuit groups that need shielding. Moreover, multilevel cans can be produced to combine different heights within a single structure, as shown in Figure 7.7 (Majr Products, 2007). This system may also be used to accommodate boards with irregular topography. Design complexity is always reflected in the cost of pressing tooling. These shields can be automatically placed with standard pick-and-place equipment, and then assembled with the frame. The multicomponent or multilevel shield can also be made in one piece, welded with a spring metal gasket on its contact interface by laser or electrobeam welding, and then pinned or screwed onto the PCB.

7.3.2 METAL-COATED THERMOFORM SHIELDS

Metal-coated thermoform shields have been developed to offer light weight and improved performance over metal cans in some applications. Instead of being soldered directly to the PCB, the shields can be placed on the board after reflow and removed easily for rework, if needed. Incorporated with thermoformable plastic base materials and metallization surfaces, these shields provide the requisite EMI shielding effectiveness.

Two representative metal-coated thermoform shields have been commercialized: snapSHOT® board-level shield (W. L. Gore & Association, Inc., Newark, Delaware) and Form/Met™ shield (WaveZero, Inc., Sunnyvale, California). With the snap-SHOT shield, the plastic sheet is plated with metal tin, and then vapor deposited with a thin layer of nickel. Thereafter, the metallized plastic sheet is thermoformed into the shape of shields. The method of attachment of the snapSHOT shield to a PCB is critical. The PCB is prepared with a relatively large surface trace containing small solder spheres, which are pushed through corresponding holes located on the shield flange. With the Form/Met shield, the metallization is carried out with vapor deposition after the base plastic sheet is thermoformed and cut into shape. In this way, the metal coating will not be stretched and has less probability of being broken. Form/Met uses a conductive adhesive that can be applied automatically to the shield or board. Both shields offer additional opportunities to lighten the weight of shielding (by about 80%) for a device.

Other shielding material, like conforming formable film, can either be attached to the enclosure or attached directly onto the circuit board to provide compartmentalization and isolation of many individual components. Alternatively, the formable film may be utilized as a drop-in product within the enclosure as a "box-in-a-box" EMI solution. This formable film technology can also be designed as a product for localized or individual component shielding on the board (Bachman et al., 2001).

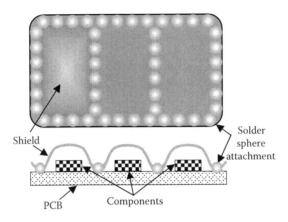

FIGURE 7.8 A snapSHOT shield with solder sphere attachment on PCB.

7.3.2.1 snapSHOT Shield

Thermoformed from lightweight, metallized plastic sheet, the snapSHOT shield usually has multicavities and individual mechanical snap features to attach with solder spheres on the PCB, as shown in Figure 7.8. Before thermoforming, the shield is premetallized with tin on the outside only, leaving the inside surface insulative. Therefore, it is less likely to short out components and traces by accident. This also allows narrower ground traces, less space between components, and reduced overall thickness when compared to existing shielding solutions. To assemble it to the PCB, standard ball grid array (BGA) solder spheres are deposited on the PCB at the locations of the snapping holes, and during soldering, capillary action makes the balls snap through the holes and makes them contact with the outer metal layer, while also retaining the shield in place (Clupper, 2006; Gore, 2008). The spheres provide both electrical connection to the shield and mechanical retention on the board. The BGA spheres can be attached to the PCB using standard SMT equipment. Solder spheres are reliably installed to ground pads on the PCB, directly after solder paste is applied, using a modified stencil printer. The rest of the SMT process, including pick-and-place and reflow, remains unaffected.

Fully integrated shielding consists of a metallized, thermoformed shell that is attached to the PCB using BGA solder balls as snap-attachment features. The shell is removable. Therefore, the snapSHOT board-level shield can be repeatedly applied to a populated board after it has gone through the solder reflow process. This enables manufacturers to inspect and rework the PCB unimpeded by the shield. When the material is formed into a shield, it can save 80% to 90% of the weight compared with the same size metal can. As the metallization is on the outside and the insulating plastic base material is in contact with the PCB, a lower profile can be achieved by virtually eliminating the component-to-shield gap. It also reduces minimum component-to-component distance by 25% to 50%, freeing up valuable space on the board (Gore, 2008).

In general, this kind of shield can be thermoformed into just about any shape, profile, or contour. Multiple heights, rounded edges, and curved shapes can all be readily incorporated into a shield design. After the shield is attached, it can be removed easily if rework or repair to the components is needed. The shielding's flexible nature makes the coplanarity of the PCB significantly less critical than with standard metal cans. Shielding effectiveness is about 60 dB through 9 GHz, approximately 30 dB better than the removable-lid metal can. Part of this difference is attributable to the fact that the metal can has apertures in its shield and seams between the lid and the metal frame. The snapSHOT shield, by contrast, is constructed from a solid sheet of metallized materials and does not have any perforations on its surface. Also, the snapSHOT shield's BGA attachment mechanism provides a robust mechanical and electrical connection along its perimeter (Clupper, 2006).

However, the potential drawbacks for snapSHOT shielding may include (a) the plating surface breaks during thermoforming; (b) the solder spheres break off or wear off during repeat removing and assembling of the shell; and (c) as the shield is plated with tin, there might be a tin whisker growth issue during application.

7.3.2.2 Form/Met Shield

Form/Met shield is comprised of a polymer film substrate (typically 0.13 mm to 0.38 mm thick) that is first thermoformed into the desired shape to facilitate its use within an electronics device and then vacuum metallized with an aluminum layer (typically, 1–6 μm thick, 99.9% pure). The materials used for the polymer film substrate usually include polycarbonate (PC), polyvinyl chloride (PVC), and polybutylene terepthalate (PBT). Using the traditional thermoforming process, the polymer film will be thermoformed into the required shape and profile before metallization (WaveZero, 2008).

Aluminum is the optimal surface metallization material from a performance and cost viewpoint. The shielding effectiveness of aluminum, as measured by its relative conductivity, is high. Aluminum is abundant, nontoxic to the environment, and readily adaptable to the vacuum metallization process. Compared with snapSHOT shields, the Form/Met shield is metallized after thermoforming, which may have the following benefits Arnold, 2002a; Arnold, 2002b:

- By first vacuum metallizing and then thermoforming, the metallized layer may severely crack and thus lose its usefulness as an EMI shielding.
- It is economical to do thermoforming first before metallizing. The shield is usually thermoformed into the shape, and the remaining materials will be cut off and recycled. No waste is created from this process.
- The thermoforming of the Form/Met shield shell is accomplished with standard processes. If the film material was metallized before thermoforming, tool design and tool wear could become an issue. Subsequent die cutting could result in the creation of metal particles that would be difficult to control in a high volume application.
- Although more ductile materials, such as tin, could be used, their shielding effectiveness is not as great as that of aluminum. The thermoforming process (when accomplished on previously metallized sheet material) would cause critical areas of the shield to become thinner and less effective for EMI shielding.

Vacuum metallization, also called physical vapor deposition (PVD), is widely used to deposit a thin aluminum or other metal layer on plastic parts. The metal is heated to the point at which it evaporates. The vapor then migrates through the chamber and condenses on the cold plastic part. The process takes place in a vacuum to allow the metal vapors to reach the plastic surface without being oxidized by oxygen in the air. The parts to be metallized are held in an appropriate fixture that can rotate in order to expose all surfaces to be metallized. Several power sources can be used for metal evaporation, such as resistance heating, induction evaporation, electron beam guns, or a vacuum arc. Resistance-heated tungsten filaments are most often used. The filaments are placed in the required position to obtain uniform coverage, and aluminum chips or staples are placed on the filaments. The metal to be evaporated can also be placed in a thermally heated boat or crucible. Evaporation coating is normally done batchwise in a cylindrical coating chamber. The coating chamber may have a diameter of up to several meters, depending on the size and number of parts to be coated. The parts may make a planetary movement around the vapor source to equally cover all sides of the parts with a metal layer. If desired, the areas not to be coated can be masked, usually by metal sheets.

Vacuum metallization is an environmentally friendly process for Form/Met shield fabrication. First, the plastic film is shaped by the thermoforming process. The thermoformed shape is then placed on fixtures that are placed in the vacuum chamber. There are no surface treatments with chemicals or mechanical abrasives used to prepare the surface for metallization. All metallizing processes take place in the chamber. In the process, thermal or evaporative deposition is used to first melt and then vaporize aluminum. Aluminum particles are attracted to the polymer substrate as a result of the potential difference between the source of aluminum (anode) and the substrate (cathode). Tungsten filaments are used to hold the aluminum cans before and during the melt process. These tungsten filaments eventually degrade in mechanical strength and are returned to the manufacturer for recycling. Aluminum deposited within the chamber is removed in a mild chemical bath and eventually the residual aluminum and chemical is recovered and recycled in an appropriate manner. No volatile organic compounds are used at any stage of the process. Aluminum deposition forms a microscopically thin surface layer of aluminum oxide. This layer is so thin that it is measured in atomic units. Upon removal from the vacuum chamber, the Form/Met shield immediately forms a self-stabilized layer. The air-formed film on new aluminum surfaces is about 2.5 nm thick, whereas the film on aluminum that is several years old may be 10 or more nm thick. This oxidizing layer does not apparently affect conductivity, surface resistivity, or shielding effectiveness even over long periods of time (Arnold, 2002b).

In a typical application, the thickness of the vacuum-deposited aluminum layer is approximately 3 μm and the polymer substrate is around 0.25 mm. The aluminum layer constitutes an insignificant fraction of the thickness of the substrate; consequently, mechanical properties are governed by the polymer and substrate thickness. Adhesion of the vacuum-metallized layer to the polymer film substrate is excellent. The surface of the polymer film is received from the film provider free of surface contaminants and vacuum deposition is accomplished without any mechanical surface modification (bead blasting) or chemical treatment. The Form/Met shield

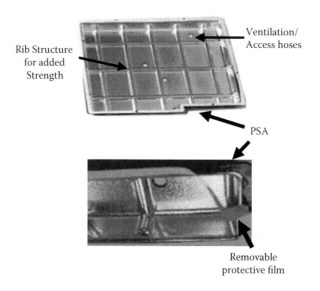

FIGURE 7.9 Form/Met board-level shields with common features. (With permission of WaveZero Company.)

remains flexible and robust after vacuum metallization and even rough handling during assembly does not scratch or otherwise affect its EMI shielding effectiveness. The ability of the Form/Met shield to withstand the rigors of handling improves the overall yields of electronic product assembly (WaveZero, 2008).

Form/Met shields have been designed and applied not only at PCB board levels, but also in enclosure level and component level. Form/Met board-level shields can be designed with complex shapes and integrated into any electronic product design. For instance, the shield can be placed on the conductive ground traces of a PCB, covering one or more active electronic components and circuits. Attachment and grounding of the shield to the circuit board is accomplished through electrically conductive adhesives, die cuts, conductive pressure sensitive adhesives (PSA), or mechanical fasteners. As shown in Figure 7.9, using a conductive PSA is one way of achieving the connection between the EMI shield and the PCB ground trace; the trace width is typically 1 to 2 mm. Alternatively, a conductive gasket can be used to achieve its ability to electrically connect the EMI shield and PCB ground plane by being compressed by the plastic housing or rib sufficiently to assure that the conductive particles in the gasket material become contiguous. A conductive gasket has two functions: electrical connectivity and z-direction stack tolerance compensation (Arnold, 2002b).

A Form/Met enclosure, as shown in the Figure 7.10, utilizes a clamshell design and a combination of snaps and periphery features that ensure ease of assembly, yet minimizes the existence of seams and slots. Designed to conform to the inner contours of the enclosure, the enclosure-level shield completely encapsulates the PCB. Ventilation and access holes for connectors and wiring can be easily incorporated into the shield's design. Integrating two or more shield features within the design achieves grounding with corresponding grounding points in the PCB. The ability to

Board-Level Shielding Materials and Components 195

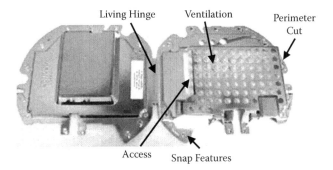

FIGURE 7.10 Form/Met enclosure with specially designed features. (With permission of WaveZero Company.)

provide a pseudo-Faraday cage results in a superior shielding solution compared to other alternatives (Arnold, 2002b; WaveZero, 2008).

Form/Met compartmentalized shielding solutions can replace one or more sheet metal cans with a single-shield component comprised of multiple compartments. EMI shielding of individual or groups of components and circuits is easily accomplished with Form/Met using a compartmentalized design such as shown in the Figure 7.11. A number of compartments can be designed into the Form/Met shield with each compartment having design features that allow its direct connection to the ground traces on the PCB. With component EMI shielding, the ground traces are typically about 1 mm in width along a periphery and 1.2 to 1.8 mm between components. The ability to create a thermoform feature between components with feature sizes around 1 mm has important applications for isolating components and circuits on a PCB. It is relatively easy to incorporate the double-sided shielding into a design. A thermoform

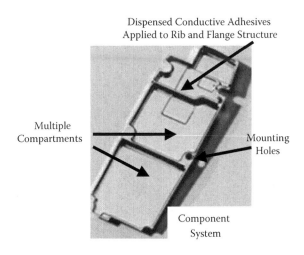

FIGURE 7.11 Form/Met compartment system. (With permission of WaveZero Company.)

design feature that allows for a multiple shield feature between compartments also provides additional EMI shielding, which reduces interference between components on the board (WaveZero, 2008).

In addition, one of many virtues of the Form/Met shield is its ability to be recycled. A thermoformed structure can be recovered and reground into a new extruded sheet that can be used in the manufacture of a new structure. The newly formed Form/Met structure will have enhanced EMI performance by virtue of the embedded conductive material distributed through the thickness. From a disposal point of view, the aluminum used as the conductive layer is nontoxic to the environment and constitutes less than 1% of the volume of the final product. Environmentally acceptable chemical dissolution processes are available if for some reason it is desirable to totally remove the aluminum from the polymer substrate (Arnold, 2002b).

7.3.2.3 Formable Shielding Film

Formable shielding film is usually thermoformed from 0.2 to 0.5 mm thick polycarbonate (PC) films, and then coated with conductive paints, such as silver-coated copper, silver + silver-coated copper, and pure silver. Vacuum metallization and plating including electroless and electroplate also can be used for the formable shielding film. Attachment of this formable shielding film to the shield can be accomplished either by attachment to the enclosure or by a direct attachment onto the circuit board (Bachman et al., 2001).

During initial assembly and testing, these materials can be designed to be removable, if required, without damage to the board. When attached to the enclosure, the formable film can be heat staked to the enclosure or can be attached using a reliable adhesive (either conductive or nonconductive depending on the application). In this case, the inclusion of a compression fit between the enclosure and the circuit board is normally part of the overall assembly design. Besides, a gasket dot can be incorporated with the formable shielding film design, which offers the possibility of eliminating the usual form-in-place or foam-type gasket while achieving effective board-level shielding and component isolation. When the formable shielding film is attached directly to the circuit board, a conductive adhesive can be used to bond the shielding film to the perimeter bonding areas. In addition, the formable shielding film is very conformable to the contours and shapes associated with individual circuit-board components. It can also be readily configured for compartmentalization of key functional areas with a single piece, and the shielding film placement can be accomplished utilizing an in-line SMT-type format during PCB assembly (Bachman et al., 2001).

7.4 CONDUCTIVE COATING METHODS USED FOR EMI SHIELDING COMPONENTS

The reliance on conductive coatings in EMI shielding systems has increased with ever-growing use of thermoformed plastic enclosures and components, the push for miniaturization, and demands for various surface finishes in EMI shields and electronic devices. Although a variety of conductive coating processes have been developed, the common methods used for EMI shielding components mainly include electroplating, conductive painting, and vapor deposition.

7.4.1 Electroplating

Electroplating protects and beautifies the surface of EMI shields; conforms the galvanic contact interface; and increases or modifies the corrosion resistance, conductivity, and solderability of metal and plastic components by depositing a variety of metallic coatings onto their surfaces. Plastics that can be electroplated include polypropylene, polysulfane, polyester, polycarbonate, and other engineering resins. Electroplated coatings have the unique ability to combine aesthetic appeal, wear and corrosion resistance with high levels of shielding effectiveness. These materials are recyclable. The plated plastic components can usually be dipped in an acid solution to remove the plated surface for recycling.

Two types of electroplating process are commonly used for EMI shielding of metal and plastic components: electroless plating and electrolytic plating.

7.4.1.1 Electroless Plating

Electroless plating is an autocatalytic chemical process in which a uniform and pure metal coating is deposited onto EMI shields (Applied Coating Technologies [ACT], 2008). Metal or plastic components are loaded into plating racks or baskets and are processed through a series of wet chemistry steps, as shown in Figure 7.12.

The coating mostly consists of a layer of electroless copper (1.0–5.0 μm), which provides the shielding effectiveness (over 80 dB in a wide frequency spectrum), with a top layer of electroless nickel (0.25–2.5 μm), which provides mechanical and corrosion protection of the copper. This process is ideal for coating of internal shields designed for board-level shielding (ACT, 2008).

Depending on the shield design requirements, either all-over electroless plating or selective electroless plating can be chosen. If the design has no cosmetic or electrical requirements that limit the area where metal plating can be applied, all-over electroless plating is the preferable choice. If the design contains areas that must remain free of metallization, or if the product design requires maintaining the molding in the texture and color of the shield component, selective electroless plating should be used. In general, both processes can produce an extremely uniform coating over the entire component, even complex shapes with deep recesses, because the electroless process is a chemical reduction process. Compared with electrolytic process, electroless plating has no line-of-sight limitations, and no external electrical current is required to sustain the process. However, the electroless deposition will only take

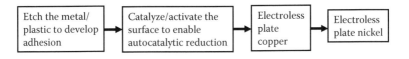

FIGURE 7.12 Electroless plating process flow.

place in the presence of a catalyst, since this process is based upon the catalytic reduction of metal ions on the surface of the substrate being plated (Skelly, 2004).

EMI shielding components are usually plated with copper first for high conductivity and then with nickel as final finish for surface protection. When using the all-over plating process, the component is chemically etched and activated to roughen the material surface and to prepare the surface for the deposition of the copper layer. The etching and activating steps impact only the outer skin of the component. The bulk material properties of the substrate are not affected. Following the etch–activation, pure copper and then nickel are deposited to the specified thickness. The same uniform, pure metal coating can also be applied via the selective plating process by spraying a catalytic primer (around 0.025 mm) on the areas to be metallized. Masking fixtures control the selective pattern. After curing the catalytic primer, the component is plated; and the plating occurs only where the catalytic primer has been applied (ACT, 2008; Skelly, 2004).

The electroless plating process has been widely used for metals and plastics, such as ABS (acrylonitrile butadiene styrene), in EMI shielding systems. This coating provides the plastic shield with apparent advantages, such as lightweight (up to 80% less than pure metal) uniform coating over all surfaces; flexible, durable structure; and environmental compliance. However, the plastic materials that can be plated with the electroless process are limited, due mainly to the poor activation of the polymer surface and the resulting poor adhesion of the metal plating. For example, PVC, polyethylene, and polypropylene cannot be plated with standard electroless plating processes (Skelly, 2004). In order to improve the activation of the polymer plating surface, a laser-supported activation and additive metallization process has been developed. The concept of the process is to set free seeds on the polymer surface by laser irradiation, and these seeds enable wet chemical reductive metal precipitation. The polymer is generally prepared and modified by incorporating dispersive organometallic complexes into the matrix, which are designed in a way that they can be activated by the laser irradiation. The laser irradiation on the one hand induces a physicochemical reaction in the form of cracking chemical bonds. On the other hand, it makes a strong adhesion of the forming metal layer possible by ablating the polymer material, namely, roughening the surface and thus providing mechanical anchoring for the metal plating (Hüske et al., 2001).

7.4.1.2 Electrolytic Plating

The electrolytic plating process can be used to deposit a multilayer coating with a total thickness up to 25 μm. Coatings produced by this process are extremely wear resistant, environmentally stable, and provide very high levels of shielding effectiveness. These coatings are widely used for EMI shields where shielding, wear and corrosion resistance, and an external decorative finish are required.

In electrolytic plating, the component being coated is generally immersed in an aqueous solution of dissolved metal ions. These ions typically come from metal salts dissolved in the process tank. Once the component is fully immersed in the solution and electrical current is passed through it into the solution, the component acts as a cathode attracting metal ions from the solution and reducing them to their metal state on the surface of the component. This process is regulated through the control of voltage, amperage, temperature, residence times, and bath purity (ACT, 2008).

Different from the electroless process, the surface of the component must be conductive for electrolytic process to occur. When plastics are to be coated, this is achieved by first electroless plating the plastic. Typically a 0.3 to 0.5 μm layer of electroless nickel phosphorous alloy is deposited. Varying the amount of phosphorous in the alloy can vary the properties of the nickel phosphorous coating. Typically, the phosphorous content can range from 3% up to 20%. High phosphorous coatings tend to be very ductile and corrosion resistant, but worsen the solderability. Once the plastic component has been made electrically conducting, it can be electroplated by the standard electrolytic process. For example, for plating the copper and nickel on the plastic surface, the copper layer is plated first and then overcoated with electrolytically deposited nickel phosphorous to provide corrosion protection and wear resistance. For enhancing the aesthetic appeal, an additional layer of chrome or tin can be applied if required. The process flow for electrolytic plating on plastic components is shown in Figure 7.13 (ACT, 2008).

Electroplating is an ideal process for EMI shields where the design allows for the component to be coated with all-over plating. Selective plating or masking of small areas is difficult. ABS is the most commonly used plastic substrate, although several other engineering polymers can be coated (Skelly, 2004). Electrolytic plating coatings have been widely used for metallic components.

7.4.1.3 Tin Plating and Tin Whisker Growth in EMI Shielding Systems

Tin plating is an electrodeposit process that offers excellent solderability; improves corrosion resistance; provides a low contact-resistance surface and galvanic compatibility with contact interface; and prevents galling of metal surfaces. Material with a

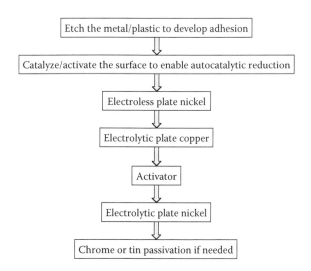

FIGURE 7.13 Electrolytic plating process flow.

tin-plated surface can be mechanically formed and shaped into various EMI shields without surface breaking as the tin plating is soft and ductile. Furthermore, tin plating is environmentally compliant and has relatively low costs; therefore, it is quite commonly used in EMI shielding. For instance, tin plating is widely used for board-level shields, as all board-level shields must be terminated by soldering to the PCB for best shielding performance.

Tin-plating surfaces are usually divided into matte tin and bright tin. Matte tin is produced from tin-plating baths that do not contain any brightener agents, whereas bright-tin baths do have brightening agents. Because bright tin is more prone to forming tin whisker, matte tin is more acceptable for most EMI shielding applications. For example, cold-rolled steel cans are usually plated by copper strike plus matte tin plating. To improve its solderability, the stainless steel is plated with Wood's nickel strike (0.1–0.50 μm thick) and followed with matte tin (1.25–5.0 μm). Here the copper strike will not help, as stainless steel quickly forms a passive tarnish, and it is not possible to plate with good adhesion onto this passive tarnish. The Wood's nickel strike is a low pH nickel and used to simultaneously dissolve the passive tarnish and providing a fresh surface for hereafter depositing the matte tin coating on it.

However, even with matte tin plating, whisker formation is still a major concern, especially for some critical applications. Tin whisker is an electrically conductive single-crystal structure, which spontaneously grows from the tin plating. Over time the whisker may grow to be several millimeters long, as shown in Figure 7.14. Whisker formation is usually identified as one of two mechanisms: spontaneous whisker formation or squeeze whisker formation. Spontaneous whisker formation occurs when the metal atoms forming the whisker diffuse through the deposit layer to the location from which the whisker then grows. In this case, they do not diffuse through any other phase layer. Squeeze whisker formation is due to externally applied compressive stress. Whereas the growth rate of spontaneous whiskers can be up to 1 cm/year under the most favorable conditions, squeeze whiskers can grow at up to 1 m/year (Kanani, 2004).

The growth of both types of whiskers can be satisfactorily modeled in terms of diffusion processes. Setting out from the observation that irregularities found at the whisker tip often remain unchanged over a period of time, it can be deduced that the metal

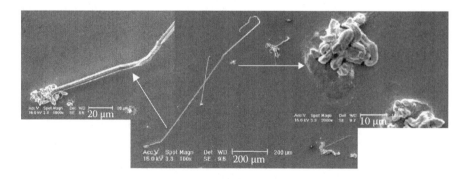

FIGURE 7.14 Tin whiskers.

atoms involved in the formation and growth of whiskers are sourced by solid state diffusion from adjacent areas to the growth point. Two diffusion mechanisms can operate in this case, namely self-diffusion and grain-boundary diffusion. Which of these diffusion mechanisms is of primary importance can only be determined on a case-by-case basis, in which diffusion coefficients are measured. Comparison of values for diffusion coefficient, not only at ambient conditions but also at elevated temperatures, indicates that the contribution from self-diffusion is negligible, as compared with that from grain-boundary diffusion. Diffusion of metal atoms along the grain boundaries in the deposited layer is the sole mechanism for whisker growth (Evans, 1972; Kanani, 2004).

However, the diffusion mechanism is not the only process involved in whisker growth. For example, the volume of a whisker of 4 μm diameter and 1 mm length, growing out of a film 10 μm thicker, represents a volume of 12.6 μm^3. This corresponds to an area of deposit of 40 μm diameter from which the whisker metal atoms must be sourced. Should several such whiskers form close to one another, as is often the case, there must be some forces responsible for providing the metal atoms necessary for further whisker growth. Experiments suggest that compressive stress, present in the deposit layer, is the primary driving force for mass transport of metal atoms. These operate not only through lattice defects in the crystal, but also result from lattice dislocations, the latter being due to the presence of foreign species in the lattice. These impurities frequently derive from the deposition electrolyte, in which they are inadvertently present (Kanani, 2004).

A further mechanism involves a spiral growth mechanism along the screw dislocations, which can penetrate through to the outer surface of the deposited layer, where they form an atom-scale step at the point of their emergence. These steps are energetically favored growth locations, at which metal atoms can readily accrete. One can thus hypothesize a whisker growth process in which the metal atoms diffuse to the surface, using vacancy-hopping, after which, by surface diffusion, they congregate at these growth sites. In this way, step defects are built up around the axis of rotation of screw dislocations. Migration of metal atoms along the axis of the screw dislocation is favored by reduced displacement energies, which provides the driving force for spiral grain growth. A characteristic of whiskers formed by this mechanism is their sharp tip. This mechanism is also invoked to explain the extremely high mechanical strength of such whiskers, in many cases close to the theoretical value. However, the tensile strength of a whisker falls off rapidly with increasing diameter (Evans, 1972; Kanani, 2004).

Tin whiskers are capable of causing electrical failures ranging from parametric deviations to sustained plasma arcing that can result in catastrophic short circuits. The general risks include the following (Richardson and Lasley, 1992):

- In low voltage, high impedance circuits, there may be insufficient current available to fuse the whisker open and a stable short circuit results. Depending on the diameter and length of the whisker, it can take more than 50 milliamps (mA) to fuse one open. More typical is ~10 mA.
- At atmospheric pressure, if the available current exceeds the fusing current of the whisker, the circuit may only experience a transient glitch as the whisker fuses open. That is, transient short circuits can result.

- If a tin whisker initiates a short in an application environment possessing high levels of current and voltage, then a very destructive phenomenon known as a *metal vapor arc* can occur. The ambient pressure, temperature, and the presence of arc suppressing materials also affect metal vapor arc formation. In a metal vapor arc, the solid metal whisker is vaporized into plasma of highly conductive metal ions, which are more conductive than the solid whisker itself. This plasma can form an arc capable of carrying hundreds of amperes. Such arcs can be sustained for several seconds until interrupted by circuit protection devices, such as fuses, circuit breakers, or until other arc extinguishing processes occur.
- Whiskers or parts of whiskers may break loose and bridge isolated conductors or interfere with optical surfaces to cause debris and contamination.

A variety of research and recommendations have been made to mitigate the effects of tin whisker growth (JEDEC Standard 22A121, 2005; Zhang et al., 2000). At this time, the only sure way of avoiding tin whiskers is not to use parts plated with pure tin for some critical applications. Studies have shown that alloying tin with a second metal can reduce the propensity for whisker growth. Generally, alloys of tin and lead (Pb) are acceptable where the alloy contains a minimum of 3% Pb by weight. The effects of tin whisker formation on alloying materials other than Pb are not well understood. In the event pure tin-plated parts cannot be avoided, there are some additional processing techniques that may be used to reduce but not eliminate the risks associated with tin whiskers. The effectiveness of these approaches varies and most require further evaluation to determine their suitability for long duration missions. Examples of these approaches include:

- Solder dipping pure tin-plated terminations and leads using a leaded solder. The effectiveness of this approach at covering all pure-tin-plated surfaces varies.
- Application of conformal coat material to pure-tin-plated surfaces. Conformal coat appears to reduce the growth rate of tin whiskers, but whiskers are still capable of growing through some conformal coat materials such as polyurethanes.
- For some device types, replate surfaces using finishes such as thick matte tin (usually not less than 5 μm), tin/lead, nickel/tin, or nickel, which are substantially less prone to whisker formation.

7.4.1.4 Immersion Surface Finishes

Other alternative coating or plating processes have been attempted, such as various organic solderability preservatives (OSPs); immersion tin; immersion silver; electroless nickel/immersion gold (ENIG); ENIG with an optional layer of thicker electroless gold (ENEG); electroless nickel/electroless palladium/immersion gold (ENEPIG); and gold direct on copper (Walsh et al., 2006).

OSPs are azole-related compounds such as imidazole, benzotriazole (thin films), and substituted benzimidazole (thick films). Thick OSPs incorporate copper in the structure, which creates an oganic-metallic coating. This provides a significant improvement in protection against copper oxidation at elevated temperature and humidity (Brian et al., 2008). OSPs are relatively inexpensive and are easily applied but limited in the number of heat cycles that can be used in subsequent assembly. Some OSPs require a nitrogen atmosphere at assembly. They are also not suitable on a contacting surface. High-temperature OSPs may cause solder voids (Walsh et al., 2006).

Immersion tins have had some early success as a solderable surface but may have a limited future. Their main drawbacks are their use of the carcinogenic ingredient thiourea, and the evidence of occasional whiskering as well as intermetallic formation. Whiskers are especially a concern with fine lines and spaces where they could be knocked out of a hole during part insertion, thus increasing the possibility of a subsequent electrical short. A copper/tin intermetallic can form for a copper or copper alloy substrate during deposition and continue to grow, limiting the useful shelf life of the stored parts. In addition, they are almost as expensive as electroless nickel/immersion gold but without the additional benefits (Walsh et al., 2006).

Immersion silver seems more attractive than immersion tin. It is less expensive and a more user-friendly process than gold and palladium. However, the silver surface is quite reactive and can form tarnish film, which significantly reduces the solderability of the coating. To prevent tarnishing, the processes include antitarnish either as an ingredient within the silver bath or applied in a subsequent step. Other process concerns are the possible inclusion of voids in the solder joint and the desire for better thickness uniformity per part (Brian et al., 2008; Walsh et al., 2006).

An ENIG finish provides a solderable flat surface that does not oxidize or discolor. It has a long shelf life, and the precious metal layer provides excellent electrical conductivity. The nickel underlayer serves as a barrier against formation of unsolderable gold copper intermetallics for a copper or copper alloy substrate for instance (Brian et al., 2008). However, ENIG is the most expensive of the final finishes and also requires the most steps. Parts must be clean, and have or be precoated with a smooth copper surface on which to build. Electroless nickel is an autocatalytic process that deposits nickel on the palladium-catalyzed copper surface. The process requires continuous replenishment of the nickel ion and the reducing agent. Good process control (constituent concentration, temperature, and pH) is the key to a consistent reproducible deposit. It is very important that the nickel be able to plate a surface with consistent phosphorus levels. Most prefer a middle range of 6% to 8% phosphorus; phosphorus levels too low would easily corrode, but levels too high make subsequent soldering of parts more difficult. Immersion golds are processed by attaching themselves to the nickel by replacing atoms of nickel with atoms of gold. The purpose of the immersion gold layer is to protect the nickel surface until it is soldered. The recommended gold thickness is 2 to 4 μin (0.0508–0.1016 μm). As the purpose of the gold layer is to maintain the solderability of the nickel surface, it is necessary that it be thin (2 to 4 μin are preferred) and pore free (Walsh et al., 2006). The most successful ENIG chemistries have several design strengths: low palladium/no chloride activators for solder mask compatibility and avoiding background and skip plating; a mid-phosphorus nickel that runs at lower temperatures, requires infrequent dummy plating, holds its ability to perform even after repeated heatings, and gives level plating around the shoulders of pads and lines; and an immersion gold that has gentle chemistry and is self-limiting in thickness (Kwok et al., 2005; Walsh et al., 2006). Typical failures observed with immersion gold coatings are associated with poorly formed joints at the solder/nickel interface. The interfacial failure has been attributed to the presence of brittle gold/tin intermetallic compounds. In other cases, the problem is caused by absence of a reliable metallurgical bond formed between tin and nickel, which results from "black nickel" phenomena. Another possible failure might occur due to the oxidation of the nickel through the pores in the coating causing a higher concentration of nickel oxides on the surface of the coating (Brian et al., 2008). ENEPIG is a very versatile surface finish with good conductivity and solderability, and may prove to be an optimal soldering surface for EMI shields.

Direct gold over copper (also called DIG [direct immersion gold]) for a copper or copper alloy substrate was specifically developed for parts where nickel could create RF interference. The planarity is exceptional, again, dependent on the quality of the copper surface. To avoid the inherent problems of copper migration through the thin gold surface, it is necessary for these parts to go to final assembly within four months. Even that period is only possible with a combined replacement/autocatalytic process. Normal immersion gold will have trouble properly attaching to the copper surface and providing the necessary pore-free layer (Walsh et al., 2006).

7.4.2 CONDUCTIVE PAINTS

Conductive paints are available in a variety of materials, which offer a range of shielding effectiveness. The conductive metals in the paint generally include nickel, copper, silver–copper hybrids, and pure silver. The coating conductivity and costs are directly related to the constituents. For instance, nickel-loaded paints offer the cheapest solution but provide the lowest level of shielding. Silver paints offer the highest degree of shielding but at high costs. Traditionally, all paints contained harsh solvents (usually methylethylketone, MEK) to improve adhesion to the substrate. However, the trend toward thin-walled moldings involving blends of polycarbonate and ABS has amplified the need to minimize the impact of the coating on the substrate. These paints are formulated using mild solvents and can be mechanically removed to allow the plastics to be recycled (Molyneux-Child, 1997).

Similar to the selective plating process, a masking fixture is used to control the location of the conductive paint. The paint can be applied using two methods: in a manual paint booth where an operator applies the paint with a paint gun; or by a PLC (programmable logic controller)-controlled paint robot. The paint robot offers advantages over manual spray methods, especially in higher volume applications where cost is critical. With the robot, the spray pattern can be programmed and then dried to apply the optimal amount of conductive paint across the entire shielded surface. Typically, a manual paint application has lower setup costs than those associated with robotic painting and is a good match for lower volume applications.

In general, conductive paint can provide good to very good shielding effectiveness, good design versatility, fair to good environmental reliability, and poor to good abrasion/scratch resistance.

7.4.3 Vapor Deposition

Vapor deposition used in EMI shielding systems mainly refers to physical vapor deposition (PVD), which comprises a group of methods to deposit thin films by the condensation of a vaporized form of the material onto various shielding surfaces. It is fundamentally a vaporization coating process in which the basic mechanism is an atom-by-atom transfer of material from the solid phase to the vapor phase and back to the solid phase, gradually building a film on the surface to be coated.

The coating method involves purely physical processes such as high temperature vacuum evaporation or plasma sputter bombardment rather than involving a chemical reaction at the surface to be coated as in chemical deposition. The PVD methods that have been developed and utilized mainly include: (a) evaporative deposition—the material to be deposited is heated to a high vapor pressure by electrically resistive heating in high vacuum; (b) electron beam physical vapor deposition—the material to be deposited is heated to a high vapor pressure by electron bombardment in high vacuum; (c) sputter deposition—a glow plasma discharge, which is usually localized around the target by a magnet, bombards the material sputtering some away as a vapor; (d) cathodic arc deposition—a high power arc directed at the target material blasts away some into a vapor; and (e) pulsed laser deposition—a high power laser ablates material from the target into a vapor.

PVD is a desirable alternative to electroplating and possibly some painting applications. PVD can be applied using a wide variety of materials to coat an equally diverse number of substrates using any of the basic PVD technologies to deposit a number of desired finishes of variable thickness with specific characteristics. The application of PVD surface coating technologies at large scale, high volume operations will result in the reduction of hazardous waste generated when compared to electroplating and other metal finishing processes that use large quantities of toxic and hazardous materials.

PVD has been specifically developed for coating plastic components with a layer of pure aluminum to provide high levels of EMI shielding (Arnold, 2002b; WaveZero, 2008). In the vapor deposition process, components are loaded onto a planetary fixture that is positioned inside the vacuum chamber, and then a vacuum is created within the chamber. A wire of the material like aluminum to be deposited is fed into the chamber and is vaporized. The components are rotated on the planetary inside the chamber, and the metal vapor condenses onto the cooler plastic components. If the design requires selective metallization, a mask can be incorporated into the fixture holding the component on the planetary. Typically, vapor deposition applies very thin conductive coatings and therefore offers a relatively lower level of shielding effectiveness, but it can be improved by increasing the coating thickness. Moreover, vapor deposition offers to designers materials that are not always available with plating or conductive paints. These materials include but are not limited to aluminum, stainless steel, Nichrome, and copper.

Consequently, vapor deposition is the most environmentally friendly coating process for EMI shields, requiring no solvents and producing no harmful by-products. The coatings have a smooth metallic appearance and are highly ductile and adhere to most engineering plastics. The process deposits a pure layer of aluminum, for instance and as aluminum dissolves in sodium hydroxide, the components are easy to recycle. The coatings produced offer considerable weight saving and high shielding effectiveness. And the components can be coated selectively using multi-impression masking tools.

7.4.4 Conductive Coating Application and Design Consideration for EMI Shielding

To choose the optimum conductive coating for an EMI shielding application, a thorough comprehension of the requirements for each shielding system is vital. The conductivity and shielding effectiveness are essential, and they are both a function of the coating thickness. Table 7.3 shows the shielding effectiveness of different coatings (Skelly, 2004). For thin coatings (the thickness <<2.5 μm), the shielding effectiveness can be estimated by (Armstrong, 2004; Carter, 1999):

$$SE = 20 \log [7 \times 10^{11}/(fR)] \quad (7.1)$$

where f is the frequency (MHz) and R is the surface resistance (Ω). Other considerations include unit cost, nonrecurring engineering or setup costs for fixing, metallization pattern tolerance in selective applications, compatibility with candidate plastic resins, and mechanical design of the shields.

All-over plating offers the best shielding effectiveness and the lowest cost. Vapor deposition can offer low costs although some degree of shielding effectiveness may be sacrificed. Conductive paints and selective painting provide high shielding effectiveness; however, the costs are typically higher than all-over plating or vapor deposition because of the cost of the conductive paints materials or the two-step process used in selective plating, that is, applying the catalyst and then plating.

Consequently, conductive coating techniques have proven to be reliable, cost effective, and suitable for volume scale production. Therefore, conductive coating has and will continue to play an important role in EMI shielding systems.

TABLE 7.3
Shielding Effectiveness of Conductive Coatings

Conductive Coatings	Thickness	Shielding Effectiveness (dB)		
		1 GHz	5 GHz	10 GHz
All-over copper/nickel plating	1.0–1.25 μm	120	113	87
Selective copper/nickel plating	2.0–2.5 μm	71	60	62
Copper loaded paint	0.025 mm	70	81	63
Copper-silver hybrid paint	0.025 mm	69	85	70
Silver loaded paint	0.0125 mm	62	70	55
Copper/stainless steel vapor deposition	0.5 μm	48	58	62
Aluminum vapor deposition	0.5 μm	46	40	35

7.5 BOARD-LEVEL SHIELDING WITH ENHANCED HEAT DISSIPATING

As more powerful electronic components and increased packaging densities are added to electronic devices, the combination of EMI protection and thermal cooling has become a development trend in board-level shielding and will provide an optimal option for producing more reliable electronic products.

7.5.1 Minimizing EMI from Heat Sinks

Heat sinks play a vital role in enhancing circuit design reliability by keeping sensitive components within a particular operating-temperature range. However, the heat sinks can also add to overall circuit EMI by resonating the EMI that the temperature-sensitive components radiate.

To reduce this kind of effect, sources of EMI radiation should be placed away from the edges of the heat sink. If one dimension is longer than the others, then it is important to mount devices away from the ends of the longer dimension. It is even more important to avoid placing sources at the corners of the structure. Heat sinks with dimensions that are close to square or cubic yield the lowest levels of radiation enhancement. In the case of a nearly square structure, fin orientation should be configured so that the fins run longitudinally with the prime direction of excitement. Devices mounted in channels provide the lowest levels of EMI radiation, although a check should be carried out to make sure the channel will not form a resonant cavity. However, EMI generation methods in circuits are often complex, and multiple noise sources are often present. Therefore, after using these rules in initial design, designers should be sure to also create a model that is specific to their particular application (Brench, 2000; Wang, 2002).

There are a number of commercially available software modeling packages with different strengths and weaknesses. Using one of the packages, designers can predict not only the EMI effects of heat sinks, but also that of any structure, chassis, or casing of any piece of equipment. Using these software models can lead to a significant reduction in design costs by pointing out trouble areas early in the design stage.

7.5.2 Combination of Board-Level Shielding and Heat Dissipation

There have been a variety of designs and products developed to combine EMI shielding and thermal cooling (English, 2007; Wang et al., 2002). For instance, Figure 7.15 shows that the ventilating shield encloses the PCB and heat dissipating system including the thermal interface material (TIM) and heat sink. This design can be used in board-level shielding to increase heat transfer while maintaining shielding effectiveness, allowing heat sinks to be used inside shielding cans to eliminate EMI radiation from the heat sink. As shown in Figure 7.16, a unique combination of board-level EMI shielding and thermal dissipation was designed to utilize a heat sink lid, which snaps into a board-level shielding frame with spring fingers. Similar to the design shown in Figure 7.15, contact is made with the heat sink through the TIM, which then conducts heat up through the heat sink. The four sides of the snap-in heat sink are flexible in order to snap into an aperture or multiple apertures within a shield lid.

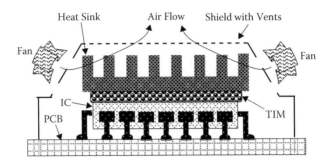

FIGURE 7.15 A ventilating EMI shield that encloses the PCB and heat dissipating system.

Multiple apertures can be made into a single lid to cool multiple components. The heat sink may also snap into a dedicated shield housing a single, discrete, hot component. The snap-in heat sink acts as a lid over and is removable for component maintenance (Armstrong, 2004). As shown in Figure 7.17, by placing a TIM on the inside of the shielding can, the air gap between the shield and electronic power can be eliminated. A heat spreader, heat sink, or heat pipe with additional TIM can be included to further enhance the thermal performance (English, 2007).

7.5.3 Thermal Interface Materials (TIMs) for Thermal-Enhanced Board-Level Shielding

In an electronic packaging with a EMI shielding system, heat generated within a semiconductor component, like the *Si* chip, must be removed to the ambient environment to maintain the junction temperature of the component within safe operating limits. This heat removal process often involves conduction or convection from the component package surface to a heat sink, heat pipe heat spreader, or EMI shield that can more efficiently transfer the heat to the ambient environment. The attachment between heat sink/spreader or combined EMI shield has to minimize the thermal resistance of this newly formed thermal joint or interface with an intimate contact. These contact surfaces are usually characterized by a microscopic surface roughness

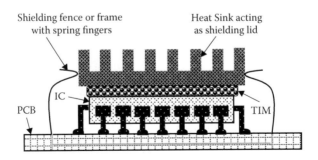

FIGURE 7.16 A heat sink snaps into the shielding frame and acts as the shielding lid.

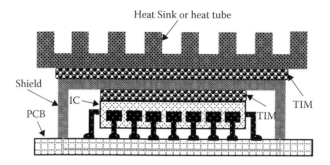

FIGURE 7.17 Combination of board-level shield, heat sink or heat tube, and thermal interface materials (TIMs).

superimposed on a macroscopic nonplanarity that can give the surfaces a concave, convex, or twisted shape. When two such surfaces are joined, intimate contact occurs only at the high points. The low points form air-filled voids; the typical contact area can consist of more than 80% air voids, which creates a significant resistance to heat flow (Mahahan et al., 2004).

TIMs are used to eliminate these air gaps from the interface by conforming to the rough and uneven mating surfaces. Because TIMs have significantly higher thermal conductivity than the air being replaced, the resistance across the interface decreases, and the component junction temperature will be reduced.

Table 7.4 shows some typical TIMs that have been developed and utilized in electronic packaging (Mahahan et al., 2004). Thermal conductive greases, also called thermal pounds, are typically made up of silicone oil with a heavily loaded suspension of thermally conductive ceramic or metal fillers. The emulsion fillers increase thermal capacity and provide body to minimize flow out of the interface. The actual bulk thermal conductivity of the resulting material is low; however, since the pastelike consistency of the thermal grease allows the final bond line to be very thin, the resulting thermal resistance across the interface can be quite low. As a two-part suspension, thermal greases can separate to form voids and dry out over time especially in service with high temperatures or extensive thermal cycling. Therefore, the pastelike nature can lead to inconsistent and messy applications.

Thermal conductive gels are similar to thermal grease with a new difference. The gel material is cured to form cross-linked polymer chains. This provides lateral stability to minimize the problem associated with liquid TIMs. Gels are reusable; however, they generally have a slightly lower thermal conductivity than thermal grease.

Thermal interface pads are preformed elastomer materials and can be compressed between the heat generator and the heat receiver. Pads are either electrically conductive or nonconductive, based on application. Pads are generally thicker than tapes and adhesives because the added thickness helps to fill larger gaps. Tapes and adhesives are used to provide a better thermal interface while also securing the component and heat/spreader/shield. However, this bond can be unreliable and may also need to be supplemented by mechanical attachment.

TABLE 7.4
Properties of Typical TIMs

Material Type	Typical Composition	Advantages	Drawbacks	Bond Line Thickness (mm)	Thermal Conductivity (W/m-K)
Grease	AlN, Ag, ZnO, silicon oil	High bulk thermal conductivity, conforms to surface irregularities no cure needed, reusable	Pump out and phase separation, migration	0.05	3 to 5
Gel	Al, Ag, silicon oil, olefin	Good bulk thermal conductivity, conforms to surface irregularities before cure, no pump out or migration, reusable	Cure needed, lower thermal conductivity than grease	0.025–0.04	3 to 4
Phase change materials	Polyolefins, epoxies, polyesters, acrylics, BN, alumina, Al, carbon nanotubes	Conforms to surface irregularities, no cure needed, no delamination, easy handling, reusable	Lower thermal conductivity than grease, no uniform bold line thickness	.05–0.05	0.5 to 5
Phase change metallic alloy	Pure In, In/Ag, Sn/Ag/Cu, In/Sn/Bi	High thermal conductivity, reusable	Complete melt possible, voiding	0.05–0.125	30–50
Solder	Pure In, In/Ag, Sn/Ag/Cu, In/Sn/Bi	High thermal conductivity, easy handling, no pump out	Reflow needed, stress crack, delamination, voiding possible, not reusable	0.05–0.125	30–86

Source: Mahahan 2004.

Phase change materials provide a combination of greaselike thermal performance with padlike handling and installation convenience. They have compositions that transform from a solid at room temperature to a mixture of solid and liquid phases at various operating temperatures. The liquid from the mixture makes intimate contact with the contact surfaces, while the solid retains the integrity of the gap. At this operating temperature, the phase change material is able to act like thermal grease, allowing it to form a thin bond line. Sometimes the installation process requires some compressive force to bring the two surfaces together and cause the material to flow, until the two surfaces come into contact at a minimum three points or the joint becomes so thin that the viscosity of the material prevents further flow. Excess liquid should then flow to the perimeter of the interface and solidify so it remains away from other components.

Phase change metallic alloys, also called low melt alloys (LMAs), are typically alloys of indium, bismuth, gallium, tin, or silver. They are in a liquid phase at operating temperatures; typical phase change occurs between 60°C and 80°C. Because LMAs are all metal, they contain no organic materials and hence require no curing during application. The superior performance of LMAs as TIMs is due to the thermally conductive nature of metals and the high degree of wetting they enjoy. Good wetting means low contact resistance and allows relatively thin bond lines. These alloys are well suited as a thermal interface between materials of dissimilar coefficients of thermal expansion (CTEs), however, reliability problems in production implementations, resulting in catastrophic failure, have retarded their adoption (Hill and Strader, 2006; Macris et al., 2004).

Thermal solders are very attractive because they have high thermal conductivities. The key to using solder TIMs is to reduce voiding during reflow. Unlike other materials used for thermal interfacing, solder voids will not propagate later during use due to pump out or migration. Thermal solders provide the best physical connection of all TIMs. Using solder eliminates the issue of bleed out from the thermal greases, while providing very good adhesion. This mechanical attachment is a particular advantage when external clamping is not feasible. To take advantage of these properties, interfaces and assembly are designed to ensure proper wetting of the contact surfaces, which minimize voiding.

Among the solders, indium may be the best as a thermal interface material. Pure indium provides a unique combination of high conductivity (86 W/m-K) and compliance (414 MPa shear flow stress and no work hardening). Since indium will wet to nonmetallic substrates, it provides unique thermal interface solutions with or without reflowing. Indium can also be cold welded in applications where normal soldering temperatures are detrimental to the electronic packaging (Mahahan et al., 2004).

Many other TIMs have been developed and more will appear because of wide demand.

7.6 SUMMARY

The board-level shielding technique has primarily employed solder-attached perforated metal cans with and without removable lids. Cans are extremely difficult to remove and replace if the shielded components need to be reworked, because of the complicated desoldering and resoldering process. Metallized thermoformable shields offer the promise of overcoming the problems associated with can types of shielding, and have improved performance over some metal cans. In addition, electrical components that need to be shielded often generate heat while operating, therefore, efforts have been made to develop a board-level shield that is simultaneously removable, compatible with single or multicompartment designs, sometimes thin in profile and lightweight, and allows for thermal dissipation of shielded components.

REFERENCES

Applied Coating Technologies (ACT). 2008. *EMI/RFI shielding by electroplating techniques.* http://www.applicoat.com/pdfs/emi-rfi-electroplating.pdf

Armstrong, M. K. 2004. Advanced PCB design and layout for EMC–Part 2: Segregation and interface suppression. *Keith Armstrong, EMC & Compliance Journal* 3: 32–42.

Arnold, R. R. 2002a. Vacuum-metalized shielding. *Compliance Engineering.* http://www.ce-mag.com/archive/02/09/arnold.html

Arnold, R. R. 2002b. *Vacuum metalized thermoformable structures for EMI shielding of printed circuit boards and electronic devices.* http://www.emcsoltech.com/pdf/EMI%20Shield%20using%20Vacuum%20metalized%20thermoformable.pdf

Bachman, B., R. Bjorlin, and G. Shawhan. 2001. New technologies for board-level and enclosure-level EMI shielding. *Interference Technology.* http://www.interferencetechnology.com/ArchivedArticles/Conductive_Coatings/New%20technologies%20for%20board-level%20and%20enclosure-level%20EMI%20shielding.pdf?regid=

Brench, C. 2000. Apparatus to minimize integrated circuit heatsink EMI radiation. US Patent 6,011,299.

Brian, P., M. Pavlov, and G. Chalyt. 2008. *Immersion gold coatings, surface analysis and solderability prediction.* http://www.ecitechnology.com/articles/Immersion_Gold_Coatings,_Surface_Analysis_and_Solderability_Prediction.pdf

Carter, D. 1999. *A look at metal barriers with discontinuities for related Tecknit products.* Tecknit Europe Ltd. http://www.tecknit.co.uk/catalog/emcholes.pdf

Clupper, T. 2006. *Removable shielding technologies for PCBs.* http://www.gore.com/MungoBlobs/412/61/EE_clupper_Removable%20Shielding_eprint.pdf

English, G. 2007. Combined board level EMI shielding and thermal management. US Patent 7,262,369.

Evans, C. C. 1972. *Whiskers.* London: Mills & Boon.

Fenical, G., and J. English. 2006. The essentials of PCB shielding. *Interference Technology,* 126–130.

Gore. 2008. *SnapSHOT.* http://www.gore.com/MungoBlobs/712/644/snapSHOT_board-level_EMI.pdf

Hill, R. F., and J. L. Strader. 2006. Practicle utilization of low melting alloy thermal interface materials. Semiconductor Thermal Measurement and Management Symposium, 220 IEEE Twenty-Second Annual IEEE, pp. 23–27.

Hüske, M., J. Kickelhain, J. Müller, et al. 2001. *Laser supported activation and additive metallization of thermoplastics for 3D-MIDS.* Proceedings of the 3rd LANE, August 28–31, Erlangen, Germany.

Intel. 1999. *Design for EMI: Application note AP-589.* ftp://download.intel.com/design/PentiumII/applnots/24333402.PDF

JEDEC Standard 22A121. 2005. Measuring whisker growth on tin and tin alloy surface finishes.

Kanani, N. 2004. *Electroplating: Basic Principles, Processes and Practice.* Boulevard, UK: Elsevier.

Kwok, R., K. Chan, and M. Bayes. 2005. *Development of an electroless nickel immersion gold process for PCB final finishes.* http://www.rohmhaas.com/electronicmaterials/CBT/vol5/papers/electroless_nickel.pdf

Livington, R. 2004. *PCB-level shielding technology for today's medical wireless world.* http://www.devicelink.com/mem/archive/04/10/011.html

Lun, T. C. 2005. *Designing for board level electromagnetic compatibility.* http://www.freescale.com/files/microcontrollers/doc/app_note/AN2321.pdf

Macris, C., T. Sanderson, R. Ebel, and C. Leyerle. 2004. *Performance, reliability, and approaches using a low melt alloy as a thermal interface material.* http://www.enerdynesolutions.com/downloads/imaps_2004_man.pdf

Mahahan, R., C.-P. Chiu, and R. Prashed. 2004. Thermal interface materials: A brief review of design characteristics and materials. *Electronics Cooling.* http://electronics-cooling.com/articles/2004/2004_february_a1.php

Majr Products. 2007. *Custom board level shielding.* http://www.majr.com/pdfs/Custom_Board_Level_Shielding.pdf

Molyneux-Child, J. W. 1997. *EMC Shielding Materials.* Oxford: Newnes.

Quesnel, N. 2008. *Optimizing EMI shielding for wireless systems.* http://www.chomerics.com/tech/EMI_shld_%20Artcls/Wireless%20EMI%20Shielding.pdf

Richardson, J. H., and B. R. Lasley. 1992. Tin whisker initiated vacuum metal arcing in spacecraft electronics. *Government Microcircuit Applications Conference* 18: 119–122.

Skelly, J. T. 2004. *Application of conductive coatings in electronic devices today and in the future.* http://www.cybershieldinc.com/images/ConductiveCoatingApplicationArticle.pdf

Walsh, D., G. Milad, and D. Gudeczauskas. 2006. What's best for your board: Immersion silver, immersion tin or may be ENEPIG? *Printed Circuit Design & FAB.* http://pcdandf.com/cms/content/view/2347/95

Wang, D.G., J. L. Knighten, and P. K. Muller. 2002. An integrated vent, heatsink and EMI shield. *Semiconductor Thermal Measurement and Management, 2002. Eighteenth Annual IEEE Symposium,* pp. 125–131.

WaveZero. 2008. *EMC product line: Form/Met^{TM}.* http://www.wavezero.com/product/EMC/formmet.html

Zhang, Y., C. Xu, C. Fan, and J. Abys. 2000. Tin whisker growth and prevention. *Journal of Surface Mount Technology* 13: 1–9.

8 Composite Materials and Hybrid Structures for EMI Shielding

8.1 INTRODUCTION

Multifunctional electrically conductive composite materials and hybrid structures have been developed with increasing demand for innovative electromagnetic interference (EMI) shielding, especially in high-power electronics. Their multifunctional requirements, for example, are driving lightweight composite processing technology to complex, hybrid configurations that require more advanced EMI shielding approaches in electronic assemblies. Many of the enhanced composite material systems being developed for high-power electronic components are strategically integrated into a single, multifunctional assembly that must also meet electromagnetic compatibility (EMC) requirements. A great effort has been made to identify potential material systems and affordable production processes that can be integrated with the complex primary structures, geometries, and multifunctional assembly joining processes to provide structural integrity and EMI shielding up to 80 dB or higher levels.

This chapter will give a primary introduction to these materials and their process technologies, including knitted wire/elastomer hybrid gaskets, shielding tapes, metal fiber reinforced composites, hybrid flexible structures, and electroless deposition for composite shielding materials.

8.2 KNITTED WIRE/ELASTOMER GASKETS

Knitted wire mesh gasketing provides a low cost and versatile method for integral EMI shielding at joints and other discontinuities in enclosures. Compared with other EMI gasketing materials, knitted wire gaskets are most suitable for low-pressure applications since they are flexible and have a high elastic resilience. One or two layers of mesh provide excellent EMI shielding, while the elastomer or other mesh cores provide a high degree of resiliency. Metal wires can be knitted over various elastomers or mesh cores to achieve the required physical properties. For instance, elastomer core gaskets are usually used for shielding large gaps in an economical way. Typical knit shield cross-sections are rectangular, round, round-with-fin, and double-round-with-fin. The mesh consists of many interlocking loops that act as thousands of small springs, which provide a resilient all-metal conductive gasket. Shielding effectiveness ranges from 35 dB at 10 kHz, 110 dB at 10 MHz, to 100 dB at 1 GHz. When an elastomer is the core for the wire mesh, it provides excellent compressibility with a high degree of resilience, while at the same time functioning as an environmental seal (Laird Technologies, 2006; Molyneux-Child, 1997).

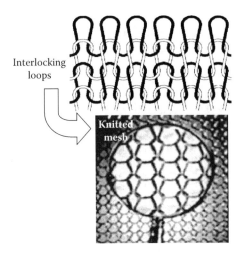

FIGURE 8.1 Knit interlocking loop and knitted mesh.

8.2.1 Fabrication Process and Materials Selection

The wire mesh is knitted using a knitting machine. After wire drawing, the metal wire is knitted into a mesh structure on a cylindrical machine in very much the same way as stockings or sweaters, as shown in Figure 8.1 (Metex, 2003). A continuous mesh strip is formed and pulled through the knitter by take-up rollers. Knitting produces a mesh of interlocking loops. These loops can move relative to one another in the same plane without distorting the mesh, giving knitted mesh a two-way stretch. From a functional point of view, it is this flexibility that gives knitted mesh its unique advantage for EMI shielding applications.

The mesh is normally knitted from wires ranging in diameter from 0.09 mm (.003 in) to 0.5 mm (.0200 in). These wires can be round or flat. A round wire when flattened has approximately twice as much surface area. Wire diameter is perhaps the most important design variable especially when ventilation and filtration are needed, because it not only determines the compression strength and resilience of the mesh gaskets, but also directly affects flow, dirt holding capacity, pressure drop, and cost (Metex, 2003). In general, larger diameter wire allows for higher strength and airflow but provides lower resilience and dirt holding capacity. Moreover, larger diameter wire is less expensive. Depending on the application requirements, a balance must be struck between using a high resilience, high density filter of high-cost fine wire or a high strength, low density filter of low-cost heavy wire.

There are several basic mesh forms that can be directly knitted. For instance, flattened mesh is sometimes run through corrugating rolls to create crimped mesh. In this form, the corrugations act like springs for resiliency and give the mesh thickness. Crimped mesh is also used to create compressed units for filtration, shock mounts, and flame arrestors. The meshes are made in densities ranging from 10% to 70% (% metal by volume). Compressed mesh is also made in a continuous strip form by running the mesh through calendaring rollers. Mesh can also be knitted directly into a hollow tube used by itself or knitted over wound-up mesh to make a cable or gasket core. It can also be rolled or padded to create filters, separators, or flame arrestors. In all forms of mesh, density and wire dispersion can be controlled closely to ensure optimum product performance and repeatability (Metex, 2003).

Knitted metal has distinct advantages over most competitive materials. Because each loop acts as a small spring when subjected to tensile or compressive stress, knitted metal has an inherent resiliency. If it is not distorted beyond its yield point, the material will resume its original shape when the stress is removed. Even when it is compressed into a special shape, a high degree of resiliency is retained. Varying the knitted structure, wire diameter, wire material, and forming pressure used to create the shielding part can control this characteristic. Knitted metal also provides high mechanical damping characteristics and nonlinear spring rates. Vibration and mechanical shock can be effectively controlled to eliminate the violent resonant conditions and provide ample protection from dynamic overloads. In a compressed form, knitted metal can handle shock loadings up to the yield strength of the material itself. The load may be applied from any direction—up, down, or from all sides. Among the most important property of knitted mesh units is the material itself. By carefully selecting the material or combinations of materials, a variety of knitted mesh gaskets can be made for different EMI shielding applications—be it corrosive, ultra-high or cryogenic temperature, high pressure, radioactive, dirty, oily, or other extreme conditions (Metex, 2003; Molyneux-Child, 1997).

The most common materials used in knitting for shield gaskets are Monel, tin-plated copper-clad steel (SnCuFe), tin-plated phosphor bronze, silver-plated copper, aluminum, stainless steel, and copper beryllium. In fact, any commercially available wire can be used to make up gasketing with dimensions varying from 2 to 30 mm or so, using wire diameters ranging from 0.05 to 0.152 mm. A wire diameter of around 0.112 mm is popular, although when aluminum wire is knitted, the wire diameter tends to be greater. Some of the more unusual materials used for wire meshes include gold, molybdenum, Inconel, pure iron, platinum, nickel, silver, and Mumetal.

Monel (Ni–Cu alloy) is the most commonly used, which combines excellent EMI shielding effectiveness, high corrosion resistance and oxidation resistance, and high tensile strength and resiliency. Monel is not suitable for applications that contact aluminum due to poor galvanic compatibility. SnCuFe has similar characteristics to Monel, with improved shielding at lower frequencies but slightly lower corrosion resistance. The typical structure of SnCuFe is shown in Figure 8.2. The 57 wt% of low carbon steel core provides high strength and permeability, the 40 wt% medium layer of copper cladding offers high conductivity, and the 3 wt% outer layer of tin coating provides low contact impedance and high corrosion resistance. Therefore, SnCuFe can offer the best shielding performance at lower frequencies, especially with respect to magnetic fields. Aluminum wire is used only when the mating surfaces are aluminum, to meet galvanic compatibility requirements. Copper beryllium knitted wire-mesh gasket probably is the only material with an excellent resilience and does not require an elastomer core to function as a gasket. Also, the copper beryllium knitted wire-mesh gasket provides generally 20 dB higher shielding effectiveness than any other material. It can also be plated for galvanic compatibility (Laird Technologies, 2006).

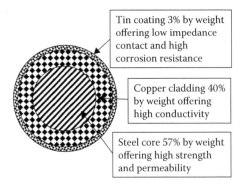

FIGURE 8.2 Typical structure of SnCuFe wire.

Choices for elastomer core materials are neoprene (sponge and solid), silicone (sponge and solid), and polyurethane foam. Core materials should be selected for temperature, chemical resistance, and mechanical characteristics. Neoprene rubber is an economical, general-purpose elastomer with good compression characteristics. It is available in solid and sponge and in hollow-core cross-sections. Silicone rubber is noted for its retention of flexibility, resilience, and tensile strength over a wide temperature range. Silicone is more resistant to cleaning solvents than neoprene. Polyurethane foam should be specified for light compression force applications. This material has good compression set characteristics and is available in rectangular and square solid cross-sections.

8.2.2 Knitted Wire Gasket Performance

Knitted wire mesh is a common and economical form of gasketing. The metal mesh construction provides excellent EMI shielding effectiveness because of the metal-to-metal contact and the ability of the mesh to cut through surface films on mating surfaces. Specifically, its resilient structure conforms to large joint unevenness, and the wiping action of the wires can break through oxides to make excellent surface contact. When a mesh gasket is pressed against a metal surface, the pressure is concentrated at the minute areas of wire contact. When the wire and metal surfaces are of comparable hardness, there will be heavy pressure on any oxide films present and these will be broken down and good electrical contact reestablished (Molyneux-Child, 1997).

The continuity of the knitted mesh wire gives the gasket high internal conductivity as well. In fact, maximum attenuation for all gaskets is achieved as compression force increases. However, care must be taken with mesh gaskets to avoid compression set. The use of elastomer cores will extend the operating range. In applications where the gasket is permanently installed between two surfaces, compression set can be tolerated (Chomerics, 2000; Laird Technologies, 2006).

The inherent resiliency of the knitted mesh makes it ideal for applications in sheet metal enclosures. The elastomer core provides a seal against rain, ventilated air, dust, and other environmental contaminants. However, wire mesh has been most commonly used in static-seal or limited-access applications because of limitations in the spring properties of the original materials and the mounting methods available. The shielding effectiveness of wire mesh gaskets ranges from 40 to 115 dB, depending on the material and the construction of the profile. Hollow-core gaskets can achieve a usable contact resistance with about 20% deflection and can be compressed up to 75%, depending on the wire material used. Solid-core mesh gaskets need less deflection to function efficiently but reach their maximum compression deflection near 40%. Compression set for CuBe wire mesh is under 1%; for other hollow- or elastomer-core meshes this value reaches 20%. Solid-core wire mesh gasket compression set can be as high as 30% (Hudak, 2000).

8.2.3 Typical Gasket Types and Mounting Methods

The basic types that knitted wire gasketing can be divided into include compacted wire sections, elastomer cored sections, seals where environmental performance is needed, and tailed gasket profiles. Figure 8.3 shows some typical profiles of knitted wire gaskets and their mounting methods (Chomerics, 2000; Laird, 2006; Molyneux-Child, 1997). Generally, rectangular and round cross-section gaskets are mounted in a channel. The channel holds the gasket in place by sidewall friction and prevents the gasket from being overcompressed. Alternatively, the gasket is held in place in tight fitting slots or bonded in place using an adhesive, such as a pressure-sensitive

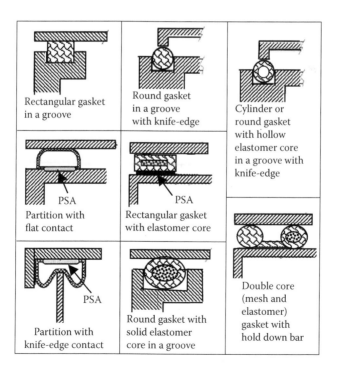

FIGURE 8.3 Some typical knitted wire gaskets and their mounting methods.

adhesive (PSA). Single-round gaskets with fin or hold down bar and double-round gaskets with fin or hold down bar are held in place by clamping the fin or hold down bar section under a metal strip; using an adhesive; or by riveting, spot welding, or adhesive bonding. Both rectangular gaskets and round gaskets with fin or hold down bar can be attached with PSA tape to ease installation.

Compacted wire provides a resilient gasket that is optimal as EMI shielding on cabinet and enclosure doors. In this construction of knitted wire mesh with elastomer core, for example, the knitted mesh is sleeved over a silicone rubber tube to give a highly compressible and semienvironmental-proof EMI gasket. Elastomer cored mesh combines the characteristics of knitted mesh including resiliency and conductivity with the excellent compression and deflection properties of silicone rubber. It is an ideal gasket material for sheet metal housings and can also be used with castings and machined surfaces. The silicone rubber tube provides a seal against dust, rain, and air. For improved EMI shielding, an additional layer of knitted wire mesh can be added. These gaskets can be retained in position in a similar manner to the all-metal gaskets. Elastomer sealed knitted mesh gaskets are manufactured with an environmental seal produced as a single element. They consist of an elastomer fluid or air sealing strip integrally bonded to an EMI shielding gasket. The standard elastomer sealing material is a closed-cell neoprene rubber, but silicone sponge and other materials have also been used successfully. The round section knitted wire gaskets have an elastomer core, whereas the rectangular section types can employ a solid resilient wire mesh, or vice versa. Circular and tailed sections are the common configurations used, but other profiles are occasionally called for. The method of joining to ensure the continuity of EMI shielding is to cut an oversized piece of the wire mesh, twist the ends and tuck them into the silicone tubing, place the connector into both ends and secure with suitable adhesive, hook stitches of cover mesh, and then twist the cover mesh together across the joint to complete the connection. Tailed mesh gaskets are used for many applications. These gasket configurations have a tail of wire mesh or hold down bar, which can be used to keep the gasket in place, either by riveting, spot welding, or by being squeezed between adjustable plates. Gaskets of a cylindrical cross-section may be formed of knitted mesh throughout or knitted wire around a core of silicone rubber. Tailed gaskets are not trapped in slots, but are attached to flat surfaces. Generally, because the metal surfaces of the lid and box to which they connect are not in contact, the shielding effectiveness is not as good as with gasketing trapped in slots. In many cases, the shielding provided by tailed gaskets is adequate, without the expense of machining slots. Furthermore, they can be used to modify existing structures, where the provision of slots would be expensive and difficult (Molyneux-Child, 1997).

It is recommended that the gasket should be compressed by at least 10% at the widest part of the gap. The shielding effectiveness can be further improved by the use of a double gasket. This is fixed to the floor using the tail or hold down bar. If it is possible to have a ledge to stop the door, then the smaller the gap between the door and ledge, the more electromagnetic leakage is minimized. In each of these cases, the gasket is rubbed by the door as it closes. This has the benefit of keeping the door surface clean, but if the door is opened frequently, the gasket may wear. For applications where there is likely to be wear, it is advisable to choose a gasket with a hard and springy wire (Instrument Specialties, 1998).

Other types, such as round cross-sectional mesh with mounting fin, can be securely crimped into the edge of an aluminum frame, forming an excellent gasketing material used where a rigid gasket and a limited compression are required. The aluminum frame can be used for mounting the gasket by spot welding, riveting, bolting, or attaching via sheet metal screws. In addition, the mounting frame may serve as a compression stop to prevent undue pressure on the metal mesh gasket. It also stiffens the surface to which it is attached, thereby serving as a load-bearing element in the structure capable of meeting exacting shock and vibration requirements. O-ring or cylindrical gaskets made from knitted wire mesh with different materials and plating surfaces, are ideal for grounding or EMI seals around shafting. PSA films are readily available with tear-off backing strips and when attached to sponge elastomer sealing sections can provide a low-cost attachment system. But it is vital to ascertain whether the adhesive is required to hold the gasket in position for many years or just to locate it initially with the actual long term fixing being achieved by other means (Chomerics, 2000; Laird Technologies, 2006).

For a mitered joint in a knitted gasket, the corner can be sewn or the wire, in this case a tailed section, crocheted at the corner to provide the desired angle. It is evident that it is not possible to produce a perfect right-angled corner with knitted wire and advice should be sought from the gasket manufacturer as to the appropriate design radii to be allowed for at the corners. Silicone systems are ideal for joining continuous gaskets with silicone cores as an alternative to the fairly complicated crocheting operation (Instrument Specialties, 1998).

8.3 EMI SHIELDING TAPES

Electronically conductive tapes are an economical shielding solution for a wide variety of EMC applications. As shown in Figure 8.4, constructed of metal foil, metallized fabric, knitted wire mesh, or carbon materials with various conductive adhesives, these tapes provide excellent conductivity and conformability, superior abrasion and corrosion resistance, strong adhesion, and durability in a thin, lightweight, and flexible EMI shielding design. As a result, EMI shielding tapes have been widely used in shielding cables; testing laboratories; and lightweight, highly conformable, and dynamic conditions.

EMI shielding tape has a multitude of uses, especially in electronic design and test laboratories for prototyping, design, and troubleshooting. Tapes with foil/fabric/mesh/cloth laminates satisfy many requirements for effective grounding, bonding, and EMI shielding. Figure 8.5 illustrates the construction of some EMI shielding tapes.

1. Self-adhesive metal tape. This tape is constructed of metal tape laminated with a layer of conductive adhesive. Metal tapes can be metal foil, such as aluminum foil, copper foil, and tin-plated copper foil; metallized fabric, like Ni–Cu-plated fabric; knitted wire mesh; or metallized carbon fiber or cloth. These tapes are often used for electrical connection surfaces, die cuts, mounting transparent foils, shielding in housings, shielding cables, and temporary shielding during tests.

222 Advanced Materials and Design for Electromagnetic Interference Shielding

FIGURE 8.4 EMI shielding tapes.

2. Galvanic compatible tape. This is a metal tape with a highly conductive self-adhesive on one side, and a masking tape on the other side. This contact surface tape is used to improve the corrosion resistance of construction metals like untreated steel plates, or to improve the galvanic compatibility

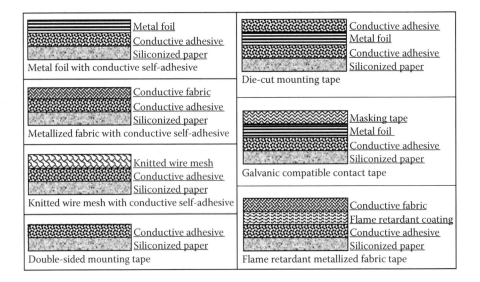

FIGURE 8.5 The construction of typical conductive tapes.

when two metal parts are connected with a gasket. After the parts have been coated, the paint overlaps the metal tape to increase the bonding.
3. Double-sided mounting tape. This double-sided self-adhesive tape is mainly used for mounting purposes. It can be applied much faster than conductive glue and can be applied very accurately. The pressure-sensitive version is not required for curing.
4. Die-cut mounting tape. The typical construction of this tape is a soft metal tape laminated on two sides with conductive self-adhesive.
5. Knitted wire mesh tape. This tape, made of a double-layered strip of knitted wire mesh, can provide effective EMI shielding for electrical and electronic cable assemblies. The knitted construction of the tape maximizes conformability and flexibility while minimizing bulk and weight. SnCuFe wire is mostly used for shielding tape to provide greater physical strength and shielding effectiveness than may be achieved with other tape materials.
6. Carbon conductive tape. The carbon conductive tape is constructed of a conductive polycarbonate base coated with carbon-filled acrylic glue, free of solvents.

In all these EMI shielding tapes, the conductive adhesive is usually made of a PSA matrix filled with conductive particles. Depending on application requirements, the conductive adhesive can be made with (a) anisotropic electrical conductivity, which allows interconnection between contact surfaces through adhesive thickness but the conductive parties are spaced far enough apart for the tape to be electrically insulating in the plane of the adhesive; or (b) isotropic electrical conductivity, in which the PSA matrix is filled with conductive fillers that allow interconnection between contact surfaces through adhesive thickness and also provide electrical conducting in the plane of the adhesive.

Light pressure can cause PSAs to stick readily to most surfaces. They typically consist of four component layers, as shown in Figure 8.6: (1) the adhesive mass, which is usually composed of a synthetic rubber, natural rubber, or acrylic polymer, may contain a variety of softeners, antioxidants, plasticizers, and curing agents; (2) the primer, which is used between the adhesive and backing to insure good adhesion between the two, may be based on natural or synthetic elastomers and may contain some tackifiers; (3) the carrier, which may be foil, crepe paper, fabric, cellophane,

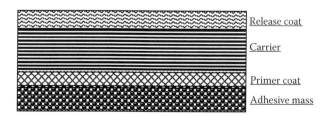

FIGURE 8.6 Four components of a typical pressure-sensitive tape.

cellulose acetate, plasticized polyvinyl chloride, or any of a number of other flexible materials, and may be reinforced with glass or other fibers; and (4) the release coat, which is applied to the side of the backing that is away from the adhesive mass so that the roll can be unwound without leaving any residual adhesive (Bemmels, 1962; Smith et al., 1984). As in all conductive PSA applications, stable EMI shielding performance in any electrical connection application may require added mechanical reinforcement like clamping or compressing in the bond area.

In addition, EMI shielding tapes can provide a low impedance connection between the braided cable shield and the metal connector backshell in moulded cables. By using EMI shielding tape, an effective EMI shielded cable assembly can be achieved without soldering the tape to the braid or backshell. The tape provides electrical continuity in seams or apertures of EMI shielded rooms. It is useful in trouble shooting for EMI radiation measurement. The tape can provide electrical contact to surfaces that cannot be soldered, such as conductive plastic or aluminium. It can help drain static charges, provides corrosion protected ground contact points, bus bar capabilities, and is a reliable ground surface.

Consequently, the EMI shielding tapes are suitable to use in a variety of application, especially for (a) shielding cables on notebook computers, copiers, or other electronic equipment, especially where cables are under dynamic action or stress; (b) shielding or grounding in weight-sensitive applications; (c) shielding over a component in which high conformability is essential; (d) shielding or grounding for electronic equipment where vibration may be present during operation; (e) EMI measurement troubleshooting and fix-it applications in test laboratories; and (f) shielding of flexible circuits.

8.4 CONDUCTIVE FIBER/WHISKER REINFORCED COMPOSITES

Traditionally, metals like copper or aluminum and their composites have been considered to be the best conventional EMI shielding materials with high shielding effectiveness because of their high dielectric constant and electric conductivity. However, heavy weight and easy corrosion restrict their wide use for shielding materials. Therefore, electrically conductive polymer composites have been considered as EMI shielding materials due to their light weight, good resistance to corrosion, flexibility, and processing advantages. These composites generally consist of an electrically nonconductive polymer matrix or substrate and into that has been mixed or layered onto an electrically conducting material. Usually the conductive reinforcements or fillers include carbon black, carbon fibers, metal flakes, or metal fibers, and carbon nanotubes (CNTs), that have been mixed into the matrix material in the melt state or layered onto the surface of a polymer substrate. In this kind of composite, the EMI shielding effectiveness depends on the intrinsic conductivity, permeability and permittivity, aspect ratio of the fillers, and the contact or percolation between the filler particles. Usually a 10 to 50 wt% well-dispersed filler is needed to achieve composites with high conductivity. There are, however, many improvements for this kind of composite that need to be done to meet increasing mechanical and electrical requirements. For instance, the mechanical properties of the composite sometimes decrease as the portion of filler is increased, the electrical conductivity is somewhat difficult to control, and homogenous dispersion of the filler in the matrix is difficult to obtain. Attempts have been made to modify and develop innovative elastic conductive materials. For example, the high electrical conductivity, high aspect ratio, and high mechanical strength of CNTs make them an excellent option to fabricate conductive composites for high-performance EMI shielding materials at low filling concentrations.

8.4.1 Materials Selection and Process of Conductive Composites

The conductive reinforcements used to synthesize elastomer composites can be classified as (a) metal fillers, including aluminum, copper, iron, stainless steel, and silver; (b) nonmetallic fillers, such as carbon black, graphite, carbon nanotubes, ferrous oxide, salts of copper and aluminum with and without binders, solid electrolytes, and semiconductors; and (c) metal-coated fabrics like metal-coated glass, nickel-coated fiberglass, and graphite.

The matrix materials used in the fabrication of conductive composites are selected on the basis of minimum degradation of the matrix due to the catalyzation and oxidation of conductive fillers; wetting of metal fillers under heavy-current operations; morphology of the polymer material including amorphous, semicrystalline, or polycrystalline; and compatibility with the processing techniques involved. Moreover, a variety of electrical and nonelectrical properties should be evaluated to meet EMI shielding requirements. Such electrical properties include bulk conductivity; surface resistivity; effective permittivity; effective permeability (if the conductive filler is a magnetic material); degradation of electrical parameters with shelf life and aging; and variation of electrical characteristics with ambient conditions, shapes, and resiliency of the end products. The nonelectrical properties also should be considered, such as cost effectiveness, density, thermal endurance, corrosion resistance, mechanical strength, hardness or durability, brittleness and flexibility, specific strength, aesthetic appearance in terms of color and surface finish, and compatibility for use with other contacting materials with adhesive or fastening feasibilities. The choice of polymers that meet these requirements is rather limited. Polycarbonate is a common matrix material used for conductive composites (Neelakanta, 1995).

The fabrication processes for conductive composites are essentially decided by the constituents of the composite. The type of conductive fillers, in terms of shape, size, and electrical conductivity, and the properties of the matrix, such as its complex permittivity, determine not only the electrical and nonelectrical characteristics of the composite but also the processing methods that are required.

Spherical or irregular-shaped filler reinforced composites can be fabricated by conventional mixing methods, and the mixtures are processed with extrusion-compounding techniques. The processing of a composite becomes more complex with shaped particles, such as flakes, whiskers, or fibers. When flakes are used, the processing could significantly affect the effective conductivity of the composite. Compared to compression molding injection molding may align the flakes in the flow direction creating a directional anisotropy of conductivity. Synthesizing with conducting flakes as part of mixing fillers, in general, would facilitate more stable conducting composites (Neelakanta, 1995).

8.4.2 CARBON FIBER/WHISKER REINFORCED MATERIALS

Carbon has been long used as a conductive filler in polymer matrices, both in crystalline (graphite) and amorphous forms (carbon black). Electrically and thermally conductive carbon fibers, including chopped strands, whiskers, or fabrics are widely utilized to enhance the structural as well as the conductive properties of the composite materials. These fibers are generally developed from pitch or polyacrylonitrile (PAN)-based precursors, the choice of which plays an important role in the final properties of the fibers. And other factors such as the volume fraction and inherent conductivity of the fibers, the aspect ratio, and their orientation play major roles in determining the conductive properties of such composites.

To further improve the EMI shielding effectiveness of the composite products, a combination of metallized glass fibers and carbon fibers, for instance, can be added to thermoplastic resin matrix to modify the mechanical and physical properties of the composites. Glass fibers have also been employed in compositions used in electrical applications, such as insulating enamels for electrical conductors. However, the use of glass fibers in conductive composites has been for the provision of physical properties such as resistance to plastic flow and to prevent abrasion of the enamel. The combination of these two fibrous materials with a thermoplastic resin enables the realization of a composite product with excellent EMI shielding effectiveness in a wide variety of applications such as radios, transmitters, computers, and other similar products.

The composite comprising the thermoplastic resin, glass fibers, and carbon fibers can be prepared by thermoform molding processes. In the composite, the fibers are commingled in the resin matrix and the mixture, molded is preferably by injection molding. The proportions of the components in the final blend can vary over a range of total fiber reinforcement to resin from about 10 wt% to about 45 wt%, with a preferred range of about 15 wt% to about 35 wt%. The proportion of metallized glass fiber to carbon fiber within the fiber reinforcing component can range from about 20 wt% to about 80 wt%, with a preferred range of about 50 wt% to about 80 wt%. Thermoplastic matrices that are generally employed include nylon 6/6, polysulfone, polyester, polyphenylene sulfide, polyetherimide, polyetheretherketone, and ETFE and PFA fluorocarbons.

Carbon fiber reinforced composites, usually at two to four times the cost of comparable glass-reinforced thermoplastics, offer up to 250 MPa tensile strength, stiffness, and other mechanical properties. Compared to the glass-reinforced composites, carbon reinforced composites (10 to 40 wt% carbon), they have a lower coefficient of expansion and mold shrinkage, higher specific strength, and improved resistance to creep and wear. More important, carbon-fiber reinforcement makes the composite more conductive. In composites containing small amounts of carbon, this characteristic is useful for applications where static charges cannot be tolerated. Composites containing higher percentages of carbon can be used for EMI shields in applications such as business-machine housings. Attenuation of electromagnetic radiation in carbon fiber reinforced nylon, for example, can reach 36 to 40 dB a higher in the frequency range from 50 kHz to 1 GHz.

PAN-derived fibers are generally selected for their high strength and efficient property translation into the composite. The pitch-based fibers are not as strong as the low-modulus PAN fibers. The ease with which they are processed into high-modulus components makes them attractive for stiffness-critical and thermally sensitive applications.

High specific stiffness, great wear resistance, and low volume and surface resistivities allow these composites to compete against many metals. Especially, long-carbon fiber reinforced composites are excellent candidates for applications requiring electrostatic dissipation and EMI shielding (GE Plastics, n.d.).

8.4.3 METAL FIBER/WHISKER REINFORCED COMPOSITES

As mentioned earlier, metal is an effective shield against EMI or static charges. Metallized polymer materials have been used for EMI shielding application. Continuous metal coatings on the polymer surface are typically deposited by vapor deposition, conductive painting, and plating. To obtain coatings thick enough to provide good electrical conductivity, these methods have their limitations due mainly to low production efficiency and relatively high cost.

As an alternative to continuous metal coatings, metal fibers and whiskers have been used for reinforcing polymer matrices to form EMI shielding composites. These materials exhibit considerably larger electrical and thermal conductivities in comparison with pure polymers. They are also light weight and have the aesthetic appearance of plastics. Therefore, they can provide adequate EM shielding effectiveness and prevent electrostatic buildup. Compared with metallized polymers, they are cost effective and damage resistant. They also have better strength, impact resistance, and color options than carbon fiber reinforced plastic composites (Koskenmaki et al., 1992).

Metal reinforcements in polymer matrices include chunky fragments, flicks, near-spherical particulates, or fibers and fibrous whiskers. Of these, metal fibers or wires are most effective in controlling electrical resistivity and providing thermal conductivity. The electrical conductivity of metal reinforced composites depends on the volume fraction of the reinforcements such that the average number of contacts with neighboring particles (percolation) reaches a threshold value. The threshold volume fraction at which the conductivity of the composite begins to increase depends on the particulate shapes, such as chunky, fibrous, spheroidal, or near spherical. Invariably, a low volume fraction is required with fibers to achieve a specified conductivity because the geometry of the fibers provides better percolation even at lower concentrations of the fibers. In general, controlling the conductivity of metal reinforced polymers is quite stringent because resistivity of such materials alters drastically even with small changes in the volume fraction of metallic reinforcements (Koskenmaki et al., 1992; Wetherhold and Patra, 2004).

Low-cost, high-conductivity metal reinforced polymer composites can be fabricated by injection molding and extrusion, with metal fibers of 20 to 30 mm length dispersed in a polymer matrix. The wide availability of polymer materials and conventional plastic-making technologies provides flexibility in designing end products with metal fiber reinforced composites. However, uniformity of fiber dispersion, length, and concentration, and methods of interlinking the fibers should be stringently controlled in fabricating metal reinforced polymer composites. Figure 8.7 illustrates

Fiber End-shape Condition	Shielding Effectiveness (dB) at 1.0GHz	Shielding Effectiveness (dB) at 1.5GHz
Straight	28	27
Acid roughened	44	41
Rippled (0.8mm dia)	40	36
Rippled (1.0mm dia)	37	33
Ripped, acid rough	46	43
Flat end-impacted	42	38

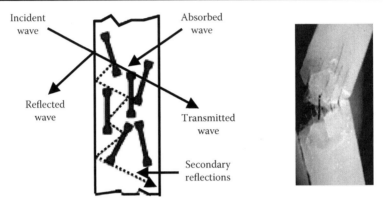

FIGURE 8.7 Metal fiber reinforced polymer composites. (Modified from Wetherhold and Patra, 2004.)

epoxy composites reinforced with ~15 vol%, 10 mm long copper fibers: straight, acid roughened, flat-end impacted, or rippled. The reinforcement of short-shaped copper fiber was used to improve fracture toughness, impact toughness, and EMI shielding effectiveness of the composites while retaining stiffness and strength (Wetherhold and Patra, 2004). End shaping ductile fibers allow anchoring and prevent pullout after debond. Possible bonding mechanisms between metal fiber and epoxy matrix includes mechanical interlocking with surface oxides or other surface features; and chemical interaction mechanism—physical absorption (van der Waals) and molecular interactions (acid-base interactions).

Variables influential to EMI shielding include fiber material, end shape, diameter, surface treatment, and skin depth. For example, acid roughening enhances EMI shielding performance by increasing the fiber surface area, which increases the number of secondary reflections. Smaller diameters perform substantially better than larger diameters due mainly to larger skin depth volume and increased fiber–fiber contact (Wetherhold and Patra, 2004).

Potential applications of metal reinforced polymer composites are high frequency EMI shielding and electromagnetic radiation absorption, such as EMI shielding for both small cable products and large-surface architectural units; electrostatic control media; plastic housings for cables to provide good thermal dissipation, electrical grounding, ESD, and specific shielding against EMI; and electrical applications in lieu of metals, which are susceptible to corrosion.

8.4.4 NANOFIBER REINFORCED POLYMER COMPOSITES

Carbon nanotube (CNT) reinforced conductive polymer composites have been explored for EMI shielding applications. The effects of the length and aggregate size of multiwall nanotubes (MWNT) on the alternate current (AC) conductance of epoxy-based composite reveals the percolation threshold for the fillers to be as low as 0.5 wt% (Toshio et al., 2004). The volume resistivity values of CNT reinforced polyimide composites for different frequencies (0 to 1 MHz) also indicate similar trends (Bai and Allaoui, 2003). The electromagnetic shielding capabilities of MWNT-filled polymethyl methacrylate (PMMA) nanocomposites are also being explored (Kim et al., 2004). These properties, along with their mechanical reinforcing/stiffening capabilities, have made CNTs extremely attractive candidates for reinforcements of conductive polymer composites (Sankaran et al., 2005).

Figure 8.8 compares surface resistivity and percolation threshold of CNT reinforced composites and other conductive composites. Compared to other conductive reinforcements like copper and aluminum fibers and vapor grown carbon fibers, single-wall nanotubes (SWNT) possess a much lower percolation threshold (1 to 2 wt%) and saturation surface resistivity value (10^2 Ω). The tendency of CNTs to form ropes provides very long conductive pathways even at ultra-low loadings, which may be exploited in EMI shielding composites, gaskets, and other uses such as electrostatic dissipation (ESD), antistatic materials, transparent conductive coatings, and radar absorbing materials (RAMs) for stealth applications. The same properties also make these composites attractive for electronics packaging and interconnection applications like adhesives, potting compounds, coaxial cables, and other types of connectors. In the long run, such multifunctional composite materials hold sufficient promise for applications in futuristic smart structures for aerospace and other critical applications.

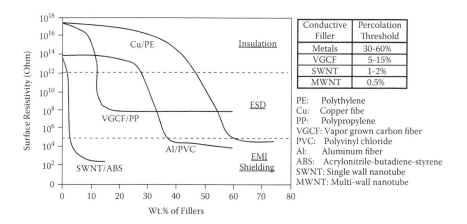

FIGURE 8.8 Surface conductivity and percolation phenomena of different conductive polymer composites. (Modified from Tour and Barrera, 2005.)

8.5 HYBRID FLEXIBLE STRUCTURES FOR EMI SHIELDING

As discussed in Chapter 7, sensitive electronic circuitry and components are susceptible to EMI emanating from other circuits and components. Integrated circuit (IC) chip carriers are specialized circuit panel structures that are frequently used to attach ICs to circuit boards. Chip carriers provide high density, complex interconnections between the IC and the circuit board. Separately attached peripheral or edge shielding is used to prevent crosswise EMI emissions. Such shielding may be in the form of conductive tape or foil. The physical size of the tape is a limiting factor for miniaturizing systems and the extra labor involved in manufacturing does not provide for an efficient manufacturing process. It is not mechanically efficient or desirable to have a separate component that takes up valuable device space and volume (Ishino et al., 1985).

In some cases, a sintered bulk ferrite body has been conventionally used as an electromagnetic shielding or absorbing material. However, the sintered ferrite material has the disadvantage that it is easily broken and is difficult to manufacture into complicated material shapes. In order to solve these problems, a flexible magnetic composite material, like gum ferrite (a composite of ferrite powder and a high-molecular compound like polymer) has been used. The advantage of the composite is that its structure is strong and it can be manufactured into complicated structures by injection molding, extrusion, and/or compression molding (Ishino et al., 1985).

In order to further improve its manufacturability and flexibility, a modified electromagnetic shielding composite structure has been developed and fabricated with a flexible conductive core and a composite ferrite material (a mixture of ferrite powder and a binder), and the composite ferrite material is deposited around the conductive core. As shown in Figure 8.9, a typical structure of flexible magnetic composite

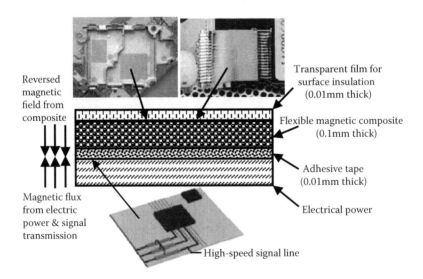

FIGURE 8.9 Typical structure of a flexible magnetic composite for EMI shielding. (Modified from TDK, 2005.)

product comprises a flexible magnetic core, a top insulation cover, and a bottom layer of double-sided adhesive tape for attachment (TDK 2005). The magnetic composite can be made up of a magnetic powder and a rubber member forming an insulator layer, for example, Fe-Si-Al alloy and polyethylene thermoplastic elastomer; or Mn/Zn ferrite and EPDM (ethylene propylene diene monomer); or Mn/Mg/Zn ferrite and soft polyvinyl chloride. The conductive magnetic composite core can be fabricated in an elongated thin shape and the completed shielding composite material is formed in a textile shape so that the magnetic material can be woven into a flexible cloth (Ishino et al., 1985; TDK, 2005).

When this hybrid flexible structure is placed in an electromagnetic field, eddy current flows in the center conductive composite structure generate reversed magnetic fields against the magnetic flux generated by electric power and signal transmission, and attenuates the EMI energy. Advanced shielding characteristics, such as decoupling effects and filtering effects, can be realized with 0.1 mm-thick magnetic layers. In addition to EMI countermeasures of small digital devices and SAR (synthetic aperture radar) countermeasures of mobile phones, a magnetic flux inducing effect that is better (where similar thick samples are compared) than conventional similar materials, where wide band characteristics were sought, is available. The hybrid flexible structures are especially ideal for thinning and weight saving of electronic units that use magnetic fields for communication.

With its electromagnetic loss mechanism and magnetic shield feature, the hybrid flexible structure offers three advantages: (1) a reduction effect (decoupling effect) can be expected against inductive coupling occurring on the facing lines and adherence units due to high frequency wave magnetic field elements arising from signal lines and the IC inside the device; (2) with the impedance addition effect for signal lines, a reduction (filter effect) of unwanted radiation elements, such as higher harmonics, can be expected; and (3) a reduction of common-mode current components superposing on flexible cables connecting high-speed circuits and so forth can be expected (TDK, 2005).

Flexible composite products have been used for electronic device and system applications, such as laptop computers, cellular phones, digital video cameras, CD drives, DVD players, copying machines, facsimile machines, printers, high-speed interface devices, car navigational systems, and optical communication devices.

8.6 ELECTROLESS METAL DEPOSITION FOR REINFORCEMENTS OF COMPOSITE SHIELDING MATERIALS

The increasing demands of optimization and application of EMI shielding in different media with broadening frequency ranges have initiated the development and utilization of universal shielding materials. Conductive composite materials, usually made out of dielectric matrix with conductive or magnetic fillers of different shapes and sizes have become an attractive approach for EMI shielding, which combines high efficiency, and flexible structure, and extended frequency range applications. Application of composite material technology allowing selection of base matrix and filler improves properties of EMI shielding and absorbing structures because of a combination of different materials properties, and introducing quite a few interfaces (Bogush, 2005). Although ferromagnetic materials, ferroelectrics, ferrites, metals, and conductive carbon particles or films have been widely used for EMI shielding and absorbing design, their applications are still limited to certain frequency ranges. A major disadvantage of ferrite-based structures, for instance, is that they are limited to resonant absorption of EMI.

Among various approaches for conductive composites, electroless metal deposition is an attractive method because it provides a low cost and low temperature process for formation of metal containing structures in the dielectric matrix. Formation of metal structures using the electroless process is based on reducing the metal ions that receive electrons generated as a result of reducing the agent oxidation reaction. Oxidation-reduction reactions occur at catalytic surfaces of substrates that typically contain nucleation sites of metals. Since the majority of electroless solutions are aqueous, the deposition temperature does not exceed 100°C, there is no limitation on the shape and size of plated substrate, and deposition is performed from soluble metal salts. The reaction rate is controlled by bath composition and it runs enough fast to provide high throughput of industrial systems.

Using polymer matrices with developed nanosize porous surfaces, for instance, allows creation of numerous metallic clusters with different sizes and shapes in the matrix that forms complex spatial conductive and magnetic structure. Application of high-volume fibers such polyacrylonitrile as a base for metallic composite material combines mechanical properties and high flexibility of fiber with high EMI absorption typical for dispersed metal structures. In this case, synthesis of material can be realized on fiber fabrication or entire linen using textile industry tools. Magnetic and high conductive metals such as nickel, cobalt, or silver have been selected for deposition due to their high stability and relative simple deposition procedure. Metallic clusters are formed from ions attached to fibers by chemical absorption with aqueous solution reduction. Such clusters serve as catalytic sites for following electroless growth or ion-molecular layering. Special types of polymer-like transformations are applied to increase chemical absorption properties of polyacrylonitrile. Figure 8.10 shows an example synthesis of metal-containing composite fibers using the electroless process (Bogush, 2005). The volume resistivity of composite fibers

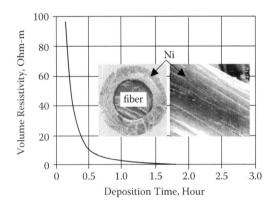

FIGURE 8.10 Volume resistivity of composite fibers fabricated by electroless Ni deposition with Pd activation as a function of deposition time. (Modified from Bogush, 2005.)

fabricated by electroless Ni deposition with Pd activation is a function of deposition time. Synthesis of composite fibers is generally carried out on the basis of modified polyacrylonitrile (PAN) after polymer-like transformation. Catalytic sites on the surface of fibers can be formed by treatment in 0.01 M $PdCl_2$ aqueous solution for 2 minutes or by sorption of nickel ions from 1 M $NiSO_4$ water solution for 3 to 4 hours at room temperature (Bogush, 2005).

Electroless deposition of metals on the fiber is usually performed from solutions containing soluble metal salts (nickel or cobalt sulphate, silver nitrate), stabilizer, surfactant and reducing agent. Growth of metal structures can be conducted using either autocatalytic deposition or ion-molecular layering with stirring and correction of main solution components: metal ions and reducing agent (Bogush, 2005).

Stabilization is the exchange of active oxygen by part of a surfactant molecule and isolation of the reaction zone from the environment to protect the formed metal clusters from oxidation. Methods of metal cluster structure stabilization include wet and gaseous drying, following treatment by aqueous and nonaqueous solutions with high-molecular surfactants. One of the best stabilization procedures with the shortest transient time is based on butanol and oil (Bogush, 2005).

Figure 8.11 shows a commercial conductive polymer composite PREMIER™ (Chomerics, Parker Hannifin Corp., Woburn, Massachusetts), which is a blend of

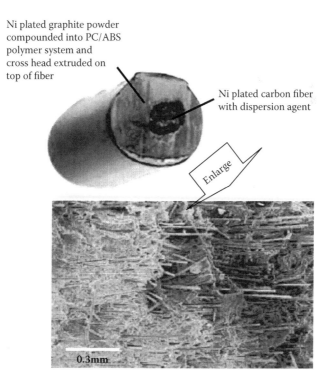

FIGURE 8.11 Ni-plated carbon fiber reinforced polymer composite. (From Chomerics, 2007. With permission.)

PC/ABS polymer and Ni-plated carbon fiber blended with Ni-plated graphite powder for enhancing performance (Chomerics, 2007). This composite can provide greater than 85 dB shielding effectiveness, high permeability (6.5), high strength and light weight. Consequently, electroless deposited fiber reinforced composites have the potential to play an important role for EMI shielding optimization and application.

8.7 SUMMARY

Many material systems and affordable production processes could be integrated with complex primary structures, geometries, and multifunctional assemblies to provide structural integrity and EMI shielding.

Knitted wire mesh gasketing provides a low cost and versatile method for integral EMI shielding at joints and other discontinuities in enclosures. Metal wires can be knitted over various elastomers or mesh cores to achieve the required physical properties and EMI shielding performance. When an elastomer is the core for the wire mesh, it provides excellent compressibility with a high degree of resilience while at the same time functioning as an environmental seal.

Electronically conductive tapes are an economical shielding solution for a wide variety of EMC applications. Constructed of metal foil, metallized fabric, knitted wire mesh, or carbon materials with various conductive adhesives, these tapes provide excellent conductivity and conformability, superior abrasion and corrosion resistance, and strong adhesion and durability in a thin, lightweight, and flexible EMI shielding design.

Conductive polymer composites offer an excellent option to mitigate EMI due to their light weight, good resistance to corrosion, flexibility, and processing advantages. Composite material technology provides tremendous potential for optimizing EMI shielding and developing universal EMI shielding materials to meet increasing EMC requirements.

REFERENCES

Bai, J. B., and A. Allaoui. 2003. Effect of the length and the aggregate size of MWNTs on the improvement efficiency of the mechanical and electrical properties of nanocomposites-experimental investigation. *Composites Part A: Applied Science and Manufacturing* 34: 689–694.

Bemmels, C. W. 1962. Adhesive tapes. In *Handbook of Adhesives,* edited by I. Skeist, pp. 584–592. Huntington, NY: Robert E. Krieger Publishing.

Bogush, V. 2005. Application of electroless metal deposition for advanced composite shielding materials. *Journal of Optoelectronics and Advanced Materials* 7: 1635–1642.

Chomerics. 2000. *EMI shielding engineering handbook.* http://www.chomerics.com/products/documents/catalog.pdf

Chomerics. 2007. *Premier™: Conductive thermoplastics for EMI shielding* http://www.chomerics.com

GE Plastics. n.d. *EMI/RFI Shielding Guide*. Pittsfield, MA: GE Plastics.

Hudak, S. 2000. A guide to EMI shielding gasket technology. *Compliance Engineering.* http://www.ce-mag.com/ARG/Hudak.html

Instrument Specialties. 1998. *Engineering Design and Shielding Product Selection Guide* (acquired by Laird Technologies). Delaware Water Gap, PA.

Ishino, K., Y. Hashimoto, and Y. Narumiya. 1985. Electromagnetic shield. US Patent 4,539,433.

Kim H. M., K. Kim, S. J. Lee, J. Joo, H. S. Yoon, S. J. Cho, S. C. Lyu, et al. 2004. Electrical conductivity and electromagnetic interference shielding of multiwalled carbon nanotubes composites containing Fe catalyst. *Applied Physics Letter* 84: 589–591.

Koskenmaki, D. C., C. D. Calhoun, P. S. Tucker, and R. L. Lambert, Jr. 1992. Metal fiber mat/polymer composite. US Patent 5,124,198.

Laird Technologies. 2006. *Knitted conductive EMI gaskets.* http://www.lairdtech.com/downloads/Knitscatalog-online1106.qxd.pdf

Metex. 2003. *Introduction to knitted wire mesh.* http://www.metexcorp.com/kwm_intro.htm

Molyneux-Child, J. W. 1997. *EMC Shielding Materials.* Oxford: Newnes.

Neelakanta, P. S. 1995. *Handbook of Electromagnetic Materials: Monotonic and Composite Versions and their Applications.* Boca Raton, FL: CRC Press.

Sankaran, S., S. Dasgupta, K. Ravi Sekhar, and C. Ghosh. 2005. Carbon nanotubes reinforced syntactic composite foams. *Proceedings of ISAMPE National Conference on Composites,* 82–82XII.

Smith, M. A., N. M. M. Jones, S. L. Page, and M. P. Dirda. 1984. Pressure-sensitive tape and techniques for its removal from paper. *JAIC* 23: 101–113. http://aic.stanford.edu/jaic/articles/jaic23-02-003.html

TDK. 2005. *Flexible composite-type electromagnetic shield materials.* http://web.welldone-elec.com/pdf/english/epuf/ir105_e.pdf

Toshio, O., I. Yuichi, I. Takashi, and Y. Rikio. 2004. Characterization of multi-walled carbon nanotubes/phenylethynyl terminated polyimide composite. *Composites Part A: Applied Science and Manufacturing* 35: 67–74.

Tour, J., and E. V. Barrera. 2005. *NanoComposites.* http://www.owlnet.rice.edu/~chem597/BarreraNotes/Introduction.pdf

Wetherhold, R. C., and A. Patra. 2004. *End-modified copper fibers and their role in a multifunctional brittle matrix composite: Fracture toughening, impact toughening, and EMI shielding.* http://www.mae.buffalo.edu/people/faculty/wetherhold/durability/bpu4-2004.pdf

9 Absorber Materials

9.1 INTRODUCTION

The increasing demands of electromagnetic compatibility (EMC) for electronic devices with various electromagnetic environments have greatly augmented the number of applications, which require electromagnetic interference (EMI) absorbing materials in frequencies ranging from the kilohertz to gigahertz of micrometer and millimeter waves. Conventional conductive shielding materials, such as metal gaskets, conductive foams, and board-level shields, become less effective at increased frequency range. EMI absorbing materials differ from conductive materials. Rather than harnessing, capturing, and grounding the EMI energy, absorber materials are designed to attenuate and absorb electromagnetic energy and convert the absorbed energy into heat. In fact, the design for absorbers has been incorporated with different loss mechanisms over wide bandwidths. Therefore, absorbers come with many different shapes and structures from thick pyramidal structures to single coatings and multilayer materials.

In practical applications, electromagnetic absorbers are generally categorized as (a) those that absorb propagated microwave energy in empty space or a vacuum, termed free space absorbers; and (b) those that absorb standing waves, which exist inside waveguides, coaxial lines, and other closed volumes where microwave radiation exists. These absorbers are called load absorbers, cavity damping absorbers, or bulk loss absorbers.

A free space absorber is generally characterized as resonant at a particular frequency or narrow range of frequencies, as the material absorbs best when it is a quarter-wavelength thick. However, an absorber for a cavity resonance application needs broadband, which depends on parameters including high magnetic and/or dielectric loss over a broad range of frequencies. Some materials work better in the low frequency range, whereas others work better at high frequency range. The most effective absorbers for cavity resonance damping are magnetically loaded with high permittivity and permeability materials. Materials with only dielectric properties can also be used for cavity resonance absorbers. They are less effective than magnetic absorbers due to the property of the electric field going to zero on a conducting wall while the magnetic field is going to maximum (Dixon, 2004).

In designing absorber materials, the following equation can be used to evaluate how relative parameters affect the absorption capability of the material (Sony, 2008):

$$A = \tfrac{1}{2}\sigma E^2 + \tfrac{1}{2}\omega\varepsilon_0\varepsilon_R E^2 + \tfrac{1}{2}\omega\mu_0\mu_R H^2 \qquad (9.1)$$

where A (W/m^3) is the electromagnetic energy absorbed per unit volume; E (V/m) is the electric field strength of the incident electromagnetic radiation; H (A/m) is the magnetic field strength of the incident electromagnetic radiation; σ (S/m) is the conductivity of the material; ω (sec^{-1}) is the angular speed of the electromagnetic wave, which is equal to $2\pi f$; (f is the frequency) ε_0 (F/m) is the dielectric permittivity of the vacuum: 8.854×10^{-12} (F/m); ε_R is the complex permittivity of the material; μ_0 (A/m) is the magnetic permeability of the vacuum: 1.2566×10^{-6} (A/m); and μ_R is the complex permeability of the material.

From Equation 9.1, attenuation and absorption of microwave energy in an absorber material basically rely on the conductivity, dielectric loss, and/or magnetic loss of the absorber material. Dielectric loss is characterized as the imaginary component of the complex permittivity and acts on the electric field. Magnetic loss is characterized as the imaginary component of the permeability and acts on the magnetic field. Specifically, microwave absorbers use dielectric loss to absorb the electric field portion of an electromagnetic wave, using carbon and other electrically conductive or capacitive particles in many cases as loading to create the proper complex permittivity. However, these materials have the potential to cause short circuits in some applications where the absorbers are located near radio frequency circuits. On the other hand, microwave absorbers employ magnetic loss by filling with magnetic fillers including special irons and ferrites. In general, dielectric-loss microwave absorbers are usually thicker due to their smaller real and imaginary parts of the permittivity. Magnetic-loss microwave absorbers are physically thinner due to their higher real parts of both the permittivity and permeability. A favorable property of the magnetic microwave absorbers is that they are insulators at direct current (DC) with volume resistivities >10^8 Ω-cm. This property allows their use inside microwave circuit modules near or in contact with circuits (Gear, 2004).

This chapter will give a brief review of typical absorbers and absorbing materials, including microwave absorber materials, anechoic chambers, dielectric absorbing materials, and electromagnetic absorbers, as well as absorbing materials selection and absorber applications.

9.2 MICROWAVE ABSORBER MATERIALS

The application of microwave absorbing materials is growing in the electronic industries, in which communication technologies at microwave frequencies have driven the development and utilization of absorbers and frequency selective surfaces. Microwave absorbers are typically designed for reflectivity minimization by alternating shape, structure, and the permittivity (ε) and permeability (μ) of existing materials to allow absorption of microwave EMI energy at discrete or broadband frequencies.

There are three conditions that can minimize the EMI reflection from a surface. When an electromagnetic wave, propagating through a free space with an impedance of Z_0, happens upon a semi-infinite dielectric or magnetic dielectric material boundary of impedance Z_M, a partial reflection occurs. The reflection coefficient at the interface can be expressed as (Chatterton and Houlden, 1992):

$$R = (Z_M - Z_0)/(Z_M + Z_0) \qquad (9.2)$$

The reflection coefficient falls to zero when $Z_M = Z_0$, or, in other words, the material in the layer is impedance matched to the incident medium. This is the first condition that can minimize the reflection coefficient. The intrinsic impedance of free space is given by

$$Z_0 = (\mu_0/\varepsilon_0)^{1/2} \approx 377 \text{ ohms} \quad (9.3)$$

where μ_0 and ε_0 are permeability and permittivity of the free space, respectively. Thus a material with an impedance of 377 *ohms* will not reflect microwaves if the incident medium is free space.

Equation 9.2 can be rewritten as

$$R = (Z_M/Z_0 - 1)/(Z_M/Z_0 + 1) \quad (9.4)$$

This gives the second condition that results in a minimum reflection coefficient when the electric permittivity and the magnetic permeability are equal. The nominal intrinsic impedance is

$$Z_M/Z_0 = (\mu_R/\varepsilon_R)^{1/2} \quad (9.5)$$

where $\varepsilon_R = (\varepsilon_r - i\varepsilon_i)/\varepsilon_0$ and $\mu_R = (\mu_r - i\mu_i)/\mu_0$; ε_r and ε_i are the real and imaginary parts of the permittivity, respectively; and μ_r and $i\mu_i$ are the real and imaginary parts of the permeability, respectively. If the incident medium is free space, and both the real and imaginary parts of the permittivity and permeability are equal, that is, $\mu_R = \varepsilon_R$, the reflectivity coefficient would be zero.

The third condition is the attenuation of the wave as it propagates into an absorbing medium. The power of the wave decays exponentially with distance, x, by the factor $e^{-\alpha x}$. Here α is the attenuation constant of the material and can be expressed as (Saville, 2005):

$$\alpha = -(\mu_0 \varepsilon_0)^{1/2} \omega (a^2 + b^2)^{1/4} \sin[(1/2)\tan^{-1}(-a/b)] \quad (9.6)$$

where $a = (\varepsilon_r \mu_r - \varepsilon_i \mu_i)$ and $b = (\varepsilon_r \mu_i - \varepsilon_i \mu_r)$. To get a large amount of attenuation in a small thickness, α must be large, which implies that ε_r, μ_i, ε_i, and μ_i must be large. However, this condition must be tempered with the first condition (Equation 9.2), where large values of permittivity and permeability would cause a large reflection coefficient.

Therefore, the absorbing material must be lossy so that the EMI energy can be dissipated within the material and not reflected back. Moreover, the design of an absorber is a compromise between the front-face reflection coefficient and the loss per unit thickness. If low reflection is desired, then the material thickness will become large in wavelengths. In practice, multilayer composite structures are used to obtain the desired loss and low reflection in the absorbing material. These structures have variable properties, such that their surface impedance Z_M is as close as possible to the incident wave impedance Z_0, and then change their intrinsic impedance inside by gradually increasing their conductivity to keep the reflection coefficient at the boundary of each layer as low as possible, and allow the materials to convert the EMI energy into Joule heating for dissipating. In fact, there have been a variety of absorbers made with two basic types of materials: resonant or graded dielectric.

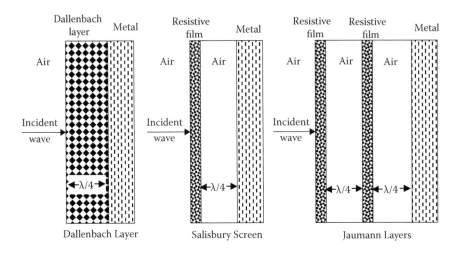

FIGURE 9.1 Resonant absorbers.

9.2.1 Resonant Absorbers

Resonant absorbers are formed through tuned or quarter-wavelength absorbing materials structured to absorb EMI energy at multiple frequencies. Resonant materials generally include Dallenbach layers, Salisbury screens, and Jaumann layers, as illustrated in Figure 9.1 (Saville 2005).

In resonant absorbers, the material is thin and the impedance is not matched between incident and absorbing media so that not all EMI energy is absorbed. Therefore, both reflection and transmission of EMI wave will occur at the first interface. And then the reflected wave will undergo a phase reversal of π, while the transmitted wave travels through the absorbing medium and is reflected from a metal backing. This second reflection also results in a phase reversal of π before the wave propagates back to the incident medium. If the optical distance traveled by the transmitted wave is an even multiple of half wavelengths, then the two reflected waves will be out of phase and destructively interface. Moreover, if the magnitude of the two reflected wave is equal, then the total reflected intensity is zero.

These resonant or tuned microwave absorbers usually exhibit high magnetic loss (permeability) or high dielectric loss (permittivity) with a quarter-wavelength of electrical thickness at the designed frequency. These absorbers require mounting on a ground plane or electrically conductive surface to achieve cancellation of the reflected EMI energy occurring at the front with the reflection occurring at the rear surface of the absorber. The phase of these two reflections is 180° or π apart due to the electrical thickness of a quarter wavelength, resulting in cancellation of the two reflections.

9.2.1.1 Dallenbach Tuned Layer Absorbers

The Dallenbach layer is a layer of homogeneous lossy material placed on a conducting substrate. The thickness, permittivity, and permeability can be adjusted so that

Absorber Materials

the reflectivity is minimized for a desired wavelength. The Dallenbach layer relies on destructive interference of the waves reflected from the first and second interfaces. To obtain a minimum reflectivity, the effectiveness impedance of the layer Z_M must equal the incident impedance Z_0.

Optimization of Dallenbach layers has shown that it is not possible to obtain a broadband with only one layer, however several layers stacked together showed increased bandwidth. Dallenbach layers have been fabricated with ferrite materials (Mayer, 1999), silicon rubber sheets filled with silicon carbide, titanium dioxide, and carbon black (Timmerman, 2000). The use of two or more layers with different absorption bands will increase the absorption bandwidth. Although Dallenbach layers can be fabricated with large bandwidths, it is not known whether the maximum bandwidth possible has been achieved. Dallenbach layers have many applications in submillimetere wavelengths, such as quasioptical beam dumps, and emitting surfaces for black body radiation sources (Saville, 2005).

9.2.1.2 Salisbury Screens

The Salisbury screen, originally patented in 1952, is also one layer of resonant absorber. Unlike the tuned absorbers, it does not rely on the permittivity and permeability of the bulk layer. The Salisbury screen consists of a resistive sheet placed on an odd multiple of quarter wavelengths in front of a metal or other conducting backing, usually separated by an air gap, or a foam or honeycomb dielectric spacer with a permittivity close to that of free space. The resistive sheet is as thin as possible with a resistance of 377 Ω matching that of free space. Therefore, the Salisbury screen works as a perfect absorber for normal incidence when the spacer thickness is an odd multiple of the quarter wavelength. Similarly, the Salisbury screen can work as a perfect reflector for the spacer thicknesses that are multiples of half wavelengths. This effect occurs only at a single frequency. For this reason, Salisbury screens in their pure form have found little practical usage (Saville, 2005).

Salisbury screens have been designed and fabricated in several ways. The initial patented structures were made of canvas on plywood frames with a colloidal graphite coating on the canvas (Neher, 1953). Conducting polymers and several other strategies have been used (Lederer, 1992). The thickness of the optimal Salisbury screen can be calculated when the sheet resistance is equal to the impedance of free space Z_0, which is given by (Lederer, 1992):

$$D = 1/(Z_0 \sigma) \tag{9.7}$$

where σ is the thermal conductivity of the sheet. The thickness of the resistivity sheet for optimum absorption has an inverse relationship to the sheet conductivity.

The bandwidth of Salisbury screen can be maximized given the maximum acceptable reflectivity. The optimum sheet resistance was calculated to be 377 Ω for the lowest reflectivity, while the optimum resistance, Rs, for a given reflectivity limit is given by (Chambers, 1994):

$$Rs = Z_0(1 - \Gamma_{cutoff})/(1 + \Gamma_{cutoff}) \tag{9.8}$$

where Γ_{cutoff} is the maximum acceptable reflectivity. Analytically, the bandwidth decreases with increasing permittivity of the spacing layer.

9.2.1.3 Jaumann Layers

Jaumann layers are a modification of the Salisbury screen that increase its bandwidth with multiple, thin, resistive layers separated with spacers on top of the metal backing (Severin, 1956). The cost of the increased bandwidth is the increased thickness of the absorber. The resistivity of the layers vary from high at the front face to low at the back.

Resistive layers have been formulated using carbon powder (25 wt%) loaded phenol-formaldehyde, cellulose, or polyvinyl acetate binder with polyethylene foams as spacers (Connolly and Luoma, 1977). Silk screening resistive layers have produced better control of thickness and resistance. A six-layer Jaumann device is capable of about a 30 dB decrease in the reflectivity from 7 to 15 GHz (Saville, 2005). Optimization of Jaumann absorbers is complicated due to the number of parameters involved, which increases as the number of layers increase. Empirical procedures and numerical optimization techniques have been developed and used for designing Jaumann absorbers (du Toit and Cloete, 1989).

Resonant or tuned microwave absorbers are usually designed to provide absorption of –20 dB (99% absorbed) of the incident microwave energy at a specific frequency with a tolerance of ±5%. The thickness of popular high permeability silicone-tuned absorbers range from 0.76 mm at 30.0 GHz to 4.06 mm at 1.0 GHz (Gear, 2004).

The elastomer-tuned resonant absorbers offer many useful properties and performance, such as minimal thickness, flexibility, a service temperature of –54°C to 163°C, and good weathering ability (Gear, 2004). These absorbers are easily formed and shaped using conventional molding and thermal forming processes. They are easy to mount with metal or other conductive backing. Applications include the reduction of narrow-frequency EMI reflections from metal surfaces around antennas, inside radar nacelles, and, in some cases, the damping of resonance occurring inside microwave modules. Tuned or resonant microwave absorbers are also effective in damping cavity resonance, although cheaper and thinner cavity damping absorbers can be specially designed.

9.2.2 Graded Dielectric Absorbers: Impedance Matching

From Equation 9.1, an EMI incident wave that impinges upon an interface will experience some reflection that is proportional to the magnitude of the impedance step between incident and transmitting media. Accordingly, three kinds of impedance-matching methods—pyramidal, tapered and matching—have been developed to reduce the impedance step between the incidental and absorption media, as shown in Figure 9.2 (Saville, 2005). For complete attenuation of the incident wave, material one or more wavelengths thick is required, making graded dielectric absorbers bulky and heavier.

9.2.2.1 Pyramidal Absorbers

Pyramidal absorbers are typically pyramidal or cone structures extending perpendicular to the surface of the thick absorbing material in a regularly spaced pattern

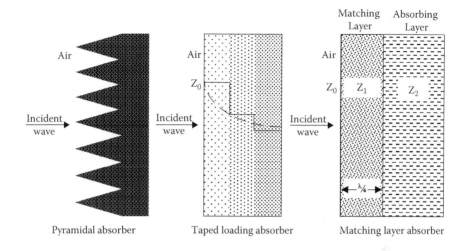

FIGURE 9.2 Graded dielectric absorbers by impedance matching.

(Emerson, 1973; Tanner, 1961). Absorption of the pyramidal absorbers is achieved by a gradual transition of impedance from that of free space to the absorber.

The height and periodicity of the pyramids are usually designed to be on the order of one wavelength. For shorter structures, or longer wavelengths, the waves are effectively absorbed by a more abrupt change in the impedance. Pyramidal absorbers can offer the best performance as they have a minimum operating frequency above which they provide high attenuation over wide frequencies and angle changes. Pyramidal absorbers are usually used for anechoic chambers, which are made with a conductive carbon in polyurethane foam. Absorption levels greater than 50 dB can be obtained with pyramids many wavelengths thick. The disadvantages of these absorbers are their thickness and tendency to be fragile. However, the method of gradual impedance transition can be applied to other materials, such as foams, honeycombs, and netting or multilayer structures for producing practical absorbers. In fact, a more robust absorber has been fabricated using multilayer resistive sheets with a pyramidal type structure (McMillan, 1958; Saville, 2005).

9.2.2.2 Tapered Loading Absorbers

The tapered loading absorber is typically a slab made with a mixture of low loss material and lossy material. The lossy component is homogeneously dispersed parallel to the incident surface, with a gradient perpendicular to the surface and increasing into the slab (Halpern et al., 1960). One type of tapered loading material consists of open-cell foam or plastic net, dipped or sprayed with lossy material from one side, or allowed to drain and dry (Jones and Wooding, 1964). It is hard to reproducibly fabricate a gradient in this manner. Another type is composed of homogeneous layers with increased loading of lossy material in the direction of EMI propagation. The advantage of these materials is that they are thinner than the pyramidal absorbers.

The disadvantage is that they have a poorer performance and it is best to vary the impedance gradient over one or more wavelengths (Saville, 2005).

9.2.2.3 Matching Layer Absorbers

As shown in Figure 9.2, the matching layer absorber places a transition absorbing layer between the incident and absorbing media to reduce the thickness required for the gradual transition materials. The transition layer has thickness and impedance values that are between the two impedances to be matched, so that the combined impedance from the first and second layers equal the impedance of the incident medium. This matching will be achieved when the thickness of the matching layer is one-quarter of a wavelength of the EMI in the layer and $Z_1 = (Z_0 Z_2)^{1/2}$. The impedance matching occurs only at the frequency that equals the optical thickness, which makes the matching layer materials narrow band absorbers. The matching layer absorber can be fabricated with an intermediate impedance transition layer and controlled quarter-wavelength thickness for absorption at microwave frequencies.

In general, graded dielectric absorbers are mostly carbon doped or impregnated urethane foams. The impedance taper or gradient is achieved through geometric shaping such as pyramids, wedges or convolutions or is electrically tapered or graded by varying the carbon loading or doping of flat layers, decreasing in impedance from front to back.

Physical graded dielectric absorbers can offer high performance providing EMI absorption levels of −40 dB (99.99% absorbed) to −50 dB (99.999% absorbed). These absorbers have impedance at the front plane (for example, tips of the pyramids) close to 377 Ω/square of free space, which decreases gradually to the back surface. They are typically used in test facilities such as anechoic chambers and other types of microwave measurement enclosures (Gear, 2004; Saville, 2005).

The microwave absorber can be easily formed and shaped by conventional forming, machining, or water jet processes. Sheets and more complex shapes can be fabricated with pressure sensitive adhesive (PSA) adhesive for mounting and with metal backing to improve shielding effectiveness. The absorber also can be treated with moisture-sealing coating allowing their use in high humidity or moderately wet environments. These multilayer absorbers have been used in military applications including the in-nose section of radar-guided missiles to reduce reflections around the seeker antenna; inside military aircraft nose sections to reduce reflections around radar systems; in lining of antenna caps used to terminate aircraft antennas allowing on-the-ground testing of radar systems; and on surface ships to reduce reflections around antennas (Gear, 2004).

9.2.3 CAVITY DAMPING ABSORBERS

Cavity resonance interference usually occurs when a cavity generates a standing wave due to stray radiation and the physical properties of the cavity. Cavity damping absorbers are generally designed using thin elastomer sheets loaded with high permeability materials providing broad frequency magnetic loss properties. Other absorbers, such as resonant or tuned absorbers, may also be good for cavity damping but may not be the best selection based on cost and weight. High-permeability cavity damping absorbers can be formed and shaped using a conventional forming tool, razor blade, water jet, or laser cutting, ranging in thickness from 0.25 mm to 1.0 mm. Silicone is commonly used for the binder material due to its flexibility, service temperature of −54°C to 163°C, power handling of 0.2 W/cm^2, low outgassing, and availability with a high performance peel-and-stick pressure-sensitive adhesive (Gear, 2004).

Cavity resonance problems are frequently met when the circuit board is integrated with the shielded cavity or chassis, due partly to increased circuit function, and reduction in the physical size of microwave circuit board in metallic shielding housings. Microwave cavities have certain frequencies that oscillate. The interference energy can be attenuated when lossy magnetic or dielectric materials are installed into the cavity. Specific cavity damping can be theoretically selected through complex resonance modeling. Optimized magnetic loss materials are frequently utilized for reducing microwave cavity resonance because these materials, like iron-loaded silicone, offer thin, nonconductive characteristics without the risk of shorting the circuit board. Comparatively, dielectric loss materials, such as latex-coated elastomer foam structures, are usually thicker but cost less than magnetic loss materials. However, if environmental conditions allow and if the thickness can be tolerated, this foam structure can be a viable option. Both dielectric loss and magnetic loss materials are effective for reducing cavity resonance (Dixon, 2004; Gear, 2004).

Most cavity damping absorber materials are applied to the cover of the microwave module. In multiple-cavity modules, twenty to fifty cavities can be in a module (Gear, 2004). Some cavities have no microwave circuitry, while others with house microwave circuitry require cavity damping absorbers. A mold-in-place process can be used to apply the cavity damping absorber to the resonant cavities of multiple-cavity housing.

9.3 ANECHOIC CHAMBERS

The anechoic chamber is an radio frequency (RF)-shielded room mainly consisting of an antenna system and RF absorber materials installed on the four walls, ceiling, and possibly the floor. The design of an anechoic chamber is basically established for performing EMC measurements according to a variety of different published EMC standards, involving many different fields of application, such as consumer electronics, automotive, aerospace, military, medical, and telecommunications. Anechoic chambers are primarily used for measuring radiated emissions and immunity in the frequency range of 30 to 1000 MHz, with extensions to 40 GHz. Different methods and criteria for validation chambers and performing EMC measurements for testing emissions and immunity are standardized, including test distances, field levels, emission limits, pass criteria, and equipment setup (Wiles, 2008).

9.3.1 ANTENNAS

Antennas are used for radiating signals for EMC testing in the anechoic chamber. With the progress of antenna technology, hybrid or combination antennas have taken the place of the conventional biconical and log periodic design. Hybrid or combination antennas have been adapted to the different tests required by the standards, which can cover the whole frequency band from 26 MHz to 3 GHz or from 80 MHz to 6 GHz in one sweep by combining and matching a biconical section to a log periodic section. Such combination antennas are often greater than a meter in width and length.

9.3.2 ABSORBER MATERIALS USED IN ANECHOIC CHAMBERS

The anechoic absorbing materials are fire retardant, thin, and typically a quarter wavelength at the lowest operating frequency. The multilayer or flat-sheet layer secular microwave absorber series provides excellent absorption of −20 dB (99% absorbed) over a frequency range of 600 MHz to 40 GHz. The absorber materials that line the inside surface of the shielded room can be classified as three basic types:

1. Electric loss pyramidal absorber. This is the preferred technology for high frequencies with a range of 100 MHz to 18 GHz. The losses are provided by carbon loading of a pyramid structure.
2. Magnetic loss ferrite tile. This is the preferred technology for low frequencies with a range of 30 to 1000 MHz. Ferrite tile, 5 to 6 mm, provides the magnetic losses, and is used in combination with hybrid foam to form a united absorber. The disadvantage of this material is that it is heavy and cannot be used for high frequencies.
3. Electric and magnetic loss hybrid absorber. This is preferred technology for broadband EMC testing generally with a frequency range of 30 MHz to 18 GHz. Specially formulated hybrid pyramid foam has good matching with ferrite tile at the bottom. However, its performance is not as good as electric loss pyramid of equal size at high frequencies.

Anechoic chambers are generally different upon their application and can be divided into the following groups: (a) partially lined room—the surfaces are not fully covered with absorbers; (b) semi-anechoic room—the walls and ceiling are covered with absorbers whereas the floor is a metal reflecting ground plane; and (c) fully anechoic room—all surfaces are covered with absorbers. The most common type of chamber will be a compact or full 3 m type. The compact chamber offers the advantage of being able to fit into the majority of buildings due to their limited height of 3 m. The full 3 m and larger chambers will be part of a dedicated building purposely built in many cases to house the chamber. In parallel with the transient nature of some markets like telecoms, most of these chambers offer the flexibility of being removable or modified to a different size if the requirements of the testing change or the company moves to a different location. The life cycle of the products has increased together with the quality of some of the key maintenance items like shielded doors.

Types of measurements conducted in these anechoic chambers include attenuation measurements, radar cross-section, EMI emissions, susceptibility and compatibility, and target simulation.

9.4 DIELECTRIC MATERIALS FOR ABSORBER APPLICATIONS

Dielectric materials have been in use for absorber applications since the inception of Salisbury screens and the Dallenbach layer. In dielectric materials, the dominant charges of their atoms and molecules are positive and negative, which are kept in the same position by the atomic and molecular forces, and are not free to dislocate. When applying an electric field to a dielectric material, however, the formation of several electric dipoles takes place, which align themselves according to the orientation of the applied electric field. The reciprocal influence on the electric field causes the storage of electric energy, which can be turned to heat by the Joule effect. And this phenomenon occurs with dielectric absorbing materials (Balanis, 1989).

Dielectric material absorbers are generally fabricated by the combination of semi-conductive or conductive fillers in a polymeric matrix. These fillers include silicon, graphite, carbon black, aluminum, copper, stainless steel, and various metal-coated powder or whiskers. And different types of polymeric matrices, rigid or flexible, have been used, such as epoxy, phenolic, bismaleimide, polyurethane, polyimide, and silicone resins. By adjusting the dielectric characteristics of the components during processing to the application and the frequency range, the dielectric material can change its behavior from transmitter or reflector to absorber.

In dielectric materials, the main properties that enable them to be applicable as microwave absorbers are the dielectric constant and the dissipation factor of energy (Equation 9.1). There have been many ways to change the dielectric properties. This is normally achieved by the distribution of fillers that alters the electrical properties of the material (Antar and Liu, 1998). For instance, using dielectric scaling technology, the scaled component can be fabricated from epoxy resins and silicone-based materials loaded with specific amounts of powdered filling agents (silicone, carbon, aluminum, copper, stainless steel, etc.) to achieve the full-scale dielectric constant at the scaled frequency (Gatesman et al., 2001).

There have been quite a few dielectric absorbing materials developed for special applications. One typical material is called circuit analog absorber. The design technology for this material uses transmission line analogy to match the incoming wave to the metal backing. By replacing the continuous resistive sheet in Salisbury- and Jaumann-type materials with a specially patterned surface containing capacitance and inductance, the imaginary part of the admittance can be tailored and used in impedance matching. The geometrical patterns like dipoles, crosses, triangles, and so forth are described in terms of their resistance, capacitance, and inductance. Resistive-capacitive materials have been made in the form of conducting polymer-coated fibers (Wright et al., 1994), and resistive-inductive materials have been made with helical metal coils imbedded in a dielectric layer (Kuehl et al., 1997). With a circuit analog absorber, significant improvements in both the bandwidth and absorption compared to Salisbury and Jaumann absorbers can be achieved. The drawback is that optimization of the patterns is a complex and time-consuming process (Saville, 2005).

Another application of circuit analog patterns is frequency-selective surface (FSS) absorbers. The FSS consists of metallic conductive patterns placed on a dielectric medium. By using materials with different conductivities, the FSS can be used for making absorbers, bandpasses, or bandstop filters. The FSS is designed with a multilayer grid device where resistance grids are spaced with increasing mesh in the direction of propagation (Kuhnhold, 1956); or the use of multilayered arrays of disks or squares (Kasevich et al., 1993). A conductive polymer blended with conducting materials such as carbon, metal powder, or ferrites has been used for making the

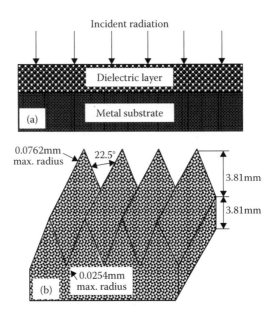

FIGURE 9.3 Far-infrared radiation absorbing material absorbers. (a) Dallenbach-like structure; (b) submillimeter-wave anechoic structure. (Modified from Giles et al., 1990.)

FSS (Narang et al., 1999). The conductivity of this material was controlled and holes punched in the material with a certain period pattern (Saville, 2005).

To meet the increasing requirements in the number of optical and quasi-optical measurement systems operating at teraherz frequencies, a variety of far-infrared radiation absorbing materials (FIRAMs) and structures have been developed (Giles et al., 1990). The FIRAM is designed as a resonant absorption structure with a quarter-wave thick artificial dielectric surface on a metal plate, as shown in Figure 9.3. One type of FIRAM consists of a metal substrate coated with an evenly thick homogeneous artificial dielectric material like a Dallenbach structure (Figure 9.3a). Realization of the Dallenbach layer as a FIRAM is achieved with a particular class of submillimeter wavelength artificial dielectric-metallic paints (Stainless Steel Coatings, 2008). These paints consist of resins such as vinyl acetate, silicone, or polyurethane uniformly loaded with stainless steel flakes. Since the metal flakes are dimensionally small compared to the wavelength, the paint exhibits the optical properties of a homogeneous media. Application of the metal-loaded vinyl acetate to form a Dallenbach layer can be performed by robotic spraying, extrusion blow molding, compression molding, casting, and spin coating. For achieving uniform thin-film materials, alternate materials such as low-density polyethylene and epoxy resin can be considered (Giles et al., 1990).

Another type of FIRAM relies on a wedge-type surface geometry fabricated from a lossy silicone-based material to achieve performance as a submilimeter-wave anechoic structure (Figure 9.3b). In each case a −25 dB reflectivity reduction is achieved at normal incidence, however, when oriented within quasi-optical measurement systems these materials reduced reflections due to unwanted stray radiation by more than −50 dB (Giles et al., 1990). The artificial dielectric layer can be optimized to operate as resonant absorbers for any frequency between 0.3 THz and 3 THz. These anechoic coatings can be designed to provide more than –25 dB of reduction in reflectivity. Since the materials are relatively inexpensive and suited to commercial manufacturing methods, the technology is practical for a variety of applications at terahertz radar frequencies (Giles et al., 1990).

In addition, carbon nanotube (CNT) composites for broadband microwave absorbing materials have been developed based on CNTs dispersed into a polymer dielectric material. The extremely high aspect ratio of CNTs and their remarkable conductive properties lead to good absorbing properties with very low concentrations. For example, a conduction level of 1 A/V-m is reached for only 0.35 wt% of CNT, whereas a 20 wt% concentration is mandatory for a classical absorbing composite based on carbon black (Saib et al., 2005).

Although a great effort has been made, dieletrical materials based on purely dielectric phenomena and electric losses are basically considered unsuitable for a broadband absorber design where the available physical space is limited, due mainly to their inability to absorb power for low frequencies. Therefore, magnetic materials seem to be better wide bandwidth absorbers although they have not been as well researched as their dielectric counterparts. However, their practical designs have been in use for a long time (Ramprecht and Sjöberg, 2007).

9.5 ELECTROMAGNETIC WAVE ABSORBERS

According to Equation 9.1, electromagnetic wave absorption materials absorb the energy in electromagnetic waves as magnetic loss and convert that energy to heat. When an absorber is located in the near field of a radiation source, the third term of Equation 9.1 will have a large influence on its absorption since the magnetic field component H is large. This indicates that moving a material with a large μ_R (magnetic loss) to a location where H is large (the vicinity of the radiation source) will make the absorption large. In other words, when suppressing EMI from a radiation source, the effect will differ widely depending on where the electromagnetic wave absorber is placed because H will differ significantly (Sony, 2008). For an electromagnetic wave absorber of the same size attached at the same position, the effect will be proportional to μ_R. How large this parameter can be made is used to evaluate the magnetic materials being developed.

Magnetic absorbers have been fabricated based on carbonyl iron and hexaferrites. These materials have absorptions in the megahertz and gigahertz ranges, and can be tuned to absorb at higher frequencies (5 to 20 GHz) based on particle size and sintering temperature. The resonance frequency is related to particle size.

The bandwidth of standard ferrite absorbers have been improved through a two-layer Dallenbach absorber design, with a ferrite layer at the air/absorber interface and a layer containing ferrite and short metal fibers at the absorber/metal interface. The fiber length is chosen to have a frequency near the required absorption frequency f_0, that is, for 8 to 13 GHz the fiber length is 1 to 4 mm for 60-μm-diameter wire. The impedance of the wires shifts from being capacitive when the frequency is less than f_0 of the fibers to inductive for $f > f_0$ the impedance is at a minimum at f_0 and the induced current is at a maximum. Between f_0 and $½f_0$ the fiber has a capacitive reactance

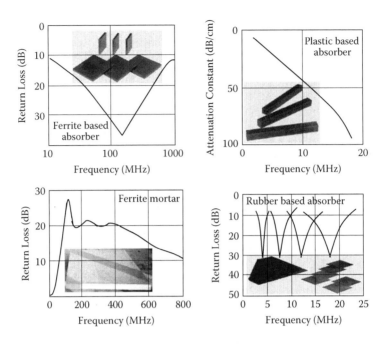

FIGURE 9.4 Some typical electromagnetic wave absorbers. (Modified from FDK, 2007.)

and for $f < \frac{1}{2}f_0$ the fiber resistance becomes small and the reactance is capacitive only. At f_0, the fiber length is nearly $\frac{1}{2}\lambda$ in the matrix. The ferrite layer acts as an impedance matching layer as well as an absorber. Performance is better than −20 dB over 8–12 GHz and up to 45 degrees of incident angle with an overall thickness of 4.6 mm (Hatakeyama and Inui, 1984).

Figure 9.4 shows some typical electromagnetic wave absorbers (FDK 2007). Ferrite-based electromagnetic wave absorbers are made of sintered ferrites, which absorb electromagnetic waves through the application of the principle of magnetic resonance loss. They are suitable for the inner walls of electromagnetic anechoic chambers and outer TV ghost suppression walls of buildings. Plastic-based electromagnetic wave absorbers are produced through the mixing and hardening of a combination of iron carbonyl and thermosetting polyester resin. They have an excellent attenuation effect in the gigahertz range and superb heat resistance. Ferrite mortars are designed to suppress TV ghost images caused by TV signal reflections from skyscrapers. They are made of ferrite mortar to reduce production costs, and they provide outstanding performance and expand the shape selection for easier and cheaper construction. The TV ghost suppression ferrite mortar is a cement-based alternative to ferrite tiles. Rubber-based electromagnetic wave absorbers consist of synthetic rubber and iron carbonyl making them flexible and easy to handle. They are suitable to use in the absorption of microwaves in the 4 to 18 GHz range (FDK 2007). In addition, the thermal electromagnetic wave absorbing materials have been developed to replace the thermal interface materials in electronic packaging, such as board-level shielding systems.

9.6 ABSORBING MATERIALS SELECTION AND ABSORBER APPLICATIONS

A wide variety of absorber materials have been developed for use in EMI absorption. When choosing absorbers in an optimal design, their electrical and physical performance, as well as application requirements should be critically analyzed and evaluated. The broader the frequency coverage is the thicker, heavier, and more expensive the absorber that is needed the lower the minimum frequency coverage is the thicker and heavier absorber is also needed. The physical parameters that need to be evaluated include environmental characteristics, temperature characteristics, mechanical properties, and lifetime expectation. Elastomer-type absorbers, like rubber absorbers, have better environmental resistance than broadband foam types. For example, hypalon is widely used in naval applications because of its superior weather resistance and color fastness. Nitrile is used for fuel and oil resistance. Fluoro-elastomers and silicones have an excellent operating temperature range (Laird Technologies, 2007).

Broadband foam materials can be used for external environments, but precautions must be taken to protect the absorber. Open-cell foams can be filled with low-loss plastics to make rigid panels for use outdoors. Broadband absorbers can be encapsulated in fiber-reinforced plastics to form flexible absorber panels that can be draped over reflectors. The useful temperature range of most absorber material is $-20°C$ to $120°C$. Some materials are available with higher maximum temperatures (Laird Technologies, 2007).

In choosing an absorber, the application requirements must be defined and clarified, such as frequency bands coverage, priority frequency bands, specular energy, high angles of incidence radiation, and surface wave attenuation.

9.6.1 Absorbing Materials and Absorber Types

Carbon powder, fiber, and fabric as well as CNT have been used for absorber fabrications. Absorbers for anechoic chambers, for instance, were originally made by coating mats of curled animal hair with carbon black impregnated neoprene. Carbon black and fibrous carbon has been incorporated into Dallenbach layers (Saville, 2005). Carbon- and CNT-loaded composite absorber materials have been developed and evaluated for current and future applications.

Metal and metal particles are also widely used for constituents of absorber materials, including iron oxide, powdered iron, powdered aluminum and copper, steel wool, evaporated metal, nickel chromium alloy, and other metal wires. Broadband absorbers have been made from solid aluminum metallic particles or dielectric-filled metallic shells in the shape of spheroids dispersed in a matrix (Janos, 1994).

Conducting polymer and conductive polymer composites have become common absorber materials. In a polymer such as polypyrrole, partial oxidation or doping causes the polymer to become conductire through the formation of polarons and bipolarons. The charge is carried along chains (Capaccioli et al., 1998). Thin, flexible elastomer absorbers are optimal for outdoor use. They can be installed and applied by means of adhesive bonding to metal substrates. The adhesives that are usually used vary with the type of elastomer chosen and generally include epoxies, urethanes, contact adhesives, and pressure-sensitive adhesives (PSAs; Laird Technologies, 2007). In general, hypalon and nitrile are the easiest elastomers to bond. In some cases, it is necessary to cover a tight radius or complex curvature. An alternative to flat sheet material is conformably molded parts. Conformal molds increase the ease of bonding and reduce the likelihood of applying any built-in stress into the material. For gasket applications, the elastomeric absorber can be extruded.

To improve weather resistance, the absorber can be painted. Typically, an epoxy- or urethane-based paint is used. To avoid gaps between each absorber sheet, absorptive gap fillers are used to minimize any impedance mismatches from sheet to sheet. This method can minimize the formation of surface waves and reflections. Other noncorrosive fillers, such as iron silicide, can also be used for corrosive environments (Laird Technologies, 2007).

Open-cell foam absorbers are normally used in a protected environment. Adhesive bonding is also typically used for installation and application. In many cases, cohesive failure of the material occurs before adhesive failure. Painting or coating the front surfaces of open-cell foam absorbers can provide them with further protection. Moreover, the broadband open-cell foam or netting absorbers are usually protected by encapsulating them in surface-coated fabrics. The bagging fabric is completely closed around the absorber, making it weatherproof.

Closed-cell foam absorbing materials are usually used to produce rigid structural absorptive panels. The rigid, closed-cell foam with surface painting will be impervious to external environments. Absorptive honeycomb can also be used as the inner core for structural panels. The face sheets of the panel can be made from fiberglass, graphite, or metal as the grounding plane. Such panels are light weight and strong, and can be used as structural absorbing materials in some applications (Laird Technologies, 2007).

9.6.2 Applications

Microwave absorbing materials are mainly used for EMI reduction in military and commercial electronics.

9.6.2.1 Military

The use of microwave absorbing materials in military applications dates back to the 1940s, spawned by the introduction of radar. With increased use of microwave electronics, such as electronic warfare (EW) systems, radar systems, identification friend- or- foe (IFF) systems, and radio frequency (RF) communications systems, EMI interference becomes a major problem. The interference causes errors in targeting, reducing system efficiencies and limiting operating capabilities. Microwave absorbing materials are used to eliminate or reduce the EMI to manageable levels. Based on the lowest frequency of the EMI interference, amount of absorption required, space available, and environmental conditions, a variety of absorbers can be chosen for each specific application. However, microwave-absorbing materials are continually being developed to keep pace with increased applications of microwave electronics, both military and commercial.

9.6.2.2 Commercial

Use of absorbers for EMI reduction in commercial electronics has been growing. High-frequency wireless devices, for instance, often have powerful transmitters and sensitive receivers in close proximity inside a cavity or housing. Moreover, spurious signals can cause leakage or system interference, which degrades performance. In these cases, magnetic absorbers are usually placed inside the cavity to reduce the EMI of the cavity and absorb unwanted reflections. Other applications of absorbers can be found in wireless local area network (LAN) devices, network servers, very small aperture terminal (VSAT) transceivers, radios, and other high-frequency devices.

As devices such as computers and cell phones move to higher frequencies and speeds, the need for absorbers or absorbing shields will increase.

9.6.2.2.1 Board-Level Shielding with Absorbers

Increasing usage of printed circuit board in complex electronics requires unique shielding solutions. When the frequency is greater than 2 GHz, the board-level shielding can be enhanced with the addition of microwave absorbers.

9.6.2.2.2 Air Filters with Absorbing Performance

Reticulated absorbing foam is used for air intake filtration on routers and other network systems. The absorbing capability of the foam will help to reduce or eliminate high-frequency EMI in the ventilation systems.

9.6.2.2.3 Thermally Conductive Absorbers

The thermal conductivity of magnetic absorbers can be achieved and enhanced by the addition of thermally conductive fillers. These fillers do not degrade the microwave absorbing property of the material while enhancing its thermal conductivity. Thermally conductive absorbers can be used as a thermal pad or any other forms of thermal interface materials.

9.7 SUMMARY

A variety of absorber materials and structures have been developed and used for military and commercial applications. Coatings in the form of Dallenbach layers, although not broadband, are useful for reducing EMI from intricate shapes. Jaumann layers are appropriate for broadband lightweight absorbers and many analytical methods have been developed for design optimization. Composite and combined electric–magnetic materials offer great potential for thin broadband absorbers. Magnetic materials are limited to carbonyl iron and ferrites with a concern about their corrosion resistance. Conducting polymers are attractive electrical materials due to the potential for controlling their permittivity and permeability through synthetic means.

The need for microwave absorbers or absorbing shields is increasing as devices such as computers, cell phones, and other electronics move to higher frequencies and speeds. To generate consistency with the electrical performance, thickness, weight, mechanical properties, and cost, the properties of magnetic or dielectric materials can be adjusted to enhance their performance and utility.

REFERENCES

Antar, Y., and H. Liu. 1998. Effect of radar absorbing materials on RCS of partially coated targets. *Microwave and Optical Technology Letters* 17(5): 281–284.

Balanis, C. A. 1989. *Advanced Engineering Electromagnetics*. New York: John Wiley & Sons.

Capaccioli, S., M. Lucchesi, P. A. Rolla, and G. Ruggeri. 1998. Dielectric response analysis of a conducting polymer dominated by the hopping charge transport. *Journal of Physics: Condensed Matter* 10: 5595–5617.

Chambers, B. 1994. On the optimum design of a Salisbury screen radar absorber. *Electronics Letters* 30(16): 1353–1354.

Chatterton, P. A., and M. A. Houlden. 1992. *EMC Electromagnetic Theory to Practical Design*. Hoboken, NJ: John Wiley & Sons.

Connolly, T. M., and E. J. Luoma. 1997. Microwave absorbers. US Patent 4,038,660.

Dixon, P. 2004, May. Dampening cavity resonance using absorber material. *RF Design*, pp. 16–20. http://rfdesign.com/mag/0405rfdf1.pdf

du Toit, L. J., and J. H. Cloete. 1989. A design process for Jaumann absorbers. *Antennas and Propagation Society International Symposium* 3: 1558–1561.

Emerson, W. H. 1973. Electromagnetic wave absorbers and anechoic chambers through the years. *IEEE Transactions on Antennas and Propagation* 21(4): 484–490.

FDK. 2007. *Electromagnetic wave absorbers & ferrite sheets for EMI suppression.* http://www.melatronik.de/Firmen/fdk/eabsorb.pdf

Gatesman, A. J., T. M. Goyette, J. C. Dickinson, J. Waldman, J. Neilson, and W. E. Nixon. 2001. *Physical scale modeling the millimeter-wave backscattering behavior of ground clutter.* http://stl.uml.edu/PubLib/Gatesman,%20Phys%20Model.pdf

Gear, J. T. 2004, August. Microwave absorbers manage military electronics RF interference. *RF Design*, pp. 6–9. http://rfdesign.com/mag/08deff1.pdf

Giles, R. H., A. J. Gatesman, and J. Waldman. 1990. A study of the far-infrared optical properties of Rexolite™. *International Journal of Infrared and Millimeter Waves* 11(11): 1299–1302.

Halpern, O., M. H. J. Johnson, and R.W. Wright. 1960. Isotropic absorbing layers. US Patent 2,951,247.

Hatakeyama, K., and T. Inui. 1984. Electromagnetic wave absorber using ferrite absorbing material dispersed with short metal fibers. *IEEE Transactions on Magnetics* 20(5): 1261–1263.

Janos, W. A. 1994. Synthetic dielectric material for broadband-selective absorption and reflection. US Patent 5,298,903.

Jones, A. K., and E. R.Wooding. 1964. A multilayer microwave absorber. *IEEE Transactions on Antennas and Propagation* 12(4): 508–509.

Kasevich, R. S., M. Kocsik, and M. Heafey. 1993. Broadband electromagnetic energy absorber. US Patent 5,214,432.

Kuehl, S. A., S. S. Grove, E. Kuehl, et al. 1997. *Manufacture of Electromagnetic Materials*. Netherlands: Kluwer Academic.

Kuhnhold, R. 1956. Absorption device for electro-magnetic waves. US Patent 2,771,602.

Laird Technologies. 2007. *Microwave absorbing materials.* http://www.lairdtech.com/downloads/absorber_1-17-07lowresforwebsite.pdf

Lederer, P. G. 1992. Modeling of practical Salisbury screen absorbers. *IEE Colloquium on Low Profile Absorbers and Scatterers*, pp. 1/1–1/4.

Mayer, F. 1999. High frequency broadband absorption structures. US Patent 5,872,534.

McMillan, E. B. 1958. Microwave radiation absorbers. US Patent 2,822,539.
Narang, S. C., A. Nigam, S. Yokoi, et al. 1999. Electromagnetic radiation absorbing devices and associated methods of manufacture and use. US Patent 5,976,666.
Neher, L. K. 1953. *Nonreflecting background for testing microwave equipment*. US patent 2,656,535.
Ramprecht J., and D. Sjöberg. 2007. Biased magnetic materials in RAM applications. *Progress in Electromagnetics Research* 75: 85–117.
Saib, A., L. Bednarz, R. Daussin, C. Bailly, X. Lou, J.-M. Thomassin, C. Pagnoulle, et al. 2005. Carbon nanotube composites for broadband microwave absorbing materials. *European Microwave Conference* 1: 4.
Salisbury, W. W. 1952. Absorbent body for electromagnetic waves. US Patent 2,599,944.
Saville, P. 2005. A *review of optimisation techniques for layered radar absorbing materials* (Technical memorandum 2005-003). Defence R&D Canada. http://pubs.drdc.gc.ca/PDFS/unc57/p523186.pdf
Sony. 2008. *Sony's electromagnetic wave absorber reduces EMC and SAR problem*. http://www.sony.net/Products/SC-HP/cx_news/vol25/pdf/emcstw.pdf
Stainless Steel Coatings, Inc. 2008. www.steel-it.com
Severin, M. 1956. Noreflecting absorbers for microwave radiation. *IRE trans. Antennas & Propagation* 4:385–392.
Tanner, H. A. 1961. *Fibrous microwave absorber*. US Patent 2,977,591.
Timmerman, A. T. 2000. Microwave-absorbing material. Canadian Patent 2,026,535.
Wiles, M. 2008. Choosing right chamber depends on the application. *Conformity*. http://www.conformity.com/PDFs/0802/0802_F5.pdf
Wright, P. V., T. C. P. Wong, B. Chambers, and A. P. Anderson. 1994. Electrical characteristics of polypyrrole composites at microwave frequencies. *Advanced Materials for Optics and Electronics* 4(4): 253–263.

10 Grounding and Cable-Level Shielding Materials

10.1 INTRODUCTION

Cables can act as antennas and radiate electromagnetic interference (EMI) energy at frequencies above 30 MHz. Interconnecting cable shielding and grounding play an important role for a successful electronic system design in meeting electromagnetic compatibility (EMC) compliance requirements. Limiting radiated EMI from a cable can be achieved by placing the cable's conductors inside a shield. The most important decision is to select the best material, right style, and wrap configuration of the shielding itself. Effective shielding is only part of the solution; terminations and grounding of the shield are also critical to solve the EMI problem and meet EMC requirements. Cable shielding and grounding have been widely applied in telecommunication, instrumentation, electronic data processing in places such as internal and external computer data cables, internal and external power cables, internal floppy disk and hard disk ribbon cables, and cables between printed circuit boards (PCBs) and data connectors.

This chapter will give a brief review of the EMI design guideline for shielded cable assemblies, cable shielding materials selection, and bonding and grounding of cable terminations.

10.2 CABLE ASSEMBLY AND ITS EMI SHIELDING DESIGN

Proper shielding design for cable assembly and its connecting terminations is crucial to obtaining the full EMC performance that a shielded cable is capable of. Depending on their application requirements, interconnecting cables can be classified as coaxial cable, twisted pair, ribbon cable, or fiber optics (Croop, 1997; Mroczkowski, 1997).

Coaxial cable is suitable for high-speed, single-ended driven applications, such as microwave, radio frequency (RF) transmission, Internet, and other applications requiring high bandwidth. Coaxial cable has zero loop area and shielding components, and good performance over controlled impedance ribbon, but it is usually bulky, difficult to terminate, and expensive.

Twisted pair cable has a small loop area, and reduced crosstalk or electromagnetic induction between pairs or wires, as two insulated conductors are twisted around each other. Twisted pair cable is usually used for low-speed data transmission, analog and digital applications, and differential drive. Shielded twisted pair cable has more reduced pair-to-pair crosstalk and additional protection against EMI, and it is mainly used for high-speed data transmission.

Ribbon cable is better than buddle for shielding, has a low cost, and high longitudinal flexibility. It is easy to shield but hard to terminate with shield. Ribbon cable is mainly used for low-speed internal applications, such as printers, scanners, and other repeated flexing applications. Controlled impedance ribbon can be used for high-speed applications with high temperature and abrasion resistance.

Fiber optics is one of the best and most promising cables for shielding, although it is relatively expensive. A fiber-optic system is similar to the copper wire system that fiber optics is replacing. The difference is that fiber optics use light pulses to transmit information down fiber lines instead electronic pulses. At one end of the fiber-optics system is a transmitter. This is the place of origin for information coming onto fiber-optic lines. The transmitter accepts coded electronic pulse information coming from copper wire. It then processes and translates that information into equivalently coded light pulses. A light-emitting diode (LED) or an injection-laser diode (ILD) can be used for generating the light pulses. Using a lens, the light pulses are funneled into the fiber-optic medium where they transmit themselves down the line. Fiber optics is steadily replacing copper wire as an appropriate means of communication signal transmission. They span the long distances between local phone systems and providing the backbone for many network systems.

In general, a shielded cable assembly consists of several layers of components (Mroczkowski, 1997):

1. Cable conductors. Cable conductors carry power, signal, or ground path currents depending on the cable assembly application. Selection is based on mechanical and electrical properties. Solid and stranded wires are available. The materials used are usually solid or stranded copper wires with coatings such as silver or tin.
2. Cable dielectric. The dielectric material, which acts as a buffer between the conductor and shielding components, allows the cable to maintain consistent electrical properties and minimizes signal loss. The material can be either solid or foamed.
3. Shielding. Shielding materials protect against signal loss and prevent EMI in the circuit. Several combinations of foils and braids are usually used to achieve various levels of shielding performance. For example, single round braid shields and flat braid combined aluminum/polyester foil shields have excellent shielding performance; and spiral wrap shields, aluminum/polyester foil shields, and helical ribbon foil shields all exhibit good shielding performance.
4. Cable jacket. The cable jacket is an additional form of insulation as well as a protectant against environmental dangers. A jacket is typically made of polyvinyl chloride (PVC), polyethylene, or fluorinated ethylene propylene (FEP). The material used will determine the jacket's temperature rating as well as minimum and maximum operating temperatures. For instance, PVC has good weathering and abrasion resistance; Teflon is able to withstand high temperatures and is chemically inert; and a cross-linked PVC irradiation vinyl jacket has superior mechanical properties and short-term heat resistance.

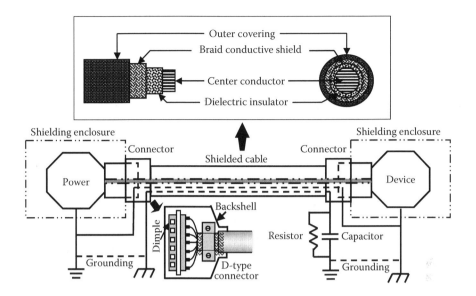

FIGURE 10.1 Cable shielding and grounding scheme.

In many systems, cable shielding is required for EMC compliance. Although proper cable and circuit design can provide benefits of low EMI, shielding is still a final defense against EMI. Together with grounding return wires, cable shielding can effectively reduce and eliminate EMI in the cable system. As shown in Figure 10.1, the coaxial cable, for instance, is connected at one end with a direct current (DC) connection to the common frame ground. The other end of the shield is typically connected with a capacitor, or a network of a capacitor and a resistor, to prevent DC current flow in the shield. In the case where connectors are involved that penetrate the system's enclosure, the cable shield must have circumferential contact to the connector with a conductive backshell to provide effective shielding. Moreover, properly terminating a shield requires 360 degrees of shielding to be maintained through all connectors or glands, to another shielded cable, or to a metal or metallized enclosure. For example, a metallized-plastic D-type shielded connector can be used for the terminating (Figure 10.1). The D-type shielded connector uses a metal saddle clamp to bond the shield to the metallized surface of the connector's backshell. The backshell then makes multiple bonds to the metal body of the connector itself, and the multiple dimples in the socket part make multiple connections to the mating connector. The saddle clamp is an approximation to 360-degree shield bonding, and has the significant advantage that braided shields need not be disturbed during assembly, which is going to provide very good EMC performance (Armstrong, 2006b).

For shielded cable assembly, as shown in Figure 10.1, the shield effectiveness is influenced by transfer impedance of the cable shield and contact impedance of the cable connector, which depends on the materials' conductivity, and the interface contact area and contact force (Nadolny, 2007).

Primarily, the shield of the cable and connector should be a good conductor having conductivity higher than 10^6 S/m (about 1.724% IACS), and therefore capable of providing a series of DC resistance of less than 0.1 milliohms for a finished product when measured from cable capture to connector termination, assuming no contact resistance. However, the contact characteristics become important material parameters at the separable interface, the braid capture region, and the connection to ground. As discussed in Chapter 2, it is critical that these connection points have a low resistance, high-durability connection, which can be achieved by proper surface metal plating. Moreover, the skin effect makes this metal plating a highly effective shield in certain frequency ranges, depending on the plating thickness and uniformity. To reduce contact impedance of the connector termination, some mechanical design approaches can be considered and applied (Hoeft and Hofstra, 1991; Nadolny, 2007):

1. As shown in Figure 10.1, adding dimples or dents to D-type connectors at the separate interface can dramatically improve electrical performance. Dimples or dents are intended to guarantee contact at several locations around the periphery of the separable interface.
2. A similar approach to item 1 can be applied to the cable braid capture, such as ferule and compression nut, staple and channel, and various forms of crimps, in an attempt to approximate the circumferential connection of the shield to the connector shell.
3. For connecting cable to chassis connectors, when a receptacle is mounted to a panel, a circumferential connection from the receptacle shield to the panel can be realized by mounting a nut plate in the back of the receptacle, then providing uniform pressure forcing the receptacle shell to contact the panel housing. Other approaches include incorporating special panel cutout shapes, which guarantee circumferential connection of the receptacle to the panel, or incorporating a conductive gasket.
4. For connecting cable to board connectors, the ground plane of a PCB can be utilized as the shield termination point. The shielded connector can be soldered directly to a surface of the ground plane to obtain the ideal circumferential shield termination, if the manufacturability is available. An alternative approach is to use ground pins, which are integral to the connector shield design and are soldered into via in the PCB. Fingerstock or other gaskets could also be applied to the PCB to contact the surface ground plane and the connector shield.
5. Numerical simulation of electrical performance would be optimal for a more exacting shielded cable, and connector design and producing process.

As any other conductive contact interface, the metal or plating used for connecting the cable, connector or gland, conductive gasket, and enclosure must be galvanically compatible. When high levels of shielding are required, or where frequencies of over 300 MHz are to be shielded, multipoint bonding and/or conductive gaskets will be required to approach or achieve 360-degree termination between the cable shield and the enclosure shield. Even if using gaskets is avoided, the design should still permit them to be employed if it is found later that they are needed (Armstrong, 2006b).

Consequently, effective cable shielding itself is only part of the solution for assembly of a shielded cable. Termination, including grounding, of the shield also is critical. For nonshielded cables or fiber-optics assembly, the way to provide cable shielding is usually by using steel conduit to house all wires and cables. The steel conduit will provide substantially higher shielding levels than the cable shields. Both cable shields and conduit connected to a shielded zone must have equal or greater shielding effectiveness than the shield. Conductive or metallic cable shielding or conduit can be used in the zonal/topological protection concept to extend the boundary formed by equipment enclosures and thus provide a way to interconnect elements while maintaining boundary continuity. Cable shielding is also used to protect a wire or wires as they travel from one boundary to another. This would be the case with a shielded EMI signal traveling from its entrance into an enclosure to the EMI receiver. Meanwhile, some form of grounding is required in any electrical or electronic system for protecting personnel from electrical shock, controlling interference, proper shunting of transient currents around sensitive electronics, and other reasons. Grounding does not directly provide protection against EMI, but it must be done properly to prevent creation of more serious EMI vulnerabilities. Ideally, grounding would keep all system components at a common potential. In practice, however, because of possible inductive loops, capacitive coupling, line and bonding impedances, antenna ringing effects, and other phenomena, large potentials may exist on grounding circuits (Herndon, 1990).

10.3 CABLE-LEVEL SHIELDING MATERIALS

Interconnecting cables including low-voltage cables, middle-voltage cables, and certain special-purpose cables usually require some form of EMI/EMP electromagnetic pulse shielding. There are some system characteristics that should be considered in selecting cable shields and shielding materials, such as functioning of overcurrent devices, required levels of fault or surge current, the manner in which the system may be grounded, desired levels of EMI/EMP protection, and environmental factors (including incidence of lightning; Voltz and Snow, 2002). Environmental considerations are important to overall cable operation and may have an effect directly or indirectly on the shielding. One of the key considerations is ambient temperature and its effect on current-carrying capacity. Indoor, outdoor, and underground locations have specific thermal characteristics that affect cable ampacity and potential shielding requirements. And shielding choices are affected by corrosion and wet locations. Several electrical system parameters should be considered, such as fault current, voltage, splicing devices and techniques, grounding of shields, shield losses, and insulating barriers in the shield.

Generally, cable-level shielding materials can be metallic, conductive heat-shrinkable, and ferrite.

10.3.1 METALLIC SHIELDING

A metallic shield needs to provide electrical continuity and grounding for functions (Voltz and Snow, 2002). For example, there needs to be EMI/ESP protection to prevent external electromagnetic fields from adversely affecting signals transmitted over the center conductor, and to prevent undesirable radiation of a signal into nearby or adjacent conductors. The metallic shield also needs to act as a second conductor in matched or tuned lines. And the metallic shield must provide protection from transients, such as lightning and surge or fault currents. Furthermore, a nonmetallic covering or jacket is usually used over the metallic shield for mechanical and corrosion protection. If a cable has a dielectric insulator between the center conductor and the metallic shield, as shown in Figure 10.1, it must have some form of stress relief at every splice and termination. This can be stress cones, molded devices, and heat-shrink or cold-shrink kits. All must be suitable for the voltage class and cable size.

For cables used in the middle-voltage range, there usually is one additional layer semiconducting materials to form a conductive bridge from either the conductor to the inside surface of the insulation or from the outer surface of the insulation to the metallic shield. The function of the semiconducting materials is to prevent voids and ensure an even distribution of the electrical stress. The semiconductive materials are generally formulated with conductive carbon black usually dispersed in a suitable polyolefin, polyethylene, or butyl acetate compound. Minor additives are included to aid dispersion, heat stability, and adhesion to the insulation. The properties of these materials vary depending on whether they are formulated for strand shielding or for insulation shielding (Voltz and Snow, 2002).

Nonmagnetic metallic materials, including copper and aluminum, are typically used for metallic cable shields. Aluminum requires a larger diameter of wire or a thick cross-section of tape to carry the same current as copper. At equivalent current-carrying capacity, an aluminum shield would be lighter in weight but about 40% larger in size. Aluminum shields are usually used for instrumentation, control, and communication cables, while copper shields are used with medium- and high-voltage cables (Voltz and Snow, 2002).

The metallic shield can take many forms. Braided copper is a general-purpose shield, and certain applications may allow the use of a lighter, less bulky shield construction. Some typical types include (Croop, 1997; Mroczkowski, 1997):

- Braided round or flat shield. Braiding consists of interweaving groups of strands of metal over an insulated conductor.
- Metal tapes including flat, corrugated and wire mesh tapes, and metal foils.
- Solid shield. Solid or tubular shields are applied by tube swaging or by forming interlocked, seam welded, or soldered tape around the dielectric, offering 100% shielding as well as a protective armor.
- Served shield. A served or spiral shield consists of a number of metal strands wrapped flat as a ribbon over a dielectric in one direction. The weight and size of the shielding are approximately one-half of those of braided shields since there is no strand crossover or overlap.
- Metal shield and conductive yarn. An effective combination of metal strands and conductive yarns provides effective shielding and reduces shield weight.
- Metallic powder loaded plastics. Polyvinyl chloride (PVC) and polyethylene loaded with conductive additives are normally limited to low, audio frequency ranges due to poor shielding effectiveness.
- Zipper cable shielding. A convenient, inexpensive method of EMI/EMP (electromagnetic pulse) shielding cable harness for computers, communication

equipment, and other sensitive electronic systems. This cable shielding is a knitted wire mesh with a protective cover of heavy-duty vinyl, a combination that provides flexibility and durability as well as shielding.

10.3.2 CONDUCTIVE HEAT–SHRINKABLE SHIELDING

Conductive shrinkable EMI shielding is a heat-shrinkable polyolefin material that provides effective EMI shielding for cables, connectors, and cable/connectors terminations. This kind of material offers significant weight savings over conventional metal braid shielding and can be applied easily with standard shrinkable tubing heating devices.

In fact, many efforts have been made to offer various conductive heat-shrinkable shields for wires, cables, and lines, typically in the form of a tubular, heat-shrinkable outer layer within which is received a conductive inner or outer layer. For example:

- A heat-shrinkable shield was formed of an outer layer of heat-shrinkable tubing, having a thin layer of a metal-filled polymeric matrix bonded to the inner surface (Derby, 1971). This shield may be sheathed over an insulated wire or cable, and then heated to shrink the outer layer of the shield over the insulation of the conduction wire or cable.
- An electromagnetic radiation suppression cover was made for a reflecting structure, which includes a tubular outer layer of an electromagnetic radiation absorber formed of a nonconductive composite with one or more kinds of dissipative particles dispersed in a shrinkable dielectric binder. An inner sealant layer is employed to fill any voids between the absorber and the structure. A thin metallic foil can be bonded between the sealant and the absorber as a ground plane (Lau et al., 1992).
- A magnetic shielding was made with a heat-shrinkable outer layer of a thermoplastic polymeric material and an inner layer of a magnetic shielding layer. The shielding layer can be formed of a thermoplastic material filled with a powdered ferrite (Nakamura et al., 1985).
- A heat-shrinkable EMI shielding jacket was made with a conductive inner layer, a thermoplastic interlayer, and a heat-shrinkable outer layer. The thermoplastic interlayer bonds the conductive inner layer to the heat shrinkable outer layer when the jacket is heated. The conductive inner layer can be a nonconductive fabric and a conductive metal foil formed of copper, aluminum, or another metal, or a combination or blend of conductive and nonconductive fibers (Tzeng et al., 1999).
- A cable assembly was designed and consisted of a plurality of insulated wires, an inner shrink tube lying tightly around the cable, a metal braiding lying around the inner shrink tube, and an outer shrink tube lying tightly around the metal braiding. While the outer shrink tube has an electrically conductive coating on its radius inner surface, the inner shrink tube has an electrically conductive coating on its radius outer surface. And the wire braiding is sandwiched between the conductive coatings. The two continuous conductive coatings for the heat-shrinkable tubing provided enhanced EMI shield at aircraft radio frequencies (Maleski et al., 1996).

- A heat-shrinkable EMI/RFI (radio frequency interference) shielding material has been made with the capability of conforming to the shape of the object being shielded, which comprises a heat-shrinkable woven fabric that is coated with a conductive material, such as silver, nickel, zinc, cobalt, tin, lead, platinum, gold, carbon, graphite, or other conductive coating (Millas, 2002).

A representative commercial conductive heat-shrinkable shielding is marketed under the name CHO-SHRINK® (Chomerics, 2001). The shielding is formed by coating a silver-based conductive compound onto the inside, outside, or both surfaces of a heat-shrinkable tube, which remains flexible, uniform, and intact even after maximum shrinking. The operating temperature is –54°C to 135°C, the shrink temperature is 121°C–191°C, and the typical shrink ratio is 60% to 85%. When the tubing is set, the inner conductive layer provides an electrical connection over the cable and to the outside surface of the connector that are joined by the tubing. Coaxial cable butt joints and cable to shielded connector housing joints can be made fast and easy and without the use of solder. Moreover, the conductive shrinkable connector boots can provide EMI shielding, cable shield grounding, and strain relief at connector backshell terminations. The highly conductive silver-based coating is applied to the inside surface of these molded polyolefin boots providing 60 to 80 dB attenuation above 500 MHz and in combination with a braided shield will provide levels of attenuation exceeding 80 dB above 10 GHz. The conductive shrinkable transitions provide shielding continuity for cable assemblies and harnesses.

The major advantage of the conductive heat-shrinkable cable shielding over conventional metal shielding is its ease of assembly. However, for optimal shielding performance, during assembling (a) terminate the shield at both ends with full 360° contact to a low impedance ground; (b) incorporate mechanical strain relief into the cable design itself, avoiding stretching and bending the cable excessively; and (c) transition pieces should be generous to preserve continuity at all junction points (Chomerics, 2001).

10.3.3 Ferrite Shielding Materials

Ferrite cable shields are designed to reduce EMI energy emitted from the cable without affecting the data transmitted through the cables. They are cost effective and used for suppressing any electromagnetic noise and reducing the need for other, more complicated, shielding measures. These ferrite components and EMI absorbers are likely to play an expanded role in EMI control for achieving EMC.

Figure 10.2 shows ferrite cable shields and their physical behaviors (Philips, 1997). Low frequency signals are not affected by a ferrite shield. At low frequencies (f), a ferrite core causes a low-loss inductance (L), resulting in a minor increase of impedance ($Z = 2\pi fL$). Interferences between electronic signals and the ferrite shield normally occur at elevated frequencies causing magnetic losses to increase and at the ferromagnetic resonance frequency (f_r) causing the complex permeability ($\mu'\mu''$) to drop rapidly to zero, while the impedance reaches a maximum.

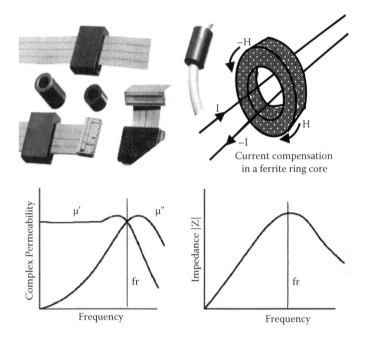

FIGURE 10.2 Ferrite cable shields and their physical behaviors. (Modified from Philips, 1997.)

This impedance is the most important parameter for suppression, and will become almost completely resistive and at very high frequencies even capacitive with losses. The impedance peaks at the resonant frequency and the ferrite is effective in a wide frequency band around it.

In general, the ferrite shield for EMI suppression is usually required on cables that are carrying DC or AC (alternating current) power. In that case, current compensation is needed to avoid saturation of the ferrite, which would result in loss of impedance. Current compensation is based on the principle that in cables passing through a ferrite core, the carried load and signal current are generally balanced. These currents generate opposed fluxes of equal magnitude that cancel out and no saturation occurs. An EMI signal, however, usually travels in the same direction on all conductors, the so-called common mode. This mode causes flux in the ferrite and will be suppressed by the increased impedance. For high frequency signals, current compensation is a beneficial effect for reasons other than saturation. In an input/output (I/O) cable, the regular RF signal could be suppressed together with the interference. Since the actual signal is differential mode, current compensation avoids this unwanted damping effect on the actual signal. A ferrite shield is mainly active against common-mode interference, although its small stray inductance will also have some effect against differential-mode interference (Philips, 1997).

Ferrite shielding materials have been widely used because they provide simple, convenient, and effective solutions for EMI problems in cables and connectors.

Both EMI attenuation and suppression of unwanted high frequency oscillations can be accomplished with no loss in DC or low frequency signal strength. The basic composition of ferrite materials is a combination of ferrous oxide and one or more other powdered metals, such as manganese, zinc, cobalt, or nickel. There are infinite varieties of formulas and performance levels of ferrite shielding materials. Each discrete ferrite formulation results in a stoichiometric ratio, which is its performance characteristic signature regarding electrical, magnetic, and mechanical relations. The operative characteristic that makes ferrite effective in EMI suppression is its variable sensitivity to frequency. With a ferrite is installed as a suppressor, lower frequencies will pass with no significant loss. However, when the frequency is high and around resonant value, the signal couples with the ferrite to create impedance that is quite high compared with the rest of the circuit. The offending EMI is thus immediately and consistently blocked out by way of impedance damping of the unwanted high frequency signals. It is this greater resistive impedance that allows the basically passive, apparently simple material to suppress multiple signals in a variety of application situations (Ferroxcube, 2007; Philips, 1997).

Typical ferrite shielding materials used include (also see Table 10.1; Ferroxcube, 2007):

TABLE 10.1
Property of Some Common Ferrite Shielding Materials

Material Type	Product Name	Permeability	Saturation B (mT)	Operating Temperature (°C)	Impedance (Ωm)
NiZn ferrites					
	4A15	1200	350	125	10^5
	4S2	700	350	125	10^6
	4B1	250	350	250	10^7
	4C65	125	350	350	10^8
MnZn ferrites					
	3E 8	18000	350	100	0.1
	3E 7	15000	400	130	0.1
	3E 6	12000	400	130	0.1
	3E 5	10000	400	120	0.5
	3E 26	7000	450	155	0.5
	3E 27	6000	400	150	0.5
	3C11	4300	400	125	1
	3S1	4000	400	125	1
	3C90	2300	450	220	5
	3S4	1700	350	110	10^3
	3B1	900	400	150	0.2
	3S3	250	350	200	10^4
Iron powder	2P90	90	1600	140 (max)	Low

Source: Modified from Ferroxcube 2007.

- NiZn ferrites. Suitable for EMI suppression up to gigahertz frequencies. They usually have low permeability but their high resistivity (10^5 Ωm) ensures that eddy current can never be induced in these ferrites. As a result, they maintain an excellent magnetic performance up to very high frequencies.
- MnZn ferrites. Low-resistivity MnZn ferrites (1 to 10 Ωm) have limited operation to a maximum of about 30 MHz. However, with precise control of material composition, the MnZn materials 3S4 has an increased resistivity of 10^3 Ωm, and have effective EMI suppression up to the gigahertz range, making it an attractive alternative to NiZn materials. Moreover, its high permeability gives it excellent low-frequency characteristics.
- Iron powder. The permeability of this material is low and its bandwidth is less than that of NiZn ferrites because of their low resistivity. Its main advantage, however, is a saturation flux density that is much higher than that of ferrites; therefore it is suitable for very high bias currents.

When selecting a ferrite shield and the shielding material, the following factors should be considered (Ferroxcube, 2007; Philips, 1997):

1. The frequency where maximum attenuation is needed will determine material requirements. The most suitable ferrite would offer the highest impedance levels at the interference frequencies, which usually cover a broad spectrum.
2. Core shape is usually defined by the cable type.
3. Installation requirements will be decided based on an entire or split core type.
4. Attenuation/impedance level influences on maximum suppression.
5. Ferrite characteristics vary with operating conditions. For example, impedance can change with temperature or DC.

10.4 BONDING AND GROUNDING

One of the most important factors in the design and maintenance of electrical systems, including cable assembly, is proper bonding and grounding. Inadequate bonding or grounding can lead to unreliable operation of systems (such as EMI), electrostatic discharge damage to sensitive electronics, personal shock hazard, or damage from a lightning strike. In fact, the concept of bonding and grounding is not the same. Bonding is the establishment of a low impedance current path between two adjacent components through an intimate interface surface. Grounding, also referred to as "earthing," is a specific form of bonding wherein a low impedance circuit or path is established to a designated ground plane or location, which may involve one or more bonds as well as an additional circuit element such as a cable or strap. Thus, proper grounding of the cable assembly or other conductors in the electromagnetic field will normally incorporate both bonds between components and a specific bond to the ground structure (earth).

10.4.1 BONDING

Bonding refers to the processes in which components or modules of an assembly, equipment, or several systems are electrically connected by means of low impedance conduction. The purpose is to make the structure homogeneous with respect to the flow of RF currents. This mitigates electrical potential differences, which could produce EMI with metal parts. A proper grounding design for an EMI shielding system should (Armstrong, 2006a; Molyneux-Child, 1997): (1) establish an effective path for EMI or other fault currents that facilitate the operation of the EMI or overcurrent protective devices; and (2) minimize shock hazard by providing a low-impedance path to ground. The design should also aim for (3) reduced inductance and controlled resonances in the component cavity. At high frequencies the impedance of a bonding conductor increases, due to inductance and, as the dimension becomes comparable with $\lambda/4$, fluctuations occur due to standing waves. Another goal should be for reducing transfer impedance. Good surface conductivity or low bonding impedance is required for all metal-to-metal contact areas. The achievement of a low value of bonding impedance is dependent on good metal-to-metal contact and the effectiveness of the contact is likely to be adversely affected by corrosion and vibration.

Galvanic corrosion may occur on metal surfaces and cause an increase in contact resistance. It is difficult to eliminate this effect entirely, since it occurs where any two galvanically incompatible metals are in contact in the presence of moisture and air. The following measures should be considered at the design stage to reduce corrosion problems (Molyneux-Child, 1997): (1) coat any exposed metal surfaces with appropriate plating or conducting paints; (2) avoid, where possible, the flow of current—especially DC—through the bonding connection, since this current will accentuate the corrosion process; (3) ensure, after the bond has been made that it is kept tight and protectively coated to exclude moisture and air; and (4) design the bonding system such that the corrosion occurs in an easily replaceable item, such as a washer.

In general, the bonding method used should not produce a DC resistance in excess of $0.0025\ \Omega$ across any bond, or more than $0.025\ \Omega$ at 150 kHz for any applied AC voltage (American Public Transportation Association [APTA], 1998). By far the best performing bond consists of a permanent, directive, metal-to-metal contact, such as is provided by welding, brazing, sweating, or swaging. Though adequate for most purposes, even the best soldered joints have appreciable contact resistance and cannot always be depended upon for the most satisfactory type of bonding. Semipermanent joints, such as provided by bolts or rivets, can provide effective bonding, but any relative motion of the joined parts will affect the bonding effectiveness through the introduction of varying impedance.

To minimize the inductance, bond wire or straps should be as short as is feasible and also be in the form of wide strips rather than circular conductors. Lock or star washers should be used with bolt lock-thread bonding nuts to ensure the continuing tightness of a semipermanent bonded joint. Star washers are particularly useful in cutting through protective or insulating coatings on metal such as anodized aluminum, residual oxides, or grease films. Joints that are press fitted or joined by screws of the self-taping or sheet metal variety cannot always be relied upon to provide low impedance RF paths. The threads of such screws may contain some residual oil from manufacture in spite of a degreasing process. Frequently, there is a need for relative motion between members that should be bonded, as in the case of shock mounts. Solid strips are preferable to braid straps as the latter have several undesirable characteristics: oxides may form on individual strands of the braid and cause the cross-sectional area of the strand to vary along its length. This variable cross-section, together with the possibility of broken strands, may result in the generation of spurious signals due to sudden changes in current distribution and intermodulation products. Where antivibration mountings for equipment are specified, it may be necessary to connect flexible bond straps across these mountings to provide bonding. Bond straps usually consist of copper and sometimes aluminum strips; copper strips are preferable, but where aluminum is used it is vital to ensure that the surfaces are free of oxide and an air-tight seal is provided (Molyneux-Child, 1997).

For PCB systems, bonding the 0V plane of the PCB to its local chassis generally has benefits for EMC, with cast or sheet metal chassis providing the best benefits. A variety of types of bonds between the PCB's 0V plane and the chassis can be used, such as direct surface contact bonding, bonding via a capacitor, and bonding via a resistor, for damping structural resonances. Because each application is different, it is often difficult to know in advance of EMC testing what type of bond will give the best overall EMC performance. Experimental EMC tests are usually used to find out which is the best configuration of DC, capacitive, or resistive bonds, and which are the best values to use (Armstrong, 2006a).

10.4.2 GROUNDING

Grounding is to provide a low impedance connection between the different elements in an electronic system and the safety ground structure (earth). The purpose of grounding is to protect circuits from faults, lightning induced surges, contact breaker operation, and fuses; to provide effective EMI shielding for enclosures, cables, and connectors; to achieve full performance of filters and capacitive suppression components; and to provide a signal reference condition. The latter case need not be earthed since a signal reference plane can function satisfactorily in isolation (APTA 1998).

In fact, such reference planes are invariably earthed. The grounding arrangements must be structured so that control is achieved over the flow of power and signal currents in such a way as to minimize the interactions and avoid the chance of interference being caused. Basically, the arrangements for protective grounding are less critical, low impedance connections at power frequencies are achievable by established practices and current carrying capacity to cater for fault conditions is likely to be of greater importance than conductor routing or length. Achievement of low impedance connections between either circuits or interconnections within systems is far more critical at radio frequency (RF) and special attention must be given to the nature and length of the conductors and the range of frequencies to be covered. Typical grounding conductors include wire-type conductors, busbars, metal raceways, the armor or sheath of certain metal cables, and cable trays. Cable assemblies and flexible cords also include equipment grounding conductors (Stauffer, 2008).

When the shielded cable is installed, its metallic shielding must be solidly grounded. Where conductors are individually shielded, each one must have its shielding grounded, and the shielding of each conductor should be carried across every joint to assure positive continuity of a shielding from one end of the cable to the other. Where grounding conductors are part of the cable assembly, they must be connected with the shielding at both ends of the cable. All grounding connections should be made to the cable shield in such a way as to provide a permanent low-resistance bond. For some cases, therefore, soldering the connection to the cable shield is usually preferable to a mechanical clamp, as there is less danger of a poor connection, loosening, or injury to the cable. In this way, the area of contact should be ample to prevent the current from heating the connection and melting the solder. For additional security, a mechanical device, such as a nut and bolt, may be used to fasten the ends of the connection together. This combination of a soldered and mechanical connection provides permanent low resistance, which will maintain contact even though the solder melts. Moreover, the wire or strap used to connect the cable shield ground connection to the permanent ground must be of ample size to carry fault currents. In addition, the metallic coverings of cables, as well as installations of shielded single conductor cables, must be evaluated to determine the best method of grounding, which is necessary as voltage is induced in the shield of a single conductor cable carrying alternating current due to the mutual induction between its shield and any other conductors in its vicinity. This induced voltage can cause metal shields bonded or grounded at more than one point to have circulating currents flowing in them, the magnitude of which depends on the mutual inductance to the other cables, the current in these conductors, and the resistance of the shield. This circulating current does not depend on the length of the cables nor the number of bonds. The only effect of this circulating current is to heat the shield and thereby reduce the effective current carrying capacity of the cable. The induced voltage can also cause shields bonded or grounded at only one point will have a voltage built up along the shield. The magnitude depends on the mutual inductance to other cables, the current in all the conductors, and the distance to the grounded point. This voltage may cause discharge or create an unsafe condition for workers. The usual safe potential is about 25 volts for cables having nonmetallic covering over the shield (Okonite, 2008).

There generally are no special materials needed for grounding. However, proper types of grounding systems should be considered and chosen when grounding a cable assembly.

10.4.2.1 Shields Grounded at One Point

In implementing single-point grounding, the objective is to avoid setting up low impedance loops in circuits and systems, which could result in a high current flow and hence generation of magnetic fields. Interaction with magnetic fields is a common cause of EMI arising at low frequencies. The implementation of single-point grounding in a system, where an earth reference plane is provided, should avoid inadvertent connections through connectors and controls; isolated balanced cables as far as possible; and in multiway cables ensure that the signal returns are via internal conductors, not through the screen (Okonite, 2008).

Grounding and Cable-Level Shielding Materials

10.4.2.2 Multigrounded Shields

Whereas the single-point ground system usually operates better at low frequencies, multiple grounding is generally required for systems operating at frequencies above 1 MHz and special measures are needed to overcome the effects produced due to differences in earth potential at various points in the system. If operating conditions permit, it is desirable to bond and ground cable shields at more than one point to improve the reliability and safety of the circuit. This decreases the reactance to fault currents, EMI effects, and increases the human safety factor. The single conductor cables carrying alternating currents may, in general, be operated with multisheath grounds. To achieve proper grounding, the following regulations should be considered for multigrounded shields (Okonite, 2008):

- Cables in AC circuits should not be installed with each phase in separate magnetic conduits under any circumstances due to the high inductance under such conditions. That is, shielded cables should be installed with all three phases in the same duct.
- Cables of any size may be installed with multishield grounds, provided allowance is made for heating due to current induced in the shield. Cables carrying direct current may always be solidly grounded at more than one point, except where insulating joints are required to isolate earth currents or to permit cathodic protection.
- Connections should be made to low impedance ground or earth plane, so that potential differences are minimized and very short connecting leads can be used.
- Attention should be paid to physical position and bonding arrangements of internal noise sources, filters, and earth conductors to minimize the signal induced voltages. It may be necessary to cut slots in the earth plane to channel currents in a suitable manner to avoid voltage of significant levels occurring in particular areas.
- The use of a balanced signal transmission arrangement with good common mode rejection at the terminations is particularly important.
- Signal cables generally should be positioned adjacent to the ground plane and earth conductors to minimize the earth loop in which unwanted induced voltages could occur. Every stage or unit of a system should be engineered to have a single-point grounding connection.

10.4.2.3 Hybrid Grounding

Basically, this is a mix of the two methods outlined earlier, that is, a grounding system that employs both single-point and multipoint earths.

10.5 SUMMARY

The selection of cable shielding, and the connection of the shield to the electronic devices and safety ground is very important in meeting EMC requirements. The cable shields and shielding materials mainly include metallic shields, conductive heat-shrinkable shielding, and ferrite shielding materials. Effective shielding for a shielded cable assembly can only be achieved by both the cable shielding itself and the proper termination shielding.

Another important factor in the design and maintenance of electrical systems, including cable assemblies, is proper bonding and grounding. Bonding is the establishment of a low impedance current path between two adjacent components through an intimate interface surface. Grounding is a specific form of bonding wherein a low impedance circuit or path is established to a designated ground plane or location.

When installing shielded cable, metallic shielding must be solidly grounded. Where conductors are individually shielded, each must have its shielding grounded, and the shielding of each conductor should be carried across every joint to assure positive continuity of a shielding from one end of the cable to the other.

REFERENCES

American Public Transportation Association (APTA). 1998. *Standard for grounding and bonding* (APTA SS-E-005-98). Washington, DC: American Public Transportation Association.

Armstrong, K. 2006a. Advanced PCB design and layout for EMC, Part 3–PCB-to-chassis bonding. *The EMC Journal.* Issue 53. http://www.compliance-club.com/journal_article.aspx?artid=90

Armstrong, K. 2006b, May/July. Design techniques for EMC, Part 2–Cables and connectors. *The EMC Journal.* www.compliance-club.com

Chomerics. 2001. *CHO-SHRINK® heat shrinkable shielding.* http://vendor.parker.com/Groups/Seal/Divisions/Chomerics/Chomerics%20Product%20Library.nsf/24eb

Croop, E. J. 1997. Wiring and cabling for electronic packaging. In *Electronic Packaging and Interconnection Handbook*, 2nd ed., edited by C. A. Harper, 4.1–4.55. New York: McGraw-Hill.

Derby, M. J. 1971. Heat shrinkable electromagnetic shield for electrical conductors. US Patent 3,576,387.

Ferroxcube. 2007. *The use of soft ferrites for interference suppression.* http://www.ferroxcube.com/appl/info/sfemisup.pdf

Herndon, R. 1990. *Electromagnetic pulse (EMP) and tempest protection for facilities.* Washington, DC: U.S. Army Corps of Engineers.

Hoeft, L. O., and J. S. Hofstra. 1991. Electromagnetic shielding provided by DB-25 subminiature connectors. *IEEE 1991 International Symposium on Electromagnetic Compatibility*, 50–53.

Lau, F. P., D. M. Yenni, Jr., R. W. Seemann, et al. 1992. Electromagnetic radiation suppression cover. US Patent 5,106,437.

Maleski, H. R., M. D. Beadell, and K. A. Kerfoot. 1996. Lightweight shielded cable assembly. US Patent 5,571,992.

Millas, E. E. 2002. Conductive heat-shrink tubing for EMI/RFI cable shielding. http://www.interferencetechnology.com/ArchivedArticles/Cables_and_Connectors/i_02_20.pdf?regid=

Molyneux-Child, J. W. 1997. *EMC Shielding Materials.* Oxford: Newnes.

Mroczkowski, R. S. 1997. Connector and interconnection technology. In *Electronic Packaging and Interconnection Handbook*, 2nd ed., edited by C. A. Harper, 3.1–3.82. New York: McGraw-Hill.

Nadolny, J. 2007. EMI *design of shielded cable assemblies.* http: www.connectorwizard.com/reference/articles/pdfs/Shielded_cable_assemblies_Samtec.pdf

Nakamura, K., M. Ishibashi, H. Kameda, et al. 1985. Heat shrinkable magnetic shielding article. US Patent 4,555,422.
Okonite. 2008. *Shielding.* http://www.okonite.com/engineering/shielding.html
Philips. 1997. *Cable shielding.* http://www.elnamagnetics.com/library/cableshi.pdf
Stauffer, B. 2008. *Understanding grounding and bonding basics.* http://www.necplus.org/Articles/Grounding%20and%20Bonding%20Basics.aspx
Tzeng, W. V., R. Saccuzzo, and J. E. Mitchell. 1999. *Heat-shrinkable jacket for EMI shielding.* US Patent 6,005,191.
Voltz, D. A., and J. H. Snow. 2002. Guide for specifying and selecting cable for petrochemical plants. *IEEE Industry Applications Magazine* 8: 60–69.

11 Special Shielding Materials in Aerospace and Nuclear Industries

11.1 INTRODUCTION

Aircraft and aerospace power systems depend heavily upon electronic systems, which must be shielded against electromagnetic interference (EMI). EMI may come in the form of lightning strikes, interference from radio emitters, nuclear electromagnetic pulses, or even high-power microwave threats. Also, humans must be protected from radiation to safely explore space. The predominant sources of extraterrestrial ionizing radiation, namely, galactic cosmic rays, consist primarily of nuclei of atoms (up to Fe) and solar energetic particles, which include mainly high-energy protons. In addition, neutrons that are formed due to breakdown of the incoming radiation flux in the shielding material have to be accounted for.

In fact, many fields have benefited from radiation shielding, including space-related or high-altitude applications and nuclear applications. Traditional radiation shielding materials include boron, tungsten, titanium, tantalum, gadolinium, hafnium, osmium, platinum, gold, silver, or palladium or some combination of these materials.

However, the need is increasing for lightweight EMI and structural radiation shielding materials to shield humans and reduce EMI in aerospace transportation vehicles and space structures, such as space stations, orbiters, landing vehicles, rovers, and habitats.

In addition, special materials for nuclear radiation shielding have been developed and used in many field applications, such as particle accelerators, nuclear reactors, radioactive biological and nuclear waste containment vessels, satellite hardware shielding, radiation shielding on high-altitude fighter planes, radiation protection for passengers and crew on high-altitude commercial airliners and military vehicles, and patient shielding for medical devices.

This chapter will address lightweight shielding materials for aerospace applications, radiation shielding for space power systems, and nuclear shielding materials and their applications.

11.2 LIGHTWEIGHT SHIELDING MATERIALS FOR AEROSPACE APPLICATIONS

Lightweight shielding materials and structure technologies are needed to mitigate EMI from electronic systems and protect humans from hazards of space radiation in aerospace vehicles. These radiation species include particulate radiation (electrons, protons, neutrons, alpha particles, light ions, heavy ions, etc.) and electromagnetic radiation (ultraviolet, x-rays, gamma rays, etc.). Although some conventional EMI shielding materials provide adequate shielding, many have weight restrictions. On the other hand, conventional spacecraft materials, aluminum, and higher atomic number structural alloys also provide relatively little shielding and, under certain conditions, could increase radiation risk. As a result, exploration of lightweight shielding materials has focused on nonparasitic radiation shielding materials and multifunctional materials, which function as EMI and radiation shielding, and support the structural integrity. Innovative processing methods have been developed to produce various forms of lightweight shielding materials, such as resins, fibers, fabrics, fiber-reinforced composites, nanomaterials, foams, light alloys, and hybrid materials.

11.2.1 Lightweight Radiation Shielding Material

The reduction of radiation dose aboard spacecraft and aircraft by appropriate shielding measures plays an essential role in the future development of space exploration and air travel. The design of novel shielding materials has involved hydrogenous composites, as liquid hydrogen is most effective in attenuating charged particle radiation. The shielding properties of hydrogen-rich polymers and rare earth-doped high-density rubber have been evaluated in various ground-based neutron and heavy ion fields. Compared with conventional aluminum shielding, hydrogen-rich plastics and rare-earth-doped rubber are effective in attenuating cosmic rays by up to 10%. The appropriate adaptation of shielding thicknesses might thus allow reducing of the biologically relevant dose. Moreover, owing to the lower density of plastic composites, tremendous mass savings can be achieved and can result in a significant reduction of launch costs (Vana et al., 2006).

A variety of lightweight and flexible shields have also been fabricated, some of which used boron, titanium, iron, cobalt, nickel, and tungsten, or other heavy metal particles suspended in a polymeric flexible medium, such as silicone. These shields can be reusable, sterilizable, or disposable (Bray et al., 2000; Cadwalader, 2001; Cadwalader and Zheng, 2004; Orrison, 1990). Weight savings as high as 80% could be achieved by simply switching from aluminum EMI shielding covers for spacecraft power systems to EMI covers made from intercalated graphite fiber composites. Because the EMI covers typically make up about one-fifth of the power system mass, this change would decrease the mass of a spacecraft power system by more than 15% (Gaier, 1992). Intercalated graphite fibers are made by diffusing guest atoms or molecules, such as bromine, between the carbon planes of the graphite fibers. The resulting bromine-intercalated fibers have mechanical and thermal properties nearly identical to pristine graphite fibers, but their resistivity is lower by a factor of five to ten, giving them better electrical conductivity than stainless steel and making these composites suitable for EMI shielding.

EMI shields in spacecraft, however, must do more than just shield against EMI. They must also protect electrical components from mechanical, thermal, and radioactive disturbances. Intercalated graphite fiber composites have a much higher specific strength than the aluminum they would replace, so achieving sufficient mechanical strength and stiffness would not be difficult. Also, the composites absorb and reradiate electromagnetic radiation in the infrared region, so they are actually considerably better at rejecting heat than the very reflective aluminum covers. The shielding of high-energy radiation generally increases with the total number of electrons in an atom. Thus, carbon, which has 6 electrons per atom, is a poorer shield than aluminum, which has 13. Intercalation with a small percentage of bromine (35 electrons/atom) or iodine (53 electrons/atom) has been implemented to improve the radiation shielding, though the average number of electrons per atom would still be below that of aluminum. The test results indicate a substantial improvement in the radiation shielding: not only by a factor of eight over the pristine graphite composite, but actually by a factor of three over aluminum (Gaier, 1992; Jaworske et al., 1987). Intercalated graphite composites may well be the EMI cover of choice for spacecraft electronics operating in a high-radiation environment.

A commercial lightweight radiation shielding product is the Xenolite™ (APC Cardiovascular, Crewe, Cheshire, England) garment, which is a polymer-based composite material, designed to offer full radiation protection with up to 30% reduction in weight (Cadwalader and Zheng, 2004). Xenolite greatly relieves back and shoulder stress when radiation protective garments must be worn for long periods of time. Very flexible and durable, with styles that offer maximum support at various stress points, Xenolite is a environmentally friendly protective material, and is 100% recyclable, which removes disposal problems (Lite Tech, 2004).

Consequently, specially designed composite materials, which can serve as both effective shielding materials against cosmic ray and energetic solar particles in deep space, as well as structural materials for habitat and spacecraft, could be used for critical lightweight radiation shielding materials of aerospace vehicles. For instance, low-density polyethylene, coupled with a high hydrogen content and other fillers, is known to have excellent shielding properties. And some specialized polyethylene-fiber reinforced composites, which combine shielding effectiveness with the required mechanical properties of structural materials, are promising.

11.2.2 Fiber Reinforced Composite Materials

Ever since the mechanical properties of fiber reinforced composite materials, such as glass, polymers, or graphite fibers in a polymer matrix, have surpassed those of aluminum, it has been a trend to use these materials in the aerospace industry. Their light weight combined with high strength naturally lends them to use in aircraft and spacecraft. However, the electrical conductivity of these composite materials is usually not high enough to withstand the effects of EMI, lightning strike, and space radiation. Therefore, there has been no widespread development of all composite airplanes yet.

As a solution to improve the electrical conductivity of the composite structure, metal has been added to the composites. This can take the form of metal surfaces such as metal foils or expanded metal foils, embedded metal wires, and metal filaments, either on the individual composite fibers or on the complete structure. Metal fibers of copper, stainless steel or aluminum provide conductivity in plastic components for static electricity dissipation or, with higher loadings, shielding against EMI. Because the percentage of needed metal fibers is quite small, base-resin or polymer properties remain relatively unchanged. For example, a graphite fiber reinforced laminate structure was made incorporating metal wire or metal coated filament, which is woven in at least the outer ply of the graphite reinforced laminate structure. The metal provides the graphite reinforced polymer with EMI shielding and lightning protection (McClenahan and Plumer, 1984). A similar idea is to fabricate fabric sheets of carbon filaments and fibers with metal coating (Ebneth et al., 1984). The inclusion of metal in composite structures, however, might cause problems in most cases due to the CTE (coefficient of thermal expansion) difference between the metal and matrix. In instances where metal coatings are used, there is a particular problem with the coatings adhering to the surfaces. As temperatures change, metal layers expand much more rapidly than the glass or carbon fiber layers below. This causes internal stress, which is relieved by the separation of the metal layer from the adjacent fiber layer. Additionally, since metals are denser than fibers, as more EMI protection is required, more weight is added to the structure. Thus, at some point the advantages of using composite structures are lost because of the added weight. Moreover, the processes by which the metals are included in or on the composite structures are often expensive because adding the metal introduces manufacturing complexities. There also may be complex electrical and chemical interactions between the metal and the organic composite. Often this requires the rigorous exclusion of water or moisture from the manufacturing system (Ebneth et al., 1984; McClenahan and Plumer, 1984).

In order to eliminate the disadvantages of using metals in composites, the use of other materials that are good conductors of electricity is needed. Carbon fibers currently used in the aerospace industry are reasonably good conductors (resistance of 2000 $\mu\Omega$-cm); however, it is still nearly three orders of magnitude greater than aluminum (2.7 $\mu\Omega$-cm). Graphitic carbon is substantially more conductive than carbon fibers, and graphitic fibers have been made to have resistances as low as 70 $\mu\Omega$-cm. Furthermore, the resistance of graphitic carbon fibers has been lowered to values less than that of silver through the process of intercalation, the addition of guest atoms or molecules into the graphite lattice, forming a compound that consists of carbon layers and intercalate layers stacked on top of each other. The incorporation of intercalated carbon or graphite into an organic composite matrix has formed composites with electrical conductivity higher than similar ones made with nonintercalated graphite (Vogel and Zeller, 1983). A high-grade graphite fiber has been fabricated by heat treating specific polymer films and intercalating them in a vacuum or in an inert gas atmosphere. This intercalated graphite has improved stability over prior graphite intercalates (Murakami et al., 1988). A method for producing exfoliated graphite fibers has also been developed, by intercalating fibers and then exfoliating the intercalated fibers by heating. These intercalated graphite fibers exhibit a high degree of electrical conductivity (Chung, 1990).

Furthermore, a highly conductive lightweight hybrid composite laminate has been fabricated by weaving strands of high strength carbon or graphite fibers into a two-dimensional fabriclike structure, depositing a layer of carbon onto the fibers of the fabric structure, heat treating the fabric structure to a temperature sufficient to graphitize the layer of carbon, and intercalating the graphitic carbon layer. The final composite comprised at least one layer of the hybrid material and at least one layer of strands of high strength carbon or graphite fibers in a polymer matrix. The hybrid materials were compatible with matrix compounds, had a coefficient of thermal expansion that was the same as the underlying carbon or graphite fiber layers, and were resistant to galvanic corrosion. And the composite had an electrical resistivity that could be less than about 15 to 70 µΩ-cm (Gaier, 1993). These composite materials are especially useful for aircraft and spacecraft components, which must withstand lightning strikes. They are also useful in EMI/EMP (electromagnetic pulse) shielding, electrostatic protection, grounding planes, fault current paths, power buses and other applications where high strength, light weight, and good conductivity are desired.

In fact, a variety of innovative fiber reinforced composite materials has been developed to improve the shielding effectiveness, current-handling capability, and electromagnetic absorption of existing composites. Some high electromagnetic shielding composites, for instance, are achieved by using graphite fibers with bromine or similar mixed halogen compounds, or nickel coating. Some composite laminated panels having high mechanical performance and high EMI energy absorption capabilities have been fabricated, which have sandwich structures, and are characterized by a reflection coefficient lower than 17 dB in the frequency range of 7 to 18 GHz (Caneva et al., 1998; Holloway et al., 2005).

11.2.3 NANOMATERIALS

Nanomaterials and nanocomposites provide a potential approach to EMI shielding, lightning strike protection, and heat dissipation in aerospace vehicles. Nanomaterials, imbedded in composite structures, would eliminate the need for complex and heavy protection shielding systems and would help maximize the potential of composites for aerospace applications. For instance, nanostrands have been coated onto the resin or other polymer fibers or fabric matrixes to create or improve conductivity. The unique properties of nanomaterials set them apart in their abilities to easily form highly conductive, three-dimensional, submicron, nanostructured lattices or composites. In addition, metallike nickel nanomaterials or metal-coated nanotubes are inherently magnetic, thus they can be aligned or oriented in the composite matrix, resulting in tailored conductive and possibly improved mechanical properties. These properties are crucial for shielding integrity.

Primarily, nanomaterials can be taken as a new step in the evolution of understanding and utilization of materials for EMI shielding and radiation protection. Nanomaterial science started with the realization that chemical composition is the main factor in determining what a nanomaterial is. Hereafter, it was discovered that the fabrication process and after-fabrication steps could influence those properties substantially. Also, small additives proved to be able to modify these properties. With the arrival of nanotechnology, it was discovered that the ability to create small particles could expand the capability to create and modify materials. For example, carbon nanotechnology created nanotubes, fullerenes, and nanodiamonds. Carbon nanotubes (CNTs) are produced in many variants; the most important classification is single- or multiwall nanotubes. A single-walled carbon nanotube (SWCNT) consists of a single cylinder, whereas a multiwalled carbon nanotube (MWCNT) comprises several concentric cylinders. Nanotubes can also be made out of materials other than carbon. They provide unique properties, such as (a) they are probably the best conductors of electricity on a nanoscale; (b) thermal conductivity comparable to diamonds along the tube axis; and (c) they are probably the stiffest, strongest, and toughest nanofiber. Nanocomposites, based on polymers, have been a central area of polymer research and significant progress has been made in the formation of various types of polymer-nanocomposites. This includes an understanding of the basic principles that determine their optical, electronic, and magnetic properties. Nanocomposites typically contain 2% to 15% loadings on a weight basis, yet property improvements can equal and sometimes exceed traditional polymer composites, even containing 20% to 35% mineral or glass. The shielding effectiveness of 20 to 30 dB has been obtained in the X-band (8.2 to 12.4 GHz) range for 15% SWCNT-loaded epoxy composites (Huang et al., 2007). The EMI shielding effectiveness of a composite material depends mainly on the filler's intrinsic conductivity, permittivity, and aspect ratio. The small diameter, high aspect ratio, high conductivity, and mechanical strength of CNTs make them an excellent option for creating conductive composites for high-performance EMI shielding materials at very low filling. In addition, SWCNT-polymer composites possess high real permittivity (polarization, ε') as well as imaginary permittivity (adsorption or electric loss, ε'') in the 0.5 to 2 GHz range, indicating that such composites could also be used as the electromagnetic shielding material for cell phone electronic protection (Grimes et al., 2000).

The designs of nanocomposites containing nanoclays and nanoscale grapheme plates (NGPs) as functional reinforcements have been investigated. If successful, these nanomaterials will be available at a much lower cost and tailorable for avionic and aerospace requirements. As mentioned earlier, radiation shielding is a vitally important issue for the durability of electronics in the space environment. The electronic devices behind the nanocomposite shielding materials do not have to be radiation hardened because the nanocomposites will stop proton and electron radiation, bleed off accumulated charge buildup, provide ground continuity, and absorb slow neutrons. The nanocomposites being developed can also be applied in nanostructured components, devices, conductive and barrier coatings, and substrates. The nanofillers could be incorporated in high-temperature resins, such as bismaleimides (BMI), to form polymer nanocomposite coatings. Isolated nanofillers can impart not only impressive mechanical but also controlled thermal, dielectric, and electrical properties; enhanced impermeability to coatings; and thin films for high temperature insulation and radiation. The resins are produced to ensure low viscosity and to possess high glass transition temperatures ideal for avionics and structural applications (Wong et al., 2006).

11.2.4 FOAM STRUCTURES

Lightweight multifunctional foam structures and foam composites have been developed based on carbon foam, metal foam, polymer foam, and nanoporous structures, showing a great potential for EMI shielding and radiation protection in aerospace applications.

Typical carbon foam is produced by the thermal decomposition (coking) of a coal-based precursor under controlled conditions. Its properties can be tailed on both micro- and macroscopic scales through selection of raw material, foaming process, and heat treatment conditions (Touchstone Research Laboratory, 2004). The critical characteristics that make carbon foam uniquely suited as a core material for an EMI shielding composite structure include its electrical resistivity, which can be tailed through heat treatment over a wide range from 0.01 to 10^6 Ω. This makes it a perfect electrical conductor of wide band frequency and wide-angle incidence for radar-absorbing and electromagnetic shielding applications. Also, carbon foam can be bonded and integrated into metal and other dissimilar materials, and easily machined and shaped into complex structures. Its relatively high strength coupled with lightweight, and resistance to fire and impacts makes carbon foam an alternative EMI shielding and structural material for aircraft applications.

Nanoporous carbon foam composites have been developed and fabricated, which can be used for aerospace structural/multifunctional materials, with structural strength and radiation shielding and hydrogen fuel storage properties. A representative commercial carbon foam GRAFOAM® (GrafTech International Holdings Inc., Parma, Ohio) is a multifunctional core material that can be used in construction, defense, fire protection, transportation, and aeronautics industries (GrafTech International, 2007). This carbon foam can help reduce lives lost due to fire and storms, as well as provide rooms with improved security when used in building structures as a sandwich panel. It is lightweight but strong and provides insulation, as well as acoustical and electromagnetic shielding.

Aerogel foam composites also have great potential for EMI shielding in aerospace applications. They are produced by reacting the internal mesoporous surfaces of porous networks of inorganic nanoparticles with polymeric cross-linking agents. Reaction of the aerogel with an isocyanate monomer results in the formation of conformal polyurethane/polyurea coating around the nanoparticles. The surfaces of the silica aerogels have been modified with (1) amines that have been cross-linked with epoxies and (2) olefins, such as polystyrene. Polystyrene, because of its high hydrogen content, has been shown to be an effective radiation shielding material. It may be possible that these aerogels could afford both radiation protection and thermal insulation to astronauts on future exploration missions (Bertino et al., 2004; Zhang et al., 2004).

11.3 RADIATION SHIELDING FOR SPACE POWER SYSTEMS

Proper radiation shielding materials and structural design for space power systems are crucial to protect humans from the hazardous radiation of outer space and inside reactor power systems in space exploration. Attenuation of the radiation levels in nuclear reactor-powered spacecraft to meet mission payload radiation dose constraints can be achieved by (1) radiation shielding components in the space reactor power system (SRPS); (2) separation distance between the SPRS and mission payload; (3) radiation attenuation/scattering in SRPS/spacecraft components of the SRPS shielding; and (4) localized shielding provided within the mission payload. Radiation dose requirements weigh the relative importance of the type and size of SRPS shielding. Historically, metal hydrides, particularly lithium hydride, have been considered for neutrons. Other hydrides and other materials also have been considered for neutron and other radiation shielding. Factors related to the final selection of materials are shielding performance, unit mass/volume, material properties, and engineering and fabrication considerations (Wilson et al., 1995).

11.3.1 NEUTRON SHIELDING DESIGN AND MATERIALS SELECTION

Radiation shielding within the SRPS is defined by the radiation dose constraints of radiation-sensitive components in the SRPS power conversion systems and reactor control components of the SRPS radiation shield. The challenge for SRPS shield design is to provide a radiation shield component that provides the required radiation attenuation within the mass limitations and size envelope imposed by the overall design. Principle design requirements include: (1) accommodate the penetrations of primary heat transport coolant piping and reactor control components without compromising the shielding function, (2) accommodate the structural support of reactor subsystem components, (3) maintain the shielding function over the long-term operation of the SRPS, (4) overcome materials issues related to compatibility and performance in space reactor environments, and (5) resolve engineering and fabrication considerations that will allow conceptual designs to be reduced to flight hardware. Shield materials that provide adequate neutron attenuation performance within the mass/space constraints of an SRPS require high hydrogen densities with the resulting design challenge of maintaining the hydrogen content of the neutron shielding material over the long-term operational time.

Likely neutron shielding materials are metal hydrides because of their high mass-to-hydrogen ratio and relatively good stability in high-radiation fields. The most advantageous material is low-density lithium hydride (~0.7 g/cm^3). Hydrides yttrium hydride, zirconium hydride, and titanium hydride provide equivalent hydrogen densities but with much higher physical densities. The metallic hydrides provide the capability to retain hydrogen at elevated temperatures with each of the candidate metallic hydrides exhibiting dissociation pressures (i.e., partial pressure of hydrogen) as a function of hydrogen content (H/M atom ratio) and temperature. Because of the high dissociation pressure for stoichiometric metallic hydrides, each of candidate materials would have the potential to lose hydrogen due to permeation losses through shield vessel walls. Each of the candidate materials exhibits a plateau dissociation pressure, which would be the most likely approximation of the hydrogen content over long-term conditions. The metallic hydride composition at plateau dissociation pressure levels has been estimated as $LiH_{0.98}$, $YH_{1.8}$, $ZyH_{1.33}$, and $TiH_{1.7}$. These candidate metallic hydride shields provide equivalent shielding performance but result in significant mass penalties. Using a combination of metallic hydride materials would result in minimizing the mass penalties but would complicate the shield design configuration with vessels required for each hydride material. The use of LiH materials may require thermal management to achieve shielding operating temperatures that limit or eliminate the radiation-induced swelling of LiH materials. Use of multiple metallic hydrides has the potential to limit LiH radiation swelling concerns and limit thermal management concerns (Heinbockel et al., 2000).

11.3.2 SHIELDING MATERIALS FOR EARTH NEIGHBORHOOD INFRASTRUCTURE

The international space station (ISS) typifies the first step in establishing space infrastructure by providing a base and transportation hub during low Earth orbit (LEO). The ISS is basically an aluminum structure, a well-proven technology for that historic development. However, aluminum is a poor radiation shield material to hazardous outside of LEO applications. It has been shown that implementation of hydrogen-bearing materials is the only alternative for improved shield performance in both LEO and space operations outside the geomagnetic field. The best such materials with reasonable structural properties are aliphatic polymers such as the high-density polyethylene chosen to augment shielding within the ISS and some usages of polyethylene have now space qualified. In this case, basic structural loads and meteoroid/debris protection are provided by the aluminum structure and the polyethylene is an added mass with the sole purpose of reducing astronaut exposure. Clearly, a multifunctional wall/support structure incorporating all these functions will reduce design mass, lower costs, and improve radiation protection.

The Earth neighborhood infrastructure has been developed for operations beyond the confines of the geomagnetic field. This infrastructure has tried to make the radical change from aluminum-based structural technologies to other materials, which include inflatable polymer structures using polyurethane foams, graphite composite stiff structures, as well as the optimal aliphatic polymer systems that provide superior shielding and materials strength characteristics. The multifunctional aliphatic structural systems have apparent radiation shielding advantages over polyurethane-based structures.

Although aluminum alloys are still the primary construction material in space infrastructure, other choices of materials focus on the effects of added alloy components to their radiation-protective qualities. For example, Al–Li alloy systems have improved shielding characteristics. Aluminum metal matrix composites are favorable depending on the fiber used, and boron fibers in particular may show some advantage since they absorb low energy neutrons, which are always produced as a secondary reaction product. Neat polymers have already shown important advantages in their radiation protection properties and can be ordered in their protective abilities according to their hydrogen contents. The use of in situ polymer manufacturing and regolith composites will also be a potential cost effective means of habitat construction. Polymer matrix composites could be an important alternative to many aluminum-based structures, with anticipated improved radiation protective properties (Hou et al., 2005; Wilson et al., 2003).

11.3.3 ACTIVE ELECTROMAGNETIC SHIELDING FOR SPACE RADIATION

As particles are charged, an alternative solution to the problem of shielding crew from particulate radiation is to use active electromagnetic shields. Practical types of shields include the magnetic shield, which diverts charged particles from the crew area; and the magnetic/electrostatic plasma shield, in which an electrostatic field shields the crew from positively charged particles, while a magnetic field confines electrons from the space plasma to provide charge neutrality. The magnetic field of the Earth is a good example of a magnetic shield, and is responsible for the relatively benign radiation environment on Earth. Advances in several major technical areas are necessary to build such an electromagnetic shield, which include high-strength composite materials, high-temperature superconductors, numerical computational solutions for simulating and controlling particle transport in electromagnetic fields, and a technology base for construction and operation of large superconducting magnets.

These advanced technologies have made the electromagnetic shielding a practical alternative for space exploration missions (Landis, 1991).

The concept of using superconducting coils to produce magnetic fields for protection from energetic particle radiation was developed in the late 1950s, contemplated as a means of providing radiation protection during manned missions beyond the magnetosphere (Levy, 1961). Original development of the concept was limited to low-temperature superconductors at liquid helium temperatures. The use of such low-temperature superconductors posed a daunting set of problems: (a) Even with tile equilibrium temperatures attained by a spaceship in outer space, cooling to liquid helium temperatures would still be needed through mechanical refrigeration techniques to achieve the superconducting state. This requirement alone limited magnetic shielding designs to the use of shipboard coils, especially due to the power requirements for maintaining the cryogenic temperature. (b) Shipboard coils required high magnetic field intensities only achievable through extremely high currents in order to shield a reasonably high volume. (c) The minimum masses of the coils and the related supporting structures required for building such shield. One additional concern noted with magnetic shields is the effects that such high magnetic fields might have on living organisms, especially long-term exposure. However, with proper design the field can be reduced to values lower than normally present on Earth (Cock, 1993).

On the other hand, the mass shielding structure required high tensile strength of materials to withstand the magnetic self-force on the conductors. For the minimum mass structure, all the structural elements are in tension, and from the virial theorem, the mass required to withstand magnetic force can be estimated as (Connolly et al., 1990):

$$M = (d/S)(B^2 V)/(2 \mu_0) \qquad (11.1)$$

where d is the density of the structural material, S is the tensile strength, B is the magnetic field, V is the characteristic volume of the field, and μ_0 is the permeability of vacuum.

An alternative to the magnetic shielding design is to use an electrostatic shield. Since both solar flare protons and heavy nuclei in space radiation are positively charged, it would be thought that shielding would simply require adding to the object to be shielded a positive charge sufficient enough to repel the particles. To shield against solar flare protons would require an electrostatic potential of 10^8 volts; shielding against cosmic ray nuclei as well would require as much as 10^{10} volts. A proposed solution is a hybrid of the magnetic shield and the electrostatic shield: the plasma shield (Levy and French, 1968). An electrostatic charge is applied to the vehicle (or habitat) shell to repel the positively charged radiation; a magnetic field then prevents the plasma electrons from discharging the vehicle.

With the advent of high-temperature superconductors, a heightened interest in magnetic shields is apparent. This interest is further fueled by the planning for a manned mission to Mars, which would directly benefit from the development of magnetic shielding technologies. A number of studies have focused on the use of ceramic materials, which can achieve superconductivity at liquid nitrogen temperatures, as well as on possible configurations for a magnetic shield. One such configuration is that of a deployed torus. These new superconducting materials present a number of notable advantages, including significantly less cooling needs as well as deployed configurations for superconducting wires. The deployed configuration appears especially promising due to enormous reductions in mass and energy requirements over previous shipboard coil designs. In addition, the danger that catastrophic failure of the magnetic shield poses to a spaceship's crew is minimized by deploying the shield away from the ship as opposed to producing the necessary magnetic fields with a shipboard coil (Cock, 1993).

Computational simulation is in universal use for solving particle trajectory problems by direct numerical integration. Solving the problem of the shielding produced by any given magnetic field configuration for any energy spectrum is now straightforward, even for systems in which the magnetic field is of complicated geometry. In terms of plasma shielding, the field of particle transport in plasma is now well understood from the work done in the fusion energy program. High-temperature superconductors can be operated at 77K or higher. This is a range that will allow use of passive cooling, where the temperature is achieved by directly radiating excess heat to space, a considerable advantage over the use of superconductors requiring liquid helium temperatures (Hull, 1989). There has been a large body of experience in fabricating large superconducting magnets. Superconducting magnets are now standard technology on large particle colliders, as well as for large magnetic fusion experiments. Extremely high specific strength composite materials are available. Since the limit to magnetic field strength is produced by the tensile strength of the materials required, composite materials with strength-to-weight ratios five times (such as Kevlar 49, a structural grade aramid cloth used for composite reinforcement that requires saturation with epoxy, polyester, or vinyl ester resins, which will then create a rigid laminate) to seven times (such as PBO [polybenzoxazole] composite) higher than that of steel allow considerable weight reduction in the electromagnetic shielding structure (Landis, 1991).

Consequently, technical advances in numerical computational simulation, high-temperature superconductors, and high-strength composite materials have made the magnetic shielding concept much closer to practicality and would be an extremely attractive option for long space missions.

11.3.4 OPTIMIZATION OF SPACE RADIATION SHIELDING

As the mixture of particles in the radiation field changes with added shielding, the optimization of shield material composition should adapt to the complicated mixture of particles, the nature of their biological response, and the details of their interaction with material constituents. Shielding optimization also requires an accurate understanding of the way in which the shield material interacts with the radiation field. In addition, many of the materials must meet other mission requirements such as strength, thermal, and hardness. These materials vary widely in composition and radiation shielding properties. Moreover, a large fraction of the radiation protective properties of the shielding materials comes from materials chosen for other requirements and early entry of radiation constraints. Therefore, a multifunctional design process is required for design optimization (Wilson et al., 2001).

The shielding on spacecrafts largely depends on the basic structure and onboard systems. Their contribution to shielding as well as any additional shielding system can be optimized in one of three ways (Schimmerling et al., 1999): (1) adding bulk material around the entire habitat; (2) changing the elements constituting the structure, for example, materials with a high ratio of atomic electrons to nuclear protons and neutrons to maximize stopping relative to nuclear interactions; or (3) redistributing materials to enhance some shielding locations, such as sleeping quarters where humans will spend a substantial portion of their time. Unless radiation constraints are incorporated at an early stage of the design, shielding improvements may only be achieved at considerable cost and weight to the structure. Apparently, the solution to radiation constraints requires improved methods of design. Since the basic structure and onboard systems provide much of the shielding, the optimization of spacecraft shielding is inherently a multidisciplinary design process.

Generally, radiation shield optimization requires an evaluation of materials and making appropriate choices at each step of the design process. The shielding performance of candidate materials for each specific application needs to be characterized to allow optimum choices in the design process. Multifunctional material properties must be developed for use in system optimization procedures. Optimum protective materials are needed to finish out deficiencies in the shield design at minimum mass and costs. Novel materials for specific applications need to be developed with enhanced shielding characteristics. For example, polymeric composites are preferable alternatives to aluminum alloys. Developing sound absorbing materials that are efficient radiation shields for use in crew areas has been proposed. Advances in hydrogen storage in graphite nanofibers may have a large impact (three to six times better than aluminum) on radiation safety in future spacecraft design (Wilson et al., 2000).

11.4 NUCLEAR SHIELDING MATERIALS

The need to shield against nuclear radiation for protection of personnel and instruments is encountered in a wide variety of installations and activities. These installations include fission reactors, fusion facilities, particle accelerators, nuclear weapons, and spacecraft and space station systems. The activities include research, medical, nuclear power, military applications, space exploration, as well as transport and storage of radioactive materials.

High-energy particles from the nuclear reaction point initiate hadronic and electromagnetic showers, as they encounter the shielding materials around. If the material is thick, these cascades will continue until most of the charged particles have been absorbed. Hence, inner shielding material should be as dense as possible to shorten the showers and of sufficient depth or thickness to contain most of the deposited energy. The remaining neutrals are mostly neutrons, which can travel long distances, losing their energy gradually. These remnant neutrons lead to further photon production. Gammas result from excited decays of spallation products and from fast neutron interaction with atomic nuclei. At the lowest energies, thermal neutrons result in production of gammas by the process of nuclear capture. For shielding to be effective, most of these neutrons and gammas should be confined to the shielding volume. This implies that an optimized outer shield should be a structure made of neutron- and gamma-absorbing materials, sometimes in the form of cladding layers. These layers are selected and combined to increase moderation (slowing by elastic collisions) of unconfined fast neutrons and to attenuate gammas produced by neutron interaction processes (Baranov et al., 2005). Typical materials for shielding and shield lamination structures include copper and its alloys, iron, cast iron, steel, lead, tungsten and its alloys, concrete, carbon, aluminum, polyethylene, and other new developed materials.

The major properties to consider when selecting shielding materials against neutrons usually involve (Greenspan, 1995):

1. Reducing the speed of the neutrons and their kinetic energy by letting the neutrons undergo elastic, and sometimes also inelastic scattering collisions with the nuclei of atoms of different materials, which constitute the shield.
2. Absorbing the neutrons. As the neutron energy becomes lower, the probability generally increases that the neutron will be absorbed by the nuclei of the shield constituents, while traversing a given distance in the shield. Most of the neutron absorption reactions and neutron inelastic scattering collisions give off photons, referred to as secondary photons.
3. Absorption of the secondary photons.

As discussed earlier, hydrogen is very effective for reducing kinetic energy of neutrons. This is due to the relatively large fraction of its energy that a neutron loses when it collides elastically with a proton—the nucleus of a hydrogen atom. Therefore, hydrogen or hydrogenous materials (containing a substantial amount of hydrogen) are one of the major constituents of many shields used for neutron radiation shielding.

Among the most effective hydrogenous materials are water, and organic polymers, such as plastics and rubbers, including hydrocarbon plastics (such as polyethylene, polypropylene, and polystyrene); natural and synthetic rubber (such as silicone rubber); and other plastics or resins containing atoms in addition to hydrogen and carbon (such as acrylic, polyester, polyurethanes and vinyl resins). One of the organic polymers most widely used for radiation shielding is polyethylene. Polyethylene contains 20% (atomic density) higher hydrogen than water when both are at room temperature. One drawback of pure polyethylene as a shielding material against neutrons, however, is that the slowed-down neutrons that are absorbed by the hydrogen of the polyethylene generate a significant source of 2.2 MeV photons (i.e., photons having energy of 2.2 MeV). The neutron shielding ability of polyethylene can be improved if a good neutron absorbing material is added to it to reduce the production probability of energetic secondary photons. A neutron-absorbing material in common use is boron. One of the most commonly used polyethylene-based commercial shielding materials against neutron sources is borated polyethylene. An example is the commercial shielding material R/X® (Reactor Experiments, Inc., Sunnyvale, California), which is 5% boron-polyethylene (Evans and Blue, 1996; Greenspan, 1995). Another drawback of polyethylene as a shielding material is that it has a relatively low density. In general, the higher the density of a material, the better its shielding ability against energetic photons for a given shield thickness.

A typical density of polyethylene is 0.92 g/cm^3, while a typical density of ordinary concrete is 2.3 g/cm^3, stainless steel is 7.8 g/cm^3, lead is 11.3 g/cm^3, and tungsten is up to 19.2 g/cm^3. As a result, the neutron shielding ability of pure polyethylene and even of borated polyethylene is not very good; the relatively energetic (2.2 MeV) secondary photons are not well attenuated by the polyethylene. More specifically, pure and borated polyethylene or other low-density hydrogenous materials have superb attenuation ability for neutrons having energies up to a few million electron volts, but poor neutron shielding ability and poor photon shielding ability (Greenspan, 1995).

Consequently, when shielding against neutron sources or combined neutron and photon sources, a two-layer shield (namely, one layer of polyethylene and the other layer of a higher density material, such as concrete with high water content, or lead) are usually used. It is also customary to use a single-layer shield made of a material that features a suitable combination of hydrogen atomic density and specific weight. A novel shielding material, like cemented plastics, consisting of hydrogenous materials loaded with powder of cementitious materials has been developed. The hydrogenous material is used to physically bond medium-density, environmentally benign materials to provide improved shielding against neutrons, and against a combined radiation of neutrons and photons (Greenspan, 1995).

In nuclear reactors, for example, a reflector is generally placed around the core to reduce neutron leakage from the core and massive shields are utilized to protect critical areas from ionizing radiation dangerous to personnel or damaging to sensitive equipment. Particularly in nuclear reactors designed for use in space applications, the size and mass is severely limited by mission requirements, launch vehicle limitations, and the cost of placing massive components into orbit and then assembling them into operating systems. For a small power reactor of about 100 kWe, the required shield mass is about 25% of the reactor mass on an unmanned station. On manned stations or for reactors of greater power, the relative mass of the shield is greater and may become the limiting component. Typical space reactor designs use radiation shields external to the reactor containment with the heat developed in the shield being dissipated by thermal radiation from the surface. The problems of neutron and gamma ray shielding are typically dealt with separately with the use of lithium hydride as the neutron shield and tungsten metal as the gamma ray shield. To meet the need for compact shield designs that minimize the overall system mass, an integral reactor vessel head and radiation shadow shield have been designed and fabricated. The radiation shield is comprised of a primary shield designed to fully satisfy the gamma ray attenuation requirement and substantially satisfy the neutron attenuation requirement, and a secondary shield designed to satisfy the residual neutron attenuation requirement. The primary shield consists of a porous-packed bed of small spheres of a metal hydride, such as zirconium hydride. The secondary shield is comprised of a similar bed of borohydride spheres, like lithium borohydride. Heat deposited in the two regions by neutron and gamma ray attenuation is recovered by the reactor coolant flow and contributes to useful power output of the reactor (Pettus, 1991).

Other typical nuclear shielding materials are concrete structures and lead, which have been used as the construction materials in buildings that house radioisotopes, x-rays, and nuclear reactors. Since concrete structures are liable to absorb water, they tend to absorb radioactively contaminated water when used in installations where radioactive rays are handled; the contaminated water absorbed is extremely difficult to remove by washing. Consequently, the concrete structures are normally coated with waterproof paint or overcoated with polymer concrete to reduce or prevent water absorption. Lead is a very effective shielding material but its use is limited due to environmental considerations, its high density, and soft quality, which require much care in construction. Alternatives are being explored to make shielding products either by molding and solidifying various heavy metal oxides and their ores with cement, an organic curable binder, or by compression molding heavy metal compounds, and then calcining the molded metal compounds at high temperatures. These sorts of shielding materials have been used as auxiliary radiation shielding materials by attaching or applying them to the surface of concrete structures. Moreover, a relatively new material has been made with an inorganic melt matrix and heavy metal compound, which possesses a high shielding effect. Iron oxide is most physically stable when used in the form of a melt matrix, and is very useful as a vehicle for dissolving an inorganic material containing sulphur. The ratio by weight between sulphur, iron oxide, and lead oxide is in the range of 100:60–110:120–360. The radiation shielding material is prepared by heating 100 parts by weight of sulphur and 60–110 parts by weight of iron oxide powder to thoroughly react them together to obtain a molten matrix, then adding 120–360 parts by weight of granular lead oxide to the molten matrix to disperse the lead oxide in the matrix, then molding the resulting dispersion into a predetermined shape, and then cooling to form the shape (Anayama, 1998).

Apparently, more and more new materials are being explored for nuclear radiation shielding applications. However, current shielding materials still play an important role. In general, hydrogenous materials are preferably used for low-energy nuclear radiation shielding. Water is used for high-energy nuclear radiation due to its low cost, ready means for removing heat, and good inelastic scattering properties. However, a large amount of water is needed because it is a poor absorber of gamma radiation. Iron is used for higher and lower energies due to the large change in neutron energy after collision but it has little effect on lower energy neutrons. Iron and steel is commonly selected based on structural, temperature, and economic considerations. Concrete is a good gamma attenuator as a general shield material. Concrete is strong, inexpensive, and adaptable to different types of construction.

11.5 SUMMARY

Exploration of lightweight shielding materials has concentrated on nonparasitic radiation shielding materials and multifunctional materials for EMI and space radiation shielding, as well as structural integrity for aircraft, space transportation vehicles, and space station systems. Innovative processing methods have been developed to produce various forms of lightweight shielding materials, such as hydrogenous materials, fiber reinforced composites, nanomaterials, foams, light alloys, and hybrid materials.

The design of novel shielding materials has involved hydrogenous composites because hydrogen is most effective in attenuating charged particle radiation. Since the strength and shielding effectiveness of fiber reinforced composite materials have surpassed those of aluminum, it is common to use these lightweight materials in the aerospace industry. Nanomaterials and nanocomposites also become potential alternative materials for lightning strike protection, EMI shielding, and heat dissipation in aerospace vehicles. Nanomaterials, imbedded in composites' structure would eliminate the need for complex and heavy protection shielding systems, and would help maximize the potential of composites for aerospace applications. In addition, lightweight, multifunctional foam structures and foam composites have been developed based on carbon foam, metal foam, polymer foam, and nanoporous structures, showing a great potential for aerospace shielding applications.

Proper radiation shielding materials and structural design for space power systems are crucial to protect humans from the hazardous radiation of outer space and inside reactor power systems in space exploration. Attenuation of the radiation levels in nuclear reactor-powered spacecraft can be achieved by proper design of radiation shielding components; separation distance between the SPRS and mission payload; and localized shielding provided within the mission payload. An alternative solution to the problem of shielding crew from particulate radiation is to use active electromagnetic shields. Practical types of shields include the magnetic shield, which diverts charged particles from the crew region; and the magnetic/electrostatic plasma shield, in which an electrostatic field shields the crew from positively charged particles, while a magnetic field confines electrons from the space plasma to provide charge neutrality. Advances in high-strength composite materials, high-temperature superconductors, numerical computational solutions to simulate and control particle transport in electromagnetic fields, and a technology base for construction and operation of large superconducting magnets make electromagnetic shielding a practical alternative for space exploration missions.

As the mixture of particles in the radiation field changes with added shielding, the optimization of shield material composition should adapt to the complicated mixture of particles, the nature of their biological response, and the details of their interaction with material constituents. Shielding optimization also requires an accurate understanding of the way in which the shield material interacts with the radiation field. Moreover, many of the materials must meet other mission requirements such as strength, thermal, and hardness. A large fraction of the protective properties come from materials chosen for other requirements and early entry of radiation constraints, and a multifunctional design process is required for design optimization.

Special materials for nuclear radiation shielding have been developed, explored, and used in many field applications, such as particle accelerators, nuclear reactors, radioactive biological and nuclear waste containment vessels, satellite hardware shielding, radiation shielding on aerospace vehicles and space station systems, and patient shielding for medical devices. The nuclear radiation shield is generally comprised of a primary shield designed to fully satisfy the gamma ray attenuation requirement and substantially satisfy the neutron attenuation requirement, and a secondary shield designed to satisfy the residual neutron attenuation requirement. Typical materials for nuclear radiation shielding and shield lamination structures include copper and its alloys, iron, cast iron, steel, lead, tungsten and its alloys, concrete, carbon, aluminum, polyethylene, and other new developed materials.

REFERENCES

Anayama, Y. 1998. Radiation shielding materials. US Patent 4,753,756.
Baranov, S., M. Bosman, I. Dawson, et al. 2005. *Estimation of radiation background, impact on detectors, activation and shielding optimization in ATLAS* (ATL-GEN-2005-001). http://atlas.web.cern.ch/Atlas/GROUPS/PHYSICS/RADIATION/RadiationTF_document.html
Bertino, M. F., J. F. Hund, J. Sosa, G. Zhang, C. Sotiriou-Leventis, N. Leventis, A. T. Tokuhiro, et al. 2004. High resolution patterning of silica aerogels. *Journal of Non-Crystalline Structures* 333: 108–110.
Bray, A. V., B. A. Muskopf, and M. L. Dingus. 2000. High density composite material. US Patent 6,048,379.
Cadwalader, J. A. 2001. Light weight radiation shield system. US Patent 6,310,355.
Cadwalader, J. A., and J. Q. Zheng. 2004. Radiation attenuation system. US Patent 6,674,087.
Caneva, C., F. Nanni, and M. S. Sarto. 1998. Electromagnetic and mechanical properties of a new composite material. *Proceedings of the International Symposium on Electromagnetic Compatibility*, Rome, Italy, September 14–18.
Chung, D. D. L. 1990. Exfoliated graphite fibers and associated method. US Patent 4,915,925.
Cock, F. H. 1993. *Magnetic shielding of interplanetary spacecraft against solar flare radiation* (NASA-CR-195539). http://engineering.dartmouth.edu/~Simon_G_Shepherd/research/Shielding/docs/cocks_93.pdf
Connolly, D. J., V. O. Heinen, P. R. Aron, J. Lazar, and R. R. Romanofsky. 1990. *Aerospace applications of high temperature super-conductivity*. 41st Congress of the International Astronautical Federation, paper IAF-90-054, Dresden, Germany.
Ebneth, H., L. Preis, H. Giesecke, and G. D. Wolf. 1984. Metallized carbon fibers and composite materials containing these fibers. US Patent 4,481,249.
Evans, J. F., and T. E. Blue. 1996. Shielding design of a treatment room for an accelerator-based epithermal neutron irradiation facility for BNCT. *Health Physics* 75(5): 692–699.
Gaier, J. R. 1992. Intercalated graphite fiber composites as EMI shields in aerospace structures. *IEEE Transactions on Electromagnetic Compatibility* 34(3): 351–356.
Gaier, J. R. 1993. Intercalated hybrid graphite fiber composite. US Patent 5,260,124.
GrafTech International. 2007. *GRAFOAM® carbon foam*. http://www.graftech.com/PRODUCTS/Carbon-Foam.aspx
Greenspan, E. 1995. Medium density hydrogenous materials for shielding against nuclear radiation. US Patent 541,633.
Grimes, C. A., C. Mungle, D. Kouzoudis, et al. 2000. The 500 MHz to 5.50 GHz complex permittivity spectra of single-wall carbon nanotube-loaded polymer composites. *Chemical Physics Letters* 319(5-6): 460–464.
Heinbockel, J. H., M. S. Clowdsley, and J. W. Wilson. 2000. *An improved neutron transport algorithm for space radiation* (NASA TP-2000-209865). Hampton, VA: NASA Langley Research Center.
Holloway, C. L., M. A. Mohamed, E. F. Kuester, and A. Dienstfrey. 2005. Reflection and transmission properties of a metafilm: with an application to a controllable surface composed of resonant particles. *IEEE Transactions on Electromagnetic Compatibility*, 47(4): 853–865.
Hou, T., J. Wilson, B. Jensen, et al. 2005. *E-beam processing of polymer matrix composites for multifunctional radiation shielding* (AIAA-2005-6603). Space 2005, Long Beach, California, August 30–31.

Huang, Y., N. Li, Y. Ma, F. Du, F. Li, X. He, X. Lin, et al. 2007. The influence of single-walled carbon nanotube structure on the electromagnetic interference shielding efficiency of its epoxy composites. *Carbon* 45(8): 1614–1621.

Hull, J. R. 1989. *High temperature super-conductors for space power transmission lines.* ASME Winter Annual Meeting Conf., San Francisco, California.

Jaworske, D. A., R. D. Vannucci, and R. Zinolabedini. 1987. Mechanical and electrical properties of graphite fiber-epoxy composites made from pristine and bromine intercalated fibers. *Journal of Composite Materials* 21: 580–592.

Landis, G. A. 1991. *Magnetic radiation shielding: An idea whose time has return?* Cleveland, OH: NASA Lewis Research Center. http://www.islandone.org/Settlements/Mag-Shield.html

Levy, R. H. 1961. Radiation shielding of space vehicles by means of superconducting coils. *ARS Journal* 31: 1568–1570.

Levy, R. H., and F. W. French. 1968. Plasma radiation shield: Concept and applications to space vehicles. *Journal of Spacecraft and Rockets* 5: 570–577.

Lite Tech. 2004. *Xenolite™ radiation production garments.* http://www.apc-cardiovascular.co.uk/pdf/xenolite_brochure_06.pdf

McClenahan, D. H., and J. A. Plumer. 1984. Graphite fiber reinforced laminate structure capable of withstanding lightning strikes. US Patent 4,448,838.

Murakami, M., K. Watanabe, and S. Yoshimura. 1998. Graphite intercalation compound film and method of preparing the same. US Patent 4,749,514.

Orrison, W. W., Jr. 1990. Radiation shield. US Patent 4,938,233.

Pettus, W. G. 1991. Shield for a nuclear reactor. US Patent 4,997,619.

Schimmerling W, J. W. Wilson, F. A. Cucinotta, and M. Y. Kim. 1999. Requirements for simulating space radiation with particle accelerators. In *Risk Evaluation of Cosmic-ray Exposure in Long-term Manned Space Missions*, edited by K. Fujitaka, H. Majima, K. Ando, H. Yasuda, and M. Suzuki , 1–16. Tokyo: Kodansha Scientific.

Touchstone Research Laboratory. 2004. *CFOAM.* http://www.trl.com/developments/cfoam.html

Vana, N., M. Hajek, T. Berger, M. Fugger, and P. Hofmann. 2006. Novel shielding materials for space and air travel. *Radiation Protection Dosimetry* 120(1-4): 405–409.

Vogel, F. L., and C. Zeller. 1983. Organic matrix composites reinforced with intercalated graphite. US Patent 4,414,142.

Wilson J. W., et al. 2000. *Improved spacecraft materials for radiation protection: Shield materials optimization and testing.* Microgravity Materials Conference, Huntsville, Alabama, June 6–7, 2000.

Wilson, J., E. Glaessgen, B. Jensen, D. Humes, E. Saether, and W. Kelliher. 2003. *Next generation shielding materials for near-earth infrastructure* (AIAA-2003-6258). Hampton, VA: NASA Langley Research Center. AIAA Space 2003 Conference and Exposition, Long Beach, California, September 23–25, 2003.

Wilson, J. W., F. A. Cucinotta, M. Y. Kim, and W. Schimmerling. 2001. Optimized shielding for space radiation protection. *Physica Medica* 17: 67–71.

Wilson, J. W., M. Kim, and W. Schimmerling. 1995. Issues in space radiation protection: Galactic cosmic rays. *Health Physics* 68: 50–58.

Wong, S. E., M. Sutherland, and F. M. Uhl. 2006. Materials processes of graphite nanostructured composites using ball milling. *Materials and Manufacturing Processes* 20: 159–166.

Zhang, G., A. Dass, A.-M. Rawashdeh, J. Thomas, J. A. Counsil, C. Sotiriou-Leventis, E. F. Fabrizio, et al. 2004. Isocyanate crosslinked silica aerogel monoliths: Preparation and characterization. *Journal of Non-Crystalline Solids* 350: 152–164.

12 Perspectives and Future Trends

12.1 INTRODUCTION

Along with the growth of the electronics industry, electronic systems are getting faster, smaller, lighter and more thermally challenging. At the same time, the use of electronic equipment has been widespread in communications, computations, automations, biomedical, space, and other purposes. All these effects have led to more stringent requirements for electromagnetic compatibility (EMC) and electromagnetic interference (EMI) shielding design of electronic systems. Mechanical and electrical design interfaces are challenging, especially for new product development, in which critical and early design decisions have to be made, either assuming EMC can be achieved with good electronic design to obviate the need for an EMI shield or anticipating the inclusion of an EMI shield (Arnold, 2003). Moreover, the EMI shielding design should be optimized to meet the EMC requirements while keeping costs as low as possible. This also has increased the demand for choosing the right EMI shielding materials and for developing new materials for EMI shielding applications.

This chapter reviews the design-in concept for EMC and optimal design of EMI shielding, choosing right EMI suppression materials/devices available in the market, typical applications of EMI shielding, and future trends and perspectives of new EMI shielding materials and technology.

12.2 EARLY DESIGN-IN FOR EMC AND OPTIMAL DESIGN OF EMI SHIELDING

The design-in concept has been addressed in electronic product design because it is an efficient approach for solving EMI problems early to meet increasing stringent EMC requirements. Figures 12.1 and 12.2 show the difference between the traditional approach and design-in concept for EMC design. Traditionally, both the electrical hardware designers and mechanical designers consider EMC design. Electrical designers tend to do whatever they want to do to optimize their design, leaving unattended EMI problems for mechanical designers to fix, as shown in Figure 12.1. After the product design, a prototype is built and tested for EMC. This approach very often results in EMI problems being identified at a point too late to design in EMC compliance. Expensive fixes on the existing design are usually the only options available. Design changes generally increase as the design moves from conceptual to detailed validation. In some cases, the EMC design needs to be started over with excessive cost and a negative impact on time to market (German, 2004).

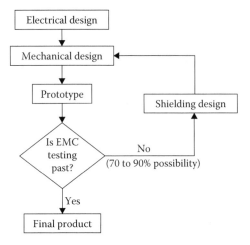

FIGURE 12.1 Traditional approach for EMC design.

With design-in as a concept, as shown in Figure 12.2, the electrical and mechanical designers usually discuss the probabilities of emissions and susceptibility issues early on, identify probable manufacturability solutions, and then design the product to support those shielding solutions. Thus, if the device shows that it does not need a shielding solution, then the shielding solution would not be purchased. On the other hand, if EMI shielding was required, it could indeed be implemented immediately without an impact on the product release. In other words, the design-in approach is used to assure that all downstream requirements are addressed in the early conceptual and design stages of electronic product development. Design-in requires inclusion of human interfaces, functional performance accomplishments, EMC, and econometric and aesthetic considerations, as well as environmental compliance (Arnold, 2003). It is a keen awareness of the potential problems, whereupon some level of design effort goes into the early product.

Consequently, addressing EMC late, as done with the traditional electronic design approach, is becoming less tenable as product complexity and densities increase and

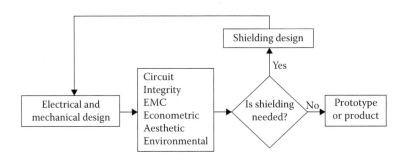

FIGURE 12.2 Design-in concept for EMC design.

design cycles continue to shrink. The traditional approach used for EMC design can break down at higher frequencies and can easily be misapplied. The resultant problem is that 70% to 90% of new designs fail first-time EMC testing, resulting in high late-stage redesign costs and often even higher lost sales costs if the product ship date is delayed. Therefore, the collaborative, design-in, conceptual analysis-based EMC design, including EMI shielding simulation, should be used early in the electronic product design process to identify and fix problems at a much lower cost (German, 2004).

Based on the design-in concept, an optimal EMI approach should be explored and implemented for electronic product development and manufacture, and include the following steps:

1. Get to know or be familiar with EMI regulations and EMC requirements. Various national and global standards limit emissions and/or immunity levels for specified products, and in some cases, require demonstrated resistance to EMI. An authorized organization or a legitimate test lab can help to determine what specifications need to be met, test for them, and help with applying for certification.
2. Distinguish and solve EMI problems in the early EMC design stage. It has become essential to include EMC design as an integrated part of the electronic product design to pass first EMC testing to ensure on-time delivery within budget. This can be achieved sometimes with a 3D solution of Maxwell's equations, which provide a mathematical representation of electromagnetic interactions. Moreover, analytical EMI simulation has become an effective tool for early EMC design. A typical EMI problem usually involves a wide range of frequency characterizations and complex geometries, such as an enclosure (which is largely relative to the profile), surface contact, slots, holes, cables, and input/output (I/O) outlets. These are important to EMC performance. EMC simulation can yield quite accurate results for EMI problems. For instance, EMC simulation is applicable for examining components and subsystems such as radiation profile versus frequency in heatsink grounding as well as assessing different grounding techniques, the impact of heatsink shape, and so forth. The shield effectiveness of different air vent sizes and shapes and metal thicknesses can also be compared. EMC simulation is also well-suited to EMC design and optimization at the system level to compute broadband shielding effectiveness; broadband radiated emissions; 3D far-field radiation patterns; cylindrical near-field radiated emissions to mimic a turntable-type measurement scenario; and current and E- and H-field distributions for visualizations that help locate EMC hot spots. Typical system-level EMC applications include: designing enclosures to ensure maximum shielding effectiveness; assessing the EMC ramifications of component location within an enclosure; computing cabling coupling both internal and external to the system; and examining the effects of radiation from the cables. EMC simulation also helps identify specific mechanisms for unwanted electromagnetic; transmissions through chassis and subsystems

such as cavity resonances; radiation through holes, slots, seams, vents and other chassis openings; conducted emissions through cables; coupling to and from heat sinks and other components; and unintentional waveguides inherent to optical components, displays, light-emitting diodes (LEDs), and other chassis-mounted components. Therefore, using simulation early in the design process makes it possible to investigate and predict key EMI phenomena, and hence optimizes electronic product design in terms of EMC requirements and shielding effectiveness before building a prototype. Modern simulation tools enable designers to evaluate more designs than it is practical to prototype, and optimize products from an EMC perspective to a level that was not possible in the past. It is also important to note that EMC design cannot be done in isolation because design changes made for EMC reasons frequently impact other design issues, such as thermal management. That is why it is significant that some EMC simulation tools enable designers to consider EMC in conjunction with other important design constraints in order to optimize overall system costs and performance (German, 2004).

3. Solving EMI problem starts on the printed circuit board (PCB). It is usually easier and less expensive to control EMI at its source. An EMC solution that works closer to the source is a PCB shield since certain board components are the source of most EMI emissions. When electrical and electronic circuits are in nonconductive enclosures, or when it is difficult or impossible to use EMI shielding gasketing, PCB shields provide the best option for EMI suppression. A properly designed and installed PCB shield can actually eliminate the entire loop area because the offending or affected circuit will be contained within the shield. For instance, board layout greatly affects EMI control. Noisy components can sometimes be moved away from sensitive areas. When the layout arrangement is not enough, some form of shielding is required to isolate the EMI source from other components and the local space. In fact, if PCB shields are considered early during the design, sections of the PCB can be used as part of the shield. PCB shields can be designed for maximum efficiency and minimum size. However, installing PCB shields late in the design can render them less effective. Moreover, it can increase costs because changes in the PCB layout are usually required. When shields are installed later, the PCB design will be less efficient and will take up more board space than necessary (Fenical, 2004). Generally, shielding on a PCB is typically some form of conductive cover mounted over one or more components. There has been a variety of board-level shielding products available, such as metal cans and conductive plastic covers with conductive paint, plating, or surface coating. In some applications, a shielding barrier separates board components to prevent crosstalk. Shielding cans, covers, and other barriers must be grounded, and this can be done using board traces and other nearby conductive surfaces. In addition, heat can be an issue when using PCB shields. Ventilation is usually an adequate way to address this problem. However, if

ventilation does not provide enough heat dissipation, PCB shields are available with integral heat sinks or other thermal dissipation systems (Quesnel, 2008).
4. Conductive enclosures are effective EMI shields. A conductive metal housing is inherently an effective EMI barrier. Plastic enclosures are readily made conductive with metallized paint, plating process, and other metal-coating processes. The best of these systems accommodates the deepest recesses in a plastic housing part where the smallest discontinuity provides a pathway for spurious emissions. An example is to enclose board with a folded box made of plastic film laminated to a layer of metal foil. The laminate shields are essentially an EMI enclosure set inside the product housing. Laminate shields must be grounded to the enclosure (Quesnel, 2008).
5. Conductive gaskets are effective and versatile seals for EMI shielding. There are many kinds of conductive EMI gaskets available and their performance in flexible, hand-carried systems is critical to a device's shielding system. EMI gaskets maintain a conductive pathway across enclosure parts. Most will provide an effective environmental seal, and they are frequently used to provide ground paths on boards and other components. Versatile gaskets, such as conductive elastomers, have cross-sections designed with finite element analysis (FEA) to optimize their mechanical performance. The result is lighter gaskets that compress so softly that they can reduce the needed number of enclosure fasteners. Elastomer gaskets can be both conductives and environmental seals, and can be added to the mounting edges of a board-mounted shielding can or cover. The gasket makes conductive contact with board traces. This design better accommodates flexible plastic-housed devices in which soldered metal cans may break loose (Quesnel, 2008).
6. Make the EMI shielding cost as low as possible. To determine the best pricing, find a working solution with the lowest installation cost. Many of shielding solutions are designed for automated or enhanced manual assembly. They can be seamlessly integrated into existing manufacturing processes (Queesnel, 2008).
7. Choose right EMI shielding materials to optimize EMI shielding design. There are a variety of EMI shielding materials and innovative technologies available or new materials in development. All these provide essential conditions for optimal EMI shielding design.

12.3 ADVANCED MATERIALS SELECTION FOR EMI SHIELDING

In choosing the most effective EMI shielding materials for products, such as telecommunications, computers, aircraft, and automotive and medical electronics, the selection can be generally clarified with several options: EMI shielding enclosure/cover/barrier, ventilation panels and windows, conductive gaskets, and other shielding products. Depending on the product's needs, these solutions provide varying EMI protection, intricacy of forms, and environmental protection.

12.3.1 Materials Selection for EMI Shielding Enclosure/Cover/Barrier

Current primary materials used for an EMI shielding enclosure/cover/barrier include the metal sheet and foil, metal-filled polymer, metal liners for housings, conductive coatings for the interior of rigid polymers, and conductive composite housings.

Several of the metals (soft magnetic/nonmagnetic) are available for an EMI shielding enclosure/cover/barrier, such as copper and its alloys, aluminium and its alloys, cold-rolled steels, nickel silver alloys, and soft magnetic Mualloys and Per-alloys. In these materials, low-cost copper, aluminium, or tinned copper foil tape backed with a highly conductive pressure-sensitive adhesive (PSA) have been used for a wide variety of commercial and military EMI shielding applications. The PSA contains a uniform dispersion of unique oxidation-resistant conductive particles that produce very low resistance through the tape.

EMI shielding through the use of metal cases for laptop computers or portable and handheld devices may not always be desired because of concerns about weight and aesthetics. The use of a metal shroud to line a plastic case provides better aesthetic and design for the outside of the housing but results in an increased assembly step and little weight minimization. Metal also lacks the ability to be formed into complex shapes, often taking up unnecessary room adjacent to the circuitry and assembled electrical components. Therefore, plastics are replacing metals as the material for electronic enclosures since plastics offer increased design flexibility and productivity with decreased cost. Because plastics are nonconductors and are transparent to electromagnetic waves, conductive coating has become an important material of providing EMI shielding to the electronic equipment enclosed in plastic enclosures. Three major types of conductive coatings have been used: electroless plating, conductive paints, and vacuum metallization.

Plating can provide the best shielding effectiveness, however, the major disadvantages of plating are its high cost and complex process cycles, and its application is limited to only certain polymer resins. Three general types of conductive metal-bearing paints are in general use. Silver paints have the best electrical properties, but they are extremely expensive. Nickel paints are used for relatively low attenuation applications and are limited by high resistance and poor stability. Passivated copper paints have moderate cost and lower resistivity, but also lack stability. All paint applications have difficulties with coating uniformly; blow back in tight areas and recesses depending on part complexity; and application problems, which can lead to flaking. Paints also fail electrostatic discharge (ESD) testing over 10 KVA (Gabower, 2003).

Conventional thin-film vacuum metallizing is not adequate to dissipate EMI. The vacuum metallized layers must have sufficient thickness to make the surfaces of the enclosure electrically conductive. The enclosure is formed in the shape and size necessary to house and shield the EMI emitter; for example, in the case of a personal computer, the enclosure may serve as a thin-walled case within the rigid outer housing of the computer. Alternatively, the enclosure may be formed to fit as an insert within a device's housing as a substitute for a metal insert shield, or the enclosure may be shaped and sized to contain only certain components that are emitters of, or susceptible to, EMI. Gangs of metallized enclosures may be devised with electrical isolation provided by gaps in the metallization layers. Different electronic devices will require varying degrees of attenuation or shielding effectiveness. The enclosure may be coated on all surfaces or selectively coated for certain applications. Many polymeric materials can be coated with vacuum metallization, such as thermoformable acrylonitrile butadiene styrene (ABS), polystyrenes, cellulose polymers, vinyl chloride polymers, polyamides, polycarbonates, polysulfones, and olefin polymers such as polyethylene, polypropylene, polyethylene terephthalate glycol (PTG), and methyl methacrylate acrylonitrile (Fajardo, 2000).

Use of these polymers with additional fillers, such as carbon black, graphite, and metal fibers, adds to the shielding effectiveness for absorbing more of the lower electromagnetic wavelengths. However, merely metal-filled resins from injection molding are usually not used for EMI shielding enclosures because they suffer from poor conductivity compared to metals. Moreover, the conductive polymer resin is very expensive and complex shape molding is difficult from flow and uniformity perspectives.

Another material process is the use of EMI shielding laminates to avoid the time consuming processes of conductive coating or vacuum metallization. Typically, a copper foil/film laminate provides from 20 to 60 dB of shielding depending on frequency, configuration, and installation.

EMI shielded vent panels and windows are typically used for open space or aperture shielding with a ventilation requirement. EMI shielded panels are usually honeycomb air vents with very high shielding performance, which are designed for wide range frequencies or specific high frequency environments and to meet related specifications. They can provide greater than 90 dB attenuation between 200 MHz to 10 GHz. EMI shielded windows can be produced in metal, glass, plastic, or a combination of them. Shielding is provided by knitted or woven wire mesh, sandwiched/laminated between the glass or plastic substrates, or by deposited conductive coatings. Shielding effectiveness is determined by the size of the wire screen openings, electrical contact between intersecting wires, and the materials and techniques employed to terminate the wire at the frame edge (Ramasamy, 1997).

12.3.2 EMI Gasketing Material Selection

An EMI gasket provides a good EMI/EMP (electromagnetic pulse) seal with effectively electrical conductive across the gasket-flange interface. The ideal gasketing surface is conductive, rigid, galvanic-compatible, and recessed to completely house the gasket. There are a variety of EMI gasketing materials available.

Metal spring gaskets, like CuBe (copper beryllium) finger stock, are the best choice for applications that need mechanical durability, such as frequent direct compression actuations or shearing forces. The metal gasket withstands lateral shearing forces as well as perpendicular forces, whereas other gasketing materials might be merely the best suited for perpendicular, direct-compression applications. Metal strips can be plated with many different metal types, offering broad galvanic compatibility with substrate metals. However, metal strips, which are available in many profiles, provide EMI shielding, contacting, or grounding, but do not supply an environmental seal.

CuBe provides the best shielding performance among the metal gaskets. In response to Environmental Protection Agency restrictions and ISO 14000 requirements regarding disposal of beryllium (Borgersen, 2000), other alloys such as CuNiSn, CuTi, phosphorus bronze, and stainless steel are gaining preference over CuBe in many applications. However, the use of these alternatives is progressing slowly because of their inferior performance and the expense of developing new tooling and designs for performance improvements.

For more demanding shielding applications, conductive silicone elastomers provide a wide range of attenuation capabilities, plus environmental and moisture sealing. Conductive elastomers are filled with filler materials from the highest conductivity silver to the lowest conductivity carbon. Carbon-filled elastomers with the lowest cost usually serve applications requiring low shielding levels. Other typical filler materials include nickel-coated graphite, silver-plated glass, silver-plated copper, and silver-plated aluminum. Nickel–coated graphite is the most popular for commercial applications because of its good combination of cost and performance attributes. Metal-filled elastomers can be extruded or molded into a wide range of form-fitting shapes and cross-sections. For gasketing very small cross-sections, a formed-in-place silicone-based silver-copper-filled or nickel-graphite-filled conductive elastomer is dispensed robotically. The robot can deposit the elastomer on a surface as narrow as 0.6 to 0.8 mm (Borgersen, 2000).

Wire mesh/strip EMI gaskets provide effective EMI shielding at the joints of seams of electronic enclosures. Any metal that can be produced in wire form can be fabricated into EMI shielding strips. They can be produced in different cross-sections, such as rectangular, round with fin, and double core. Commonly used materials include tin-plated phosphor bronze, tin-coated copper-clad steel (Sn/Cu/Fe), brass, silver-plated brass, Monel (nickel copper alloys), aluminum, and copper beryllium. Monel is used most frequently due to its good aging properties, excellent tensile strength, and spring performance. Sn/Cu/Fe offers the best EMI shielding performance, especially in H-fields, but has limited corrosion resistance. Mesh strips with elastomer cores consist of two layers of wire mesh knitted around a rectangular or round core of neoprene, silicon sponges, or a hollow solid silicone core. Knitted wire mesh provides the electrical conductivity for EMI/EMP shielding, and the core provides excellent compatibility with a high degree of corrosion resistance. A mesh strip gasket is good for applications requiring a weather seal in addition to EMI/EMP shielding. This type of gasketing is also useful in applications requiring shielding only where the adhesive backing on the elastomer would provide a convenient means of mounting (Ramasamy, 1997).

Conductive fabric-over-foam can be the lowest cost option, mainly suitable for applications where no environmental seal, complex profile, or demanding mechanical durability is needed. The covering can consist of nylon thread coated with conductive metals woven into a fabric and wrapped over soft urethane foam. Alternatively, a woven fabric may be metallized with nickel–copper or another metal coating, then wrapped around the foam core. Flexible, conformable fabric-over-foam maintains close contact with surfaces with minimal compression for low-closure-force applications. It provides snug contact over irregular surfaces and around bends and corners, making it good for electronic enclosures such as doors and access panels. The manufacturing process limits it to relatively simple cross-sectional profiles such as squares, rectangles, and D shapes. When a more complex profile is necessary, fabric-over-foam may not be the best solution (Borgersen, 2000).

12.3.3 OTHER SHIELDING MATERIALS SELECTION

Rigid epoxy adhesives are filled with silver or silver-plated particles into rigid structural bonds. They are employed to bond EMI vents, windows, or mesh gaskets to shield permanent seams and to fill cracks and gaps (Lau, 2003). Flexible conductive silicones are filled with silver-plated copper and cured into gaskets like scads when applied to properly primed metal surfaces. They are used to bond conductive silicone gaskets in place, and must be used in thin bond lines. Conductive caulks and sealants are electrically conductive silicone, epoxy, and acrylic resins. Their purpose is to provide effective EMI/EMP shielding and scaling, filling larger gaps than adhesives (Ramasamy, 1997).

Electrically conductive, transparent-coated, EMI-shielded optical display panels are produced by depositing a very thin metal film coating directly onto the surface of various optical substrate materials, like most transparent plastic/acrylic and glass sheets, to provide high EMI shielding performance coupled with good light transmission properties. (Tecknit, 2008).

Cable shielding and absorbing materials are used to redirect the electromagnetic energy, which normally would be coupled into such wiring or cables. Zipper cable shielding is a convenient, inexpensive method of EMI/EMP shielding cable harnessing for computers, communication equipment, and other sensitive electronic systems (Ramasamy, 1997). This cable shielding can be a knitted wire mesh with a protective cover of heavy-duty vinyl, a combination that provides flexibility and durability as well as shielding. EMI absorbers for cable shielding are generally discrete ferrite components designed to reduce electromagnetic energy emitted from the cable without affecting the data transmitted through the cables. These ferrite components and EMI absorbers have been expansively used in EMI control for achieving EMC. Conductive shrinkable tubing is a heat-shrinkable polyorefen material that provides effective EMI shielding for cables, connectors, and cable/connectors' terminations. It also offers significant weight savings over conventional metal braid shielding and can be applied easily with standard shrinkable tubing heating devices.

Low-pass EMI filters are used to attenuate conducted noise currents in a variety of applications, including shielded enclosures, computer racks, and other electrical/electronics subsystems. Proper attention to the filter selection can save the design time and cost. These passive low-pass filters are generally manufactured using three basic elements: inductors, capacitors, and resistors. At a given frequency, the insertion of a filter is determined as the ratio of voltages appearing across the line immediately beyond the point of insertion, before and after insertion. Low-pass EMI filters are needed to control conducted EMI on alternating current/direct current (AC/DC) power lines, data transmission, signal and control lines of systems/subsystems with electronics, and digital circuits (Ramasamy, 1997).

In summary, EMI shielding materials play an important role in controlling or mitigating EMI in equipment and systems. EMC can be achieved by predicting and analyzing the problems early in the design stage, and adapting proper hardening techniques in the system to avoid program delay and additional cost of fixing EMI after the system is designed. Some of the suppressing materials, like ferrites, lossy lines, zipper tubing, and filters can be used as quick fix techniques at any stage of system design and development. New EMI shielding products are expected with new materials continuously in development. The designer should update the EMI databank from time to time to notice the changes in product and application technology of EMI shielding and suppression materials, as more adequate standards, specifications, and guidelines are being developed by a variety of institutions (Borgersen, 2000).

12.4 FUTURE TRENDS AND APPLICATIONS

With the ongoing electrical and electronic boom, the wireless surge, and increasing power levels and high frequencies, more advanced EMI shielding materials with increased characteristics and multifunctional performance are demanded. A series of specific factors, such as increasing frequencies and different frequency bands, miniaturization, heat, cost sensitivity, aesthetic consideration, and environmental issues have to be weighed to improve EMI shielding capabilities for different application areas.

For instance, automotive, aircraft, and other transportation electronics are affected by harsh EMI environments, high reliability, and somewhat cost sensitivity. EMI can be generated from power transients, radio frequency interference, electrostatic discharge, and power line electric and magnetic fields. Since transportation vehicles can go almost anywhere, worst-case situations must be assumed. Moreover, vehicular electronics must be designed for extremely high reliability at the lowest possible cost. All these issues have to be overcome through optimal EMC design and right EMI shielding materials selection.

Therefore, EMI shielding materials have been facing great challenges to meet the increasing demands from electronic industries, linking to telecommunications, information technology, medical electronics, and automotive industry, as well as aerospace and defense industry. At the same time, growth in these industries will have a positive impact on the development and innovation of EMI shielding materials and component design techniques.

12.4.1 EMI Shielding Design Techniques

The early design-in concept with the analytical simulation method is the future direction for optimal EMI shielding design. With this, preferred shielding models and design techniques can be developed for various specific EMI problems, depending upon the particular system, its electronic design, and the type of interference source. Some techniques are only applicable for certain types of EMI threats or fields, and many times, more than one technique must be used to provide the required amount of EMI protection. The techniques can be implemented at one or more of the design levels, such as the integrated circuit (IC), PCB, module or enclosure, interconnect, and software.

Perspectives and Future Trends 303

The automotive electronic design is a good example to illustrate these techniques. EMI design techniques for power transients should provide primary transient protection on all module input power lines, plus secondary protection, such as filtering, at the circuit level. The load dump transient is usually the biggest concern. Designing noise tolerant software is also very effective in controlling susceptibility to powerline transients. The design techniques for radio frequency immunity are to keep the unwanted energy from reaching vulnerable circuits. This requires high frequency filtering on both power and I/O cables, which act as antennas, plus careful circuit layout and circuit decoupling. Cable and module shielding are also effective, but are not popular in vehicular designs due to costs. The design techniques for ESD electrostatic discharge are to limit damage by transient suppression or high frequency filtering on I/O and power lines, and to limit upsets by local filtering and decoupling, careful circuit layouts, and perhaps even shielding. Many of the design techniques for EMI emissions and immunity work equally well for the indirect ESD effects due to the transient electromagnetic fields. The design techniques for power line fields are usually instrumentation oriented, and include local shielding and filtering of the most critical circuits. The design techniques necessary for EMI emissions and immunity also minimize this threat. This is normally not a serious threat. The design techniques for radiated emissions are to suppress the emissions at the source by careful circuit layout, filtering, and grounding, or to confine the emissions by shielding. For automotive designs, the emphasis has usually been on suppression and careful circuit layout, since shielding is costly and difficult for most high-volume automotive products. Nevertheless, shielding is seeing increasing use in vehicular applications (Banyai and Gerke, 1996). In fact, effective EMI shielding solutions for vehicle construction steadily gains attention, as vehicle manufacturers are incorporating more electronics into vehicles, ranging from engine control units to adaptive cruise control to driver information and entertainment systems.

12.4.2 INTEGRATED CIRCUIT-LEVEL SHIELDING

The integrated circuit (IC) is generally the original EMI source of most electronic devices. It is common sense to suppress EMI at the source; therefore, many chip manufacturers have attempted to design their components by incorporating the EMI fixes at the IC level. However, it is not very practical to effectively use this approach for electromagnetically reducing chips. This is mainly because of constraints of chip speed, cost, performance, and wide temperature range. Lower EMI usually means slower speeds, which in turn means lower performance. Yet the trend in future ICs is toward higher performance devices with increasing clock speeds and edge rates. Nevertheless, some progress has been made to control EMI at the IC level. For example, some ICs have been designed with the capability to turn off certain high-speed control lines when not needed. Clock drivers have been redesigned to incorporate high frequency power, grounding schemes, and internal decoupling capacitors on the silicon chip to improve the EMI aspect and reduce its radiation at the IC level (Banyai and Gerke, 1996).

On the other hand, EMI is disruptive to the proper operation of IC components located near the source of the EMI. To achieve reduced EMI susceptibility, an IC package has been designed with an internal EMI shield that is incorporated into the package itself. The flip-chip multichip module (MCM) tile is metallized around the external surfaces of the substrate. The PCB into which the MCM tile is recessed (referred to as the MCM PCB), is provided with a solder wall corresponding to the peripheral shape of the MCM tile, and the MCM tile and MCM PCB are joined with the solder wall completely surrounding the IC chip. The solder wall on the MCM PCB is connected to a ground plane that surrounds the cavity into which the IC device(s) is recessed. In this manner, the MCM cavity is completely isolated from stray EMI (Low et al., 2001).

At the same time, other EMI shielding approaches and new materials have been developed to solve EMI problems at the IC level. Faraday cages in the form of wire grids have been built around the IC chip to shield the interconnections from stray fields. Various small-sized metal can shields are also provided to encapsulate IC chips and leads (Anthony and Anthony, 2006). Similar to regular board-level shields, these cans are usually fabricated using stamped metals, typically copper or aluminum. Apparently, they add cost, size, and weight to the IC package. In an effort to reduce the size and weight of the can, portions of the shielding can are perforated. This is also good for air ventilation. Although the bulk weight of the shielding can be reduced, this does not decrease the cost. Moreover, it reduces the effectiveness of the EMI shield.

Depending on the EMI frequency, absorbers can be embedded into the ICs to reduce the electromagnetical field. Absorber materials are available as elastomer and broadband absorbers. In this area, because metals and their composites have limited mechanical flexibility, heavy weight, corrosion, and difficulty of tuning the shielding efficiency, conductive polymers and polymer magnets are promising materials for absorbers because of their relatively high conductivity and dielectric constant, and ease of control of these properties through chemical processing (Lee et al., 1999). Because of their inherent low densities and high molecular masses of molecule/polymer-based magnets, their bulk applications relying on high magnetic moments either on a mass or volume basis are unlikely. In contrast, the use of polymer magnets for inductor absorbers, for example, as the core material in transformers that guide magnetic fields and for magnetic shielding of low-frequency magnetic fields, are feasible in the IC packages. The high permeability-to-mass ratio makes this class of soft (low coercive fields) magnetic materials potentially attractive for lightweight transformers, generators, and/or motors as well as DC and low frequency shielding applications (Joo et al., 1998). The high shielding efficiency (SE) of highly conducting doped polyaniline, polypyrrole, and polyacetylene has been achieved (Lee et al., 1999). Very thin conductive polymer samples have high and weakly temperature dependent shielding efficiencies. The easy tuning of intrinsic properties by chemical processing suggests that such polymers, especially polyaniline, are good candidates for low frequency shielding applications. Also, the potential for room-temperature $V(TCNE)_x$ y(solvent) polymeric magnet in the area of EMI shielding has been developed (Joo et al., 1995). [$V(TCNE)_x$ y(solvent) is magnetic at room temperature, its extrapolated T_c being about 400K (125°C) above its decomposition temperature of 350K.] The $V(TCNE)_x$ y(solvent) magnet also has the attributes of low power loss and flexible low-temperature processing (Kohlman et al., 1996).

Perspectives and Future Trends   305

Although all these efforts are being made based on incorporating IC and EMI shielding design to control EMI at the chip level, the real battle against EMI must still be waged at the PCB and module stage in the near future.

12.4.3 Printed Circuit Board (PCB)-Level Shielding

If well done, PCB-level shielding can be the most cost-efficient means of resolving EMI issues. The approaches involve proper shield selection and optimal circuit design including partitioning, board stack up, and use of isolating and routing, as well as high-frequency grounding of the board and filtering techniques. As mentioned previously, if these techniques are designed-in at the initial stage, minimal impact to schedule and cost can be accomplished.

In fact, many solutions at this level are inexpensive or even free, as long as some general rules are followed (Banyai and Gerke, 1996):

1. Plan for EMI right from the start. Identify the most critical circuits and decide what to do about them. Proper technique starts with component placement. Critical circuits, such as clock circuits, clock drivers, and their functions should be grouped together to make the shortest trace lengths between components.
2. Consider multilayer boards, and allow for plenty of filtering and decoupling in the initial designs. Use enough ground planes, design high-speed traces as transmission lines, and employ plenty of filtering and decoupling components, or place jumpers or zero-ohm resistors to hold their place, then add the real components only if testing shows a need.
3. Test early and often. It is had better to identity potential problems early in the design, when the alternates are inexpensive and plentiful. Doing early board prototype testing can produce some useful insight to potential problem areas. Board areas of high radiation and measuring of interconnect cable noise currents are indicators of potential system radiation sources. Near-field probes and current clamps can find potential trouble areas. Discovering these problem areas later in the design cycle can produce costly schedule delays, as well as costs due to scrapped boards and/or components. Additional expensive EMI solutions may be needed, including various EMI shielding approaches, as discussed in Chapter 7.

The following subsections address some future trends for PCB-level shielding and its applications.

12.4.3.1 Lightweight Shields for PCB-Level Shielding

The design and realization of lightweight shields to reduce EMI with sensitive electronic apparatus and systems is a challenge for applications such as portable electronic devices, and equipment to be used or installed onboard aircraft, satellites, and so forth. There have been some attempts to improve the shielding performances of light materials, such as plastics and composites, by inserting conducting meshes in the substrate, using conducting additives and fillers prior to injection molding, and using conducting coatings. Among these techniques, the use of conducting coatings are most promising (Sarto et al., 2001).

For example, an EMI shield is made by a polycarbonate foil having the thickness of 125 μm, which is coated by a 50 nm thick layer of nickel and a 5 μm thick layer of tin. The polycarbonate substrate is chosen due to its good thermoforming properties. Before plating the tin on the polycarbonate foil, a thin layer of nickel, having the thickness of about 50 nm, is deposited first on the polycarbonate substrate by DC magnetron sputtering. The magnetron sputtering technique is a vacuum deposition technique that has the advantage, with respect to electroplating, of allowing the realization of conductive coatings on both metallic and nonmetallic substrates. It also allows the realization of very compact coatings and permits a fine control of the thickness growth of the film. The choice of nickel was advanced by its chemical properties: in particular, it is not subjected to oxidation and it does not react with tin or polycarbonate. Moreover, it has good adhesion properties on plastic substrates. Finally, the metallized foil is thermoformed at 200°C to form lightweight shielding covers and caps of whatever shape (Sarto et al., 2001).

Similar kinds of PCB shields have been made by a thermoformed polycarbonate foil, which is coated by a thin layer of tin or other metals before or after thermoforming. Due to the great formability of polycarbonate and to the good ductility of the metal coating, the shielding foil can be used to make covers and caps with different shapes and features, to be mounted on PCBs in order to reduce EMI and radiated emissions. These kinds of shields can provide shielding effectiveness of around 60 to 80 dB in the frequency range from 30 MHz up to 2.0 GHz. These techniques have a great potential used to form low cost, lightweight shields.

12.4.3.2 Mold-in-Place Combination Gaskets for Multicompartment Shields

Mold-in-place (MIP) technology is used to create small, complex, electrically conductive elastomer gaskets with fine details, and can be used to mold conductive elastomer material with a small rib pattern onto a variety of metal and plastic substrates to create a multicompartment shielded enclosure. During installation, the shield is sandwiched between one side of the printed circuit board and the housing. As the housing is assembled, the MIP ribs are compressed, providing shielding (Fenical, 2007).

MIP gaskets can be produced utilizing injection, transfer molding, or compression molding processes to create a gasket profile that exactly meets specific application needs. EMI shielding and environmental sealing performances are improved in a number of ways when conductive gaskets are molded directly onto a substrate instead of being mechanically attached. Molding dies can be produced in an infinite number of shapes to improve mechanical and/or EMI performance, depending upon the installation requirements.

Access to components of the MIP gasket is accomplished by simple disassembly of the housing. Therefore, MIP can replace soldering multiple PCB shields with a single-piece approach, by molding the MIP gasket onto a metal PCB shield where perimeter surface requires shielding. MIP also allows access for testing, calibration, or repair, and it is ideal for handheld devices such as cellular phones, glucose monitors, inventory scanners, wireless entry transmitters, military night-viewing devices, and countless other devices where space is at a premium. The metal substrate acts as a shielded enclosure, allowing use as a perimeter and intercavity shield.

In addition, spacer gaskets feature a thin plastic retainer frame onto which a conductive elastomer is molded. The elastomer can be located inside or outside the retainer frame, as well as on its top and bottom surfaces. Customized cross-sections and spacer shapes allow for very low closure force requirements and a perfect fit in any design. Spacer gaskets provide a low impedance path between peripheral ground traces on PCBs and components. Using tools such as FEA, the gasket profiles are designed with deflection and compression ranges that provide maximum performance. A typical conductive elastomer spacer gasket provides shielding effectiveness of 100 dB at 500 MHz and 80 dB at 10 GHz (Chomerics, 2000). Some typical applications for EMI spacer gaskets include PC (personal computer) cards, GPS (global positioning systems), PCS (personal communications services) handsets, and PDAs (personal digital assistants). Spacer gaskets also offer low cost, minimal material waste, and easy installation in small enclosures. Field reliability is increased because damaged gaskets or covers become one-part replacements reducing the risk of error, making spacer gaskets ideal for high-volume applications (Fenical, 2007).

In general, MIP applications include board-level shielding and spacer gaskets utilized in wireless handsets, PDAs, enclosures, navigation systems, satellite radio receivers, and so forth. MIP is available in nonconductive compounds for environmental sealing requirements in commercial or military applications.

12.4.3.3 Rotary Form-in-Place Gaskets

Rotary head dispensing technology is a modified process of form-in-place (FIP) gaskets. The traditional method allows only for application of a round bead resulting in a final D shape. The rotary head technology provides a fourth axis dispensing capability that allows gaskets with height greater than width to be dispensed in one pass. Previously, two or three passes might have been required to create this kind of aspect ratio. Also, a gasket design under compression can be deflected to one side providing a greater surface contact area for enhanced shielding effectiveness, along with reduced compression force (Fenical, 2007).

The process uses a FIP dispenser with a head that rotates around a z axis and new types of dispensing needles designed to produce the desired finished gasket shape. The programming of the gasket trajectory, speeds, and pressures are very similar to those of traditional FIP compounds, and is capable of producing small high-aspect ratio beads that are difficult, if not impossible, with traditional FIP dispensing.

Rotary FIP dispensing also reduces labor and material usage. It is ideal in applications, such as cell phones, handheld devices, and electromedical devices with miniature electrical housings, that have very small EMI gasket footprints, thin cavity wall construction, and intricate multicavity designs. The material can also be dispensed on metal or plastic, and air cures at normal temperatures and humidity. In addition to the standard semicircular gasket profile, a variety of profile shapes can be made (Fenical, 2007).

12.4.4 EMI Shielding Modules or Enclosures

EMI module shielding can be achieved by sealing all seams and filtering all penetrations. This approach has been successfully used to provide cost effective EMI control in a highly cost-sensitive industry. There is an increasing trend toward similar shielded modules in electronic industries. EMI shielding enclosures may be more cost effective to shield and filter a module than to add a lot of components on the circuit board. This is particularly true if the module must be enclosed for environmental protection. Presently, conductive plastics and paint, as well as plating, are the typical approaches to shielding the actual enclosure material. Most EMI shielding materials can provide very high levels of protection. For example, nickel paint on plastic typically provides 60 dB or higher protection, and vacuum deposition or electroless plating on plastic typically provides 80 dB of protection. A steel or aluminum enclosure can provide well over 120 dB of high frequency shielding. These levels of shielding effectiveness are easily attainable if the leaks from seams or penetrations are sealed (Banyai and Gerke, 1996).

Research and development of novel shielding materials certainly may aid in reducing costs and weight, increasing strength, and improving shielding effectiveness of a shielded enclosure. Electrically conductive polymers with nanocomposites, aluminia-based ceramics, and electrically conducting polyphenylenes are some examples of novel materials. Use of board-level shields, conductive fabrics, and thermoformed conductively laminated plastics are some examples of cost-effective alternatives. With current shielding products, it is not imperative any more to shield the module with guaranteed shielding, as any leakages from openings such as seams and apertures could affect the final shielding effectiveness. This is why the use of conductive gaskets or vent panels becomes important for sealing the apertures and seams. The proper selection of what type of gasket, for instance, is dependent upon the environment and the enclosure design. If no environmental sealing is required, contact spring strips and wire mesh gaskets may be used. In case of a required ingress protection, depending on the classification of applications, a conductive fabric gasket or an electrical conductive silicone in various application forms, such as extruded, dispensed, mold-in-place, or form-in-place gaskets may be considered.

The following addresses some future trends in EMI shielding modules and enclosures.

12.4.4.1 Conductive Foam with Z-Axis Conductivity

Conductive foam (CF) with x-, y- and z-axis conductivity offers an innovative approach to module or enclosure shielding and grounding. CF consists of nickel-over-copper metallized open-cell plastic foam. It provides low compression load deflection, making CF an ideal EMI shielding material in vertical compression and low-cross-shear applications (Fenical, 2007).

For perimeter EMI shielding applications with low cross-shear, such as I/O panels, sliding applications that are seldom opened or applications that are only under shear during the initial assembly, the advantage of improved z-axis conductivity is excellent shielding effectiveness of more than 90 dB across a wide range of frequencies. It also provides a wide compression range up to 60% of its original uncompressed thickness.

This material provides extensive gasket design flexibility from standard I/O configurations, including D type subs, USB (universal serial bus) ports, IEEE 1394 digital interfaces, SCSIs (small computer system interfaces), and RJ-45 connectors, to die-cut shapes and rectangular strips.

12.4.4.2 Dent-Resistant Vent Panels

High-performance, dent-resistant vent panels offer an alternative to conventional metal honeycomb vent panels used to provide increased airflow along with EMI protection. A nickel-copper-plated plastic honeycomb material with a conductive foam band around the perimeter can eliminate the need for costly frames, as well as the associated attachment hardware and much of the installation labor. Vent panels with a frameless design also allow for greater airflow by as much as 10% to 20% per unit surface area of vent panel than equivalent framed vent panels. This increased percentage can be critical for cooling many of the densely packed enclosures currently being designed. Typical applications for the vent panels include communications cabinets, aeronautics and general aerospace cooling applications, medical equipment, manufacturing systems, and military applications (Fenical, 2007).

12.4.4.3 Nanocomposite Shielding Materials

Lightweight EMI shielding is needed to protect the workspace and environment from radiation coming from computers and telecommunication equipment, as well as for protection for sensitive circuits. Compared to conventional metal-based EMI shielding materials, electrically conducting polymer composites have gained popularity because of their light weight, resistance to corrosion, flexibility, and processing advantages. Moreover, with the shift toward composite materials for aircraft, spacecraft, and satellite structures for reduced weight and increased strength, there is a need for additional EMI shielding materials since carbon fiber composites are much less conductive than metals. The EMI shielding efficiency of a composite material depends on many factors, including the filler's intrinsic conductivity, permittivty, and aspect ratio. The small diameter, high aspect ratio, high conductivity, and mechanical strength of carbon nanotubes (CNTs), including single-wall nanotubes (SWNTs) and multiwall nanotubes (MWNTs), make them an excellent option for creating conductive composites for high-performance EMI shielding materials at low filling. On the current stage, the highest EMI shielding effectiveness for SWNT-epoxy composites was with 15 wt% SWNTs as the additive. The shielding effectiveness reached ~49 dB at 10 MHz, and around 15 to 20 dB in the range of 500 MHz to 1.5 GHz, indicating that SWNT-polymer composites could be used as an effective and lightweight EMI shielding material (Li et al., 2006).

A suitable method for applying a polymer nanocomposite solution has been developed to serve as EMI shielding material that could be integrated directly onto the aircraft structures. The material is the mixture of carbon nanofiber and polymer with an effective mixing process to preserve the dispersion of the particles. The best method to apply it is to spray the mixed solution directly onto the structure using a setup very similar to an automotive paint sprayer. With the right viscosity and concentration, it provides an even coating while preserving the uniform mixture of the nanoparticles required for sufficient shielding. The carbon nanofiber can be coated with silver to provide better shielding but weighs slightly more (Wang and Alexander, 2003). These materials can be used to protect ground-based structures, aircraft, spacecraft, and satellites from internal and external EMI effects. The use of polymer nanocomposites as an EMI shielding material in place of currently used aluminum systems will reduce weight and eliminate corrosion, which resulted from the hardware necessary to keep the aluminum in place. This method can be quickly and easily adapted to systems that are currently unshielded as a retrofit technology.

12.4.4.4 Ultrasoft Sculpted Fabric-over-Foam

The continuing trend of building enclosures from plastic materials to enhance design, ergonomics, weight, and costs, as well as stiffening environmental requirements, has led to an evolution in fabric-over-foam (FOF) products. Bromine-free, ultrasoft sculpted foam is a novel UL 94 V0-rated FOF shielding option. It has superior compression set values relative to traditional FOF. In many cases, the sculpted foam product has 50% better resistance to compression set, as well as improved compression force characteristics. When the gasket is compressed greater than the recommended 50% deflection, compression force is reduced by about half relative to conventional FOF. This is extremely important for use in plastic enclosures where higher compression forces cannot be tolerated. Many conventional gaskets can exhibit enough compression to strip the threaded fasteners or plastic clips from the plastic housings. This also allows for designing with thinner and lighter plastics (Fenical, 2007).

Ultrasoft sculpted FOF shielding material offers unmatched compression set performance in combination with very low compression force, and promotes better shielding performance over the life of the gasket by reducing the internal stress of the elastomeric properties of the gaskets. Their shielding effectiveness values are comparable to conventional FOF products because the conductive outer materials are the same. In combination with much lower compression forces, this product further reduces the potential for distortion of the enclosure material, thus avoiding the formation of other potential leakage points. The ability to sculpt the foam allows the EMI gasket shape to better match the desired enclosure requirement rather than the other way around (Fenical, 2007).

12.4.5 INTERCONNECTING LEVEL SHIELDING

Interconnecting cables are probably the largest source of EMI radiation after the circuit board. Use of filter connectors can solve this problem, but these are expensive. The same goes for clamp-on ferrites, which are expensive and add on weight. If filtering is required, the best approach is to add filter components at the board I/O interface. This is also a good place for transient protectors. Fiber optics may change this in the future, as they can reduce and even eliminate many cable-related EMI problems. However, fiber optics is still too expensive for most electronic applications. Even when the fiber optical cable becomes practical, the connected power wiring will still need EMI protection (Banyai and Gerke, 1996).

12.4.6 EMI Control or Immunity by Software

Software can be very effective for EMI susceptibility although normally not effective against emissions; however, there have been cases where it did make a difference. Like fault tolerance, noise tolerance can be built into the software to catch the errors before they upset the system, and then gracefully recover. It does not take much to provide this protection, and usually just a few lines of code can work wonders. For example, to detect memory errors and to tell when a memory state has changed, add checksums to blocks of data in memory. To detect program errors, add tokens to modules of code. Save the token upon entering a module, and then check it upon leaving the module. If they are not the same, an error has occurred, since the module was entered illegally. To detect I/O errors, do type and range checking on the data. A watchdog can prevent becoming lost in an endless software loop (William, 1992). This approach consists of requiring the processor to execute a specific operation regularly. It is often a separate device, although many microcontrollers incorporate an internal watchdog. If the watchdog is not reinitiated in a predetermined time, it resets the entire system, bringing it back to a known state (Banyai and Gerke, 1996).

The degree of sophistication incorporated into software to deal with errors reflects the importance attached to the correct operation of the system under design. However, some form of automatic or manual reset following errors should be incorporated in all but the most basic systems (Christopoulos, 1995).

Many of these approaches have already been used in vehicular applications with very good results. Although they are usually incorporated for safety reasons, they are also effective tools in the battle against EMI (Yarkoni and Wharton, 1979).

12.5 SUMMARY

Future electronics designs will require much more involvement of EMI shielding solutions than ever before, with increasing requirements toward miniaturization, high speeds, greater heat transfer, reduced costs, and light weight.

New advances in EMI shielding materials, mounting options, and physical designs are being made to keep up with the ever-changing electronics industries. Advanced EMI shielding designs and materials have been attempted and used in ICs, PCBs, modules and enclosures, interconnecting level shields, and EMI control by software. For example, foam-based products, such as ultrasoft sculpted FOF and offshoot CF, have been developed for lower closure force and improved high-frequency shielding. Moreover, additional innovative products are being produced on a regular basis.

Miniaturization has led to the need for many of these new developments in EMI shielding. Ever increasing frequencies also have contributed to the need for change. Electrically conductive elastomers that are produced in extremely small profiles or can be deposited onto small substrates are becoming the answer to the miniaturization dilemma.

The need to reduce costs and improve performance while increasing complexity has led to the development of versatile and higher performance EMI shielding materials and products, such as innovative absorbing material, conductive polymer composites, and nanocomposites. MIP and rotary FIP provide small, intricate gaskets with minimal footprint and excellent shielding. Frameless vent panels greatly improve cooling capabilities without a reduction in shielding while offering a more robust cost-effective solution.

Multifunctional EMI shielding materials and new technologies will be present in more products that need future applications including handheld and wireless products, computers, medical electronic equipment, processor-controlled manufacturing systems, military, aerospace, and automotive systems, and all other applications with higher frequencies.

REFERENCES

Anthony, A. A., and W. M. Anthony. 2006. Universal energy conditioning interposer with circuit architecture. US Patent 7,110,227.

Arnold, R. R. 2003. Electronic product trends drive new EMI/RFI shielding solutions. *Interference Technology*. http://www.emcsoltech.com/pdf/new%20Trends%20drive%20new%20EMI-RFI%20shielding%20solutions.pdf

Banyai, C., and D. Gerke. 1996. *EMI design techniques for microcontrollers in automotive applications* (Intel Application Note, AP-711). http://ece-www.colorado.edu/~mcclurel/iap711.pdf

Borgersen, R. 2000. *Choosing the right EMI shielding gasket*. Evaluation Engineering. http://archive.evaluation.com/archive/articles/0800deal.com

Chomerics. 2000. *EMI Shielding Engineering Handbook*. http://www.chomerics.com/products/documents/catalog.pdf

Christopoulos, C. 1995. *Principles and techniques of electromagnetic compatibility*. Boca Raton, FL: CRC Press.

Fajardo, I. 2000. RF and EMI shield. US Patent 6,088,231.

Fenical, G. 2004. Shielding for automotive EMC design. *Compliance Engineering*. http://www.ce-mag.com/archive/04/Fenical.html

Fenical, G. 2007. New development in shielding. *Conformity*. http://www.conformity.com/artman/publish/printer_174.shtml

Gabower, J. F. 2003. Electromagnetic interference shield for electronic devices. US Patent 6,570,085.

German, F. 2004, October 24. *Designing early for EMC. EDN*, pp. 93–98. http://www.edn.com/contents/images/472832.pdf

Joo, J., C. Y. Lee, H. G. Song, J. W. Kim, K. S. Jang, E. J. Oh, and A. J. Epstein. 1998. Enhancement of electromagnetic interference shielding efficiency of polyaniline through mixture and chemical doping. *Molecular Crystals, Liquid Crystals* 316: 367–370.

Joo, J., E.-J. Oh, and A. J. Epstein. 1995. Electromagnetic interference shielding capability of conducting polymers in far-field and near-field regions. *Molecular Electronics & Devices*, 107–118.

Kohlman, R. S., Y. G. Min, A. G. MacDiarmid, and A. J. Epstein. 1996. Tunability of high frequency shielding in electronic polymers, *Proceedings of the Society of Plastics Engineers Annual Technical Conference* (ANTEC 1996), Indianapolis, Indiana, pp. 1412–1416,.

Lau, S. E., D. S. Huff, R. D. Hermansen, and D. E. Johnston. 2003. Drop-resistant conductive epoxy adhesive. European Patent 0892027.

Lee, C. Y., H. G. Song, K. S. Jang, E. J. Oh, A. J. Epstein, and J. Joo. 1999. Electromagnetic interference shielding efficiency of polyaniline mixtures and multilayer films. *Synthetic Metals* 102: 1346–1349.

Li, N., Y. Huang, F. Du, X. He, X. Lin, H. Gao, Y. Ma, et al. 2006. Electromagnetic interference (EMI) shielding of single-walled carbon nanotube epoxy composites. *Nano Letters* 6(6): 1141–1145.

Low, Y. L., T. D. Dudderar, and D. P. Kossives. 2001. Integrated circuit packages with improved EMI characteristics. US Patent 6,297,551.

Quesnel, N. 2008. *Six basic steps to effective EMI control.* http://www.chomerics.com/tech/EMI_shld_%20Artcls/Six%20Basic%20Steps%20to%20Effective%20EMI%20Control.pdf

Ramasamy, S. R. 1997. A review of EMI shielding and suppression materials. *Proceedings of the International Conference on Electromagnetive Interference and Compatibility '97*, pp. 459–466. Hyderabad, India.

Sarto, M. S., S. D. Michele, P. Leerkamp, and H. Thuis. 2001. *An innovative shielding concept for EMI reduction.* http://www.ieee.org/organizations/pubs/newsletters/emcs/summer01/pp_sarto.htm

Tecknit. 2008. *Design guidelines.* http://tecknit.com/REFA_I/E-Windows.pdf

Wang, C., and M. D. Alexander. 2003. Method of forming conductive polymeric nanocomposite materials and materials produced thereby. International Patent WO/2001/016048.

William, T. 1992. *EMC for Product Engineers.* Oxford: Newnes.

Yarkoni, B., and J. Wharton. 1979. Designing reliable software for automotive applications. *Society of Automotive Engineers Transactions* (Doc. No. 790237), 88: 856–859.

Index

A

Absorber materials, 251–253
  anechoic chambers, 245–246
    absorber materials used in anechoic chambers, 246
    antennas, 245–246
  dielectric materials for absorber applications, 246–249
  electromagnetic wave absorbers, 259–250
  graded dielectric absorbers
    impedance matching, 242–244
    matching layer absorbers, 244
    pyramidal absorbers, 242–243
    tapered loading absorbers, 243–244
  microwave absorber materials, 238–245
    cavity damping absorbers, 244–245
    resonant absorbers, 240–242
  resonant absorbers
    Dallenbach tuned layer absorbers, 240–241
    Jaumann layers, 242
    Salisbury screens, 241–242
  types of absorbers, 251–252
Absorber types, 251–252
Absorbing materials, 251–253
Absorption loss, 14–15
Active electromagnetic shielding, space radiation, 283–285
Aerospace industry shielding materials, 275–292
  fiber reinforced composite materials, 277–279
  foam structures, 280–281
  lightweight radiation shielding material, 276–277
  nanomaterials, 279–280
Anechoic chambers, 245–246
  absorber materials used in, 246
  antennas, 245–246
Anisotropy coefficient, 61–62
ASTM ES7-83, dual chamber test fixture based on, 42

B

Bending formability, 59–60
Board-level shielding
  conductive coating, 196–206
    conductive paints, 204
    electromagnetic interference shielding, 206
    electroplating, 197–204
    vapor deposition, 205–206
  electroless plating, 197–198
  electrolytic plating, 198–199
  electromagnetic compatibility design, 178–181
    chassis construction, cabling, 180
    grounding technique, 180
    materials selection, 181–185
    PCB layout, 178–179
    power decoupling, 179
    trace separation, 179–180
  enhanced heat dissipating, 207–211
    board-level shielding, heat dissipation, combination of, 207–208
    minimizing electromagnetic interference from heat sinks, 207
    thermal interface materials, for thermal-enhanced board-level shielding, 208–211
  heat dissipation, combination, 207–208
  immersion surface finishes, 202–204
  manufacturing technology, 185–196
    metal cans, 186–190
    metal-coated thermoform shields, 190–196
  materials, 177–213
  metal cans
    multicompartment cans, 190
    multilevel shielding, 190
    single-piece shielding cans, 186–189
    two-piece cans, 189
  metal-coated thermoform shields
    Form/Met shield, 192–196
    formable shielding film, 196
    snapSHOT shield, 191–192
  tin plating, 199–202
  tin whisker growth, 199–202
Bonding, 268–269
Brass, 111
Butadiene-acrylonitrile, conductive elastomer fabrication, 129

C

Cable assembly, electromagnetic interference shielding design, 257–261
Cable-level shielding materials, 261–267
  conductive heat-shrinkable shielding, 263–264

ferrite shielding materials, 264–267
metallic shielding, 261–263
Cavity damping absorbers, 244–245
Cavity resonance, 23–24
Characterization methodology, materials, 37–66
Chassis construction, cabling, 180
Circular coaxial holder, 42
Clamping/compressive force, deformation range, conductive elastomer, 139–140
Composite materials, 215–235
Concerns with electromagnetic interference, 2–3
Conducted electromagnetic interference emission, 4–5
Conduction mechanism, processing optimization, conductive elastomer materials, 131–137
   appearance properties, 133–137
   conduction mechanism, 131–133
   electrical properties, 135–137
   mechanical properties, 133–135
   physical properties, 133–135
   process parameters, 131–133
   thermal properties, 137
Conductive coatings, 196–206
   conductive paints, 204
   electromagnetic interference shielding, 206
   electroplating, 197–204
      electroless plating, 197–198
      electrolytic plating, 198–199
      immersion surface finishes, 202–204
      tin plating, 199–202
      tin whisker growth, 199–202
   vapor deposition, 205–206
Conductive composite shielding materials, 86
Conductive elastomer, 127–158
   butadiene-acrylonitrile, 129
   ethylene propylene diene monomer, 129
   fabrication, raw material selection, 128–131
      base binder materials, 128–129
      conductive fillers, 129–130
   fluorocarbon, 129
   fluorosilicone, 129
   gasket design guideline, 137–146
      applied clamping/compressive force, deformation range, 139–140
      flange joint geometry, 138–139
      flange materials, joint surface treatment, 145–146
      gasket profile, materials selection, 141–145
      unevenness, flange joint geometry, 138–139
   gasket fabrication, 146–149
      extrusion, 146–147
      form-in-place, 148–149
      metal meshes, oriented wires, reinforced shielding gaskets with, 149
      molding, 147
      screen printing, 148–149
   gasket installation, 149–152
      application, 149–152
   materials, conduction mechanism, processing optimization, 131–137
      appearance properties, 133–137
      conduction mechanism, 131–133
      electrical properties, 135–137
      mechanical properties, 133–135
      physical properties, 133–135
      process parameters, 131–133
      thermal properties, 137
   natural rubber, 129
   silicone, 128–129
Conductive fiber/whisker reinforced composites, 224–229
   carbon fiber/whisker reinforced materials, 226–227
   conductive composite process, 225
   materials selection, 225
   metal fiber/whisker reinforced composites, 227–228
   nanofiber reinforced polymer composites, 229
Conductive fillers, conductive elastomer fabrication, 129–130
Conductive foam
   ventilation structure, 159–175
      conductive plastic foam, vent panels, 165–168
      fabric-over-foam, 169–175
      integral electromagnetic interference window structure, 168–169
      integral window structure, 165–169
      metallized fabrics, 169–175
      ventilation panel, 159–165
      waveguide aperture, 159–165
   Z-axis conductivity, electromagnetic interference shielding module, 308–309
Conductive heat-shrinkable shielding, 263–264
Conductive paints, 204
Conductive plastic foam, vent panels, 165–168
Connectors, 110–113
   brass, 111
   copper beryllium alloys, 91–97
      availability, 95–97
      high-conductivity alloys, 97
      high-strength alloys, 95–97
      nickel beryllium, 97
      performance, 95–97
      phase constitution, processing, 91–95

Index

copper-nickel-tin spinodal alloys, 97–107
  bending formability, 100–101
  composition, 97–99
  elastic performance, 103–104
  fatigue strength, 104–107
  heat treatment, 101–103
  mechanical properties, 100–101
  physical properties, 97–99
  plating, 103
  shielding effectiveness, 104
  soldering, 103
  spinodal decomposition, 101–103
  stress-relaxation resistance, 101
  strip gauges, 99–100
  surface cleaning, 103
  temper designations, 99–100
copper-titanium alloy, 107
design guideline, 113–118
design options, 115–116
fabrication, 118–124
  multiple-slide stamping, 120
  part profiles, 120–121
  photoetching, 120
  progressive die forming, 118–119
  roll forming, 119
gasketing design guidelines, 116–118
metal-formed, 89–125
mounting methods, 122–123
nickel silver, 111
phosphor bronze, 110–111
stainless steel, 108–110
surface mating assurance, 122–123
Contact finishes, effects of mating cycles, operating environments, 57–58
Contact force, 51–52
Contact interface compatibility, 52–58
Controlling methods, electromagnetic interference, 3–8
Copper beryllium alloys, 91–97
  availability, 95–97
  high-conductivity alloys, 97
  high-strength alloys, 95–97
  nickel beryllium, 97
  performance, 95–97
  phase constitution, processing, 91–95
Copper-nickel-tin spinodal alloys, 97–107
  bending formability, 100–101
  composition, 97–99
  elastic performance, 103–104
  fatigue strength, 104–107
  heat treatment, 101–103
  mechanical properties, 100–101
  physical properties, 97–99
  plating, 103
  shielding effectiveness, 104
  soldering, 103
  spinodal decomposition, 101–103
  stress-relaxation resistance, 101
  strip gauges, 99–100
  surface cleaning, 103
  temper designations, 99–100
Copper-titanium alloy, 107
Corrosion, 31–32, 53–55
  noble finish selection, 53–54
    gold, 53
    noble metal alloys, 53–54
    palladium, 53
  nonnoble finishes, 54–55
    nickel, 55
    silver, 55
    tin, 54–55

D

Dallenbach tuned layer absorbers, 240–241
Dent-resistant vent panels, electromagnetic interference shielding module, 309
Design principles, 113–118
  electromagnetic interference shielding, 1–36
Dielectric materials, absorber applications, 246–249
Dual mode stirred chamber, 39–40
Dust sensitivity test, 64–65

E

Earth neighborhood infrastructure, shielding materials for, 282–283
ECE gaskets. *See* Electrically conductive elastomer gaskets
Effect of strain hardening, 60–61
EIS. *See* Electromagnetic interference shielding
Electrical conductivity, 45–47
  contact resistance, 45–46
Electrically conductive elastomer gaskets, 84
Electroless metal deposition for reinforcements of composite shielding materials, 231–234
Electroless plating, 197–198
Electrolytic plating, 198–199
Electromagnetic compatibility, 1–11, 24–27, 178–181
  chassis construction, cabling, 180
  design, 293–297
  grounding technique, 180
  guidelines, 24–35
  materials selection, 181–185
  PCB layout, 178–179
  power decoupling, 179
  trace separation, 179–180

Electromagnetic interference control, 311
Electromagnetic interference gaskets, 19, 82–84
  electrically conductive elastomer gaskets, 84
  knitted metal wire gaskets, 83
  metal-plated fabric-over-foam, 84
  metal strip gaskets, 82–83
Electromagnetic interference shielding, 206, 257–261, 302–311
  absorber materials, 237–255
  absorbing materials, application, 251–253
  access, 67–87
  advanced materials selection, 297–302
  aerospace industry, shielding materials, 275–292
  anechoic chambers, 245–246
  board-level shielding components, 185–196
    manufacturing technology, 185–196
  board-level shielding design, 178–185
  board-level shielding materials, 178–185
    components, 177–213
  bonding, 267–271
  cable assembly, 257–261
  cable-level shielding materials, 257–273
  composite materials, 215–235
  conducted electromagnetic interference emission, 4–5
  conductive coating methods, components, 196–206
  conductive elastomer, 127–158
    fabrication, 128–131
  conductive elastomer materials
    conduction mechanism, 131–137
    processing optimization, 131–137
  conductive fiber/whisker reinforced composites, 224–229
  conductive foam
    integral window structure, 165–169
    ventilation structure, 159–175
  copper beryllium alloys, 91–97
  copper-nickel-tin spinodal alloys, 97–107
  copper-titanium alloy, 107
  design, 293–297, 302–303
  design guide, 1–36
  dielectric materials, absorber applications, 246–249
  effectiveness measurement, 37–45
  electrical conductivity, 45–47
  electroless metal deposition, composite shielding material reinforcement, 231–234
  electromagnetic compatibility, 1–11, 24–35
    early design-in for, 293–297
  electromagnetic compatibility design, 24–27
    guidelines, 24–35

electromagnetic wave absorbers, 259–250
emissions, electromagnetic interference, 3–8
enclosure, 67–87
  design, 67–73
  integrity, 78–82
enclosures, 308–310
enhanced heat dissipating, board-level shielding with, 207–211
environmental compliance, 32–36
  RoHS directive, 33–34
  WEEE directive, 34–35
environmental performance evaluation, 62–65
fabric-over-foam, 169–175
fabrication process, connectors, 118–124
flexible graphite gaskets, 127–158
formability, 58–62
fundamentals, 1–36
future developments, 293–313
gasket
  design, 137–146
  fabrication, 146–149
  installation, 149–152
grounding, 257–273
hybrid flexible structures for, 230–231
hybrid structures, 215–235
knitted wire/elastomer gaskets, 215–221
lightweight shielding materials, aerospace applications, 275–281
magnetic field shielding enclosure, 73–78
manufacturability, 58–62
material characterization methodology, 37–66
materials selection, 27–32, 67–73
  corrosion, 31–32
  gasketing, 29–30
  integral assembly, 30–31
  material galvanic compatibility, 31–32
  shielding housing, enclosures, 27–29
mechanical properties, 49–52
metal-formed electromagnetic interference gaskets, connectors, 89–125
metal gasket types, connectors, 118–124
metal gaskets, connectors, design guideline, 113–118
metal strip selection, performance requirement, 90–91
metalized fabrics, 169–175
microwave absorber materials, 238–245
modules, 308–310
mounting methods, 122–123
nuclear shielding materials, 286–289
permeability, 47–49
permittivity, 47–49

Index

principles, 11–24
problems with electromagnetic interference, 2–3
radiated electromagnetic interference emission, 5–8
raw material selection, 128–131
regulations, electromagnetic compatibility, 8–11
  European standard, 10
  Federal Communications Commission standard, 9–10
  international standard, 9
  military standard, 10–11
shielding effectiveness, 13–18
  absorption loss, 14–15
  multiple reflection correction factor, 16–18
  reflection loss, 15–16
shielding effects, 11–13
shielding housing, enclosures
  metal sheet, 28
  nonmetallic materials, 28–29
  ventilating panel, windows, 29
shielding modeling in practice, 18–24
  cavity resonance, 23–24
  electromagnetic interference gasket, 19
  metallic enclosure, 18–19
  shield with apertures, 20–23
space power systems, radiation shielding, 281–286
specialized materials for shielding enclosure, 82–86
stainless steel, 108–110
surface finish, contact interface compatibility, 52–58
surface mating assurance, 122–123
tapes, 221–224
thermal conductivity, 45–47
ventilation panel, 159–165
waveguide aperture, 159–165
Electromagnetic wave absorbers, 250–254
Electroplating, 197–204
  electroless plating, 197–198
  electrolytic plating, 198–199
  immersion surface finishes, 202–204
  tin plating, 199–202
  tin whisker growth, 199–202
EMI. *See* Electromagnetic interference
Emissions
  electromagnetic interference, 3–8
  radiated electromagnetic interference, 5–8
Enclosure, 67–87
  conductive composite shielding materials, 86
  design, 67–73
  electromagnetic interference gaskets, 82–84
  electrically conductive elastomer gaskets, 84
  knitted metal wire gaskets, 83
  metal-plated fabric-over-foam, 84
  metal strip gaskets, 82–83
  honeycomb materials, 85–86
  integrity, 78–82
  magnetic screening materials, 84–85
  materials, 67–73, 82–86
  painted or plated plastic enclosures, 86
  shielding tapes, 85
  thermoformable alloys, 85
Enhanced heat dissipating, 207–211
  board-level shielding, heat dissipation, combination of, 207–208
  minimizing electromagnetic interference from heat sinks, 207
  thermal interface materials, for thermal-enhanced board-level shielding, 208–211
Environmental performance evaluation, 62–65
  dust sensitivity test, 64–65
  fretting corrosion, 63–64
  gaseous testing, 63
  humidity testing, 64
  pore corrosion, 63–64
  shock test, 65
  temperature, 64
  thermal aging, 63
  thermal cycling, 63
  vibration, 65
EPDM. *See* Ethylene propylene diene monomer
Ethylene propylene diene monomer, conductive elastomer fabrication, 129
European electromagnetic compatibility standard, 10
Extrusion, conductive elastomer, 146–147

F

Fabric-over-foam, conductive foam, ventilation structure, 169–175
Fabrication of gaskets, connectors, 118–124
  multiple-slide stamping, 120
  part profiles, 120–121
  photoetching, 120
  progressive die forming, 118–119
  roll forming, 119
Federal Communications Commission electromagnetic compatibility standard, 9–10
Ferrite shielding materials, 264–267
Flange joint geometry, 138–139
  unevenness, 138–139

Flange materials, joint surface treatment, 145–146
Flexible graphite gaskets, 152–157
  application, 155–157
  fabrication, 155–157
  properties of flexible graphite, 153–155
Fluorocarbon, conductive elastomer fabrication, 129
Fluorosilicone, conductive elastomer fabrication, 129
Form-in-place, conductive elastomer, gasket fabrication, 148–149
Form/Met shield, 192–196
Formability, 58–62
  anisotropy coefficient, 61–62
  bending formability, 59–60
  effect of strain hardening, 60–61
  springback during metal strip forming, 62
Formable shielding film, 196
Fretting corrosion, 63–64
Friction force, 52
Fundamentals of electromagnetic interference shielding, 1–36

G

Galvanic compatibility, 31–32
Galvanic corrosion, contact interface compatibility, 58
Gaseous testing, 63
Gasket, material selection, 299–301
Gasket design guideline, conductive elastomer, 137–146
  applied clamping/compressive force, deformation range, 139–140
  flange joint geometry, 138–139
  flange materials, joint surface treatment, 145–146
  gasket profile, materials selection, 141–145
  unevenness, flange joint geometry, 138–139
Gasket fabrication, conductive elastomer, 146–149
  extrusion, 146–147
  form-in-place, 148–149
  metal meshes, oriented wires, reinforced shielding gaskets with, 149
  molding, 147
  screen printing, 148–149
Gasket installation, conductive elastomer, 149–152
  application, 149–152
Gasket profile, materials selection, 141–145
Gasketing, 29–30
  design, 116–118
Gaskets, 110–113
  brass, 111

copper beryllium alloys, 91–97
  availability, 95–97
  high-conductivity alloys, 97
  high-strength alloys, 95–97
  nickel beryllium, 97
  performance, 95–97
  phase constitution, processing, 91–95
copper-nickel-tin spinodal alloys, 97–107
  bending formability, 100–101
  composition, 97–99
  elastic performance, 103–104
  fatigue strength, 104–107
  heat treatment, 101–103
  mechanical properties, 100–101
  physical properties, 97–99
  plating, 103
  shielding effectiveness, 104
  soldering, 103
  spinodal decomposition, 101–103
  stress-relaxation resistance, 101
  strip gauges, 99–100
  surface cleaning, 103
  temper designations, 99–100
copper-titanium alloy, 107
  design guideline, 113–118
  design options, 115–116
  fabrication, 118–124
    multiple-slide stamping, 120
    part profiles, 120–121
    photoetching, 120
    progressive die forming, 118–119
    roll forming, 119
  gasketing design guidelines, 116–118
  metal-formed, 89–125
  mounting methods, 122–123
  nickel silver, 111
  phosphor bronze, 110–111
  stainless steel, 108–110
  surface mating assurance, 122–123
Gold noble finish, 53
Graded dielectric absorbers
  impedance matching, 242–244
  matching layer absorbers, 244
  pyramidal absorbers, 242–243
  tapered loading absorbers, 243–244
Grounding, 269–271
  hybrid grounding, 271
  multigrounded shields, 271
  shields grounded at one point, 270
  technique, 180

H

Hardness, 50
Heat dissipation, board-level shielding, combination of, 207–208

Index

Heat sinks, minimizing electromagnetic interference from, 207
Honeycomb materials, 85–86
Humidity testing, 64
Hybrid flexible structures for electromagnetic interference shielding, 230–231
Hybrid grounding, 271
Hybrid structures, 215–235

I

Immersion surface finishes, 202–204
Impedance matching, graded dielectric absorbers, 242–244
Integral assembly, 30–31
Integral electromagnetic interference window structure, 168–169
Integral window structure, conductive foam, ventilation structure, 165–169
Integrated circuit-level shielding, 303–305
Interconnecting level shielding, 310
International electromagnetic compatibility standard, 9

J

Jaumann layers, 242

K

Knitted metal wire gaskets, 83
Knitted wire/elastomer gaskets, 215–221
  fabrication process, 215–218
  gasket types, 219–221
  knitted wire gasket performance, 218–219
  materials selection, 215–218
  mounting methods, 219–221

L

Lightweight shielding materials, aerospace industry, 275–281
  fiber reinforced composite materials, 277–279
  foam structures, 280–281
  lightweight radiation shielding material, 276–277
  nanomaterials, 279–280

M

Magnetic field shielding enclosure, 73–78
  design, 76–78
  materials, 74–76
  overview, 73–74
Magnetic screening materials, 84–85
Manufacturing technology, 185–196
  metal cans, 186–190
    multicompartment cans, 190
    multilevel shielding, 190
    single-piece shielding cans, 186–189
    two-piece cans, 189
  metal-coated thermoform shields, 190–196
    Form/Met shield, 192–196
    formable shielding film, 196
    snapSHOT shield, 191–192
Matching layer absorbers, 244
Material characterization methodology, 37–66
  contact interface compatibility, 52–58
  electrical conductivity, 45–47
    contact resistance, 45–46
  environmental performance evaluation, 62–65
    dust sensitivity test, 64–65
    fretting corrosion, 63–64
    gaseous testing, 63
    humidity testing, 64
    pore corrosion, 63–64
    shock test, 65
    temperature, 64
    thermal aging, 63
    thermal cycling, 63
    vibration, 65
  formability, 58–62
    anisotropy coefficient, 61–62
    bending formability, 59–60
    effect of strain hardening, 60–61
    springback during metal strip forming, 62
  manufacturability, 58–62
  mechanical properties, 49–52
    contact force, 51–52
    friction force, 52
    hardness, 50
    permanent set, 52
    Poisson's ratio, 50
    stress relaxation, 50–51
    uniaxial tensile testing, 49–50
  noble finish selection
    gold, 53
    noble metal alloys, 53–54
    palladium, 53
  nonnoble finishes
    nickel, 55
    silver, 55
    tin, 54–55
  permeability, 47–48
    characterization of, 48–49

permittivity, 47–49
  characterization of, 48–49
shielding effectiveness measurement, 37–45
  ASTM ES7-83, dual chamber test fixture based on, 42
  circular coaxial holder, 42
  dual mode stirred chamber, 39–40
  MIL-G-83528, testing based on, 39
  MIL-STD-285, testing based on, 38–39
  transfer impedance methods, 42–45
  transverse electromagnetic cell, 40–42
surface finish, 52–58
  contact finishes, effects of mating cycles, operating environments, 57–58
  corrosion, 53–55
  galvanic corrosion, contact interface compatibility, 58
  noble finish selection, 53–54
  nonnoble finishes, 54–55
  oxidation protection, 53–55
  solderability, 55–57
thermal conductivity, 45–47
transfer impedance methods
  coaxial cables, 43
  conductive gaskets, 43–45
Material galvanic compatibility, 31–32
Mechanical properties, 49–52
  contact force, 51–52
  friction force, 52
  hardness, 50
  permanent set, 52
  Poisson's ratio, 50
  stress relaxation, 50–51
  uniaxial tensile testing, 49–50
Metal cans, 186–190
  multicompartment cans, 190
  multilevel shielding, 190
  single-piece shielding cans, 186–189
  two-piece cans, 189
Metal-coated thermoform shields, 190–196
  Form/Met shield, 192–196
  formable shielding film, 196
  snapSHOT shield, 191–192
Metal-formed connectors, 89–125
Metal-formed gaskets, 89–125
  connectors, 89–125
Metal meshes, reinforced shielding gaskets with, 149
Metal-plated fabric-over-foam, 84
Metal sheet, 28
Metal strip gaskets, 82–83
Metal strip selection, performance requirement, 90–91

Metallic enclosure, 18–19
Metallic shielding, 261–263
Metallized fabrics, conductive foam, ventilation structure, 169–175
Microwave absorber materials, 238–245
  cavity damping absorbers, 244–245
  graded dielectric absorbers
    impedance matching, 242–244
    matching layer absorbers, 244
    pyramidal absorbers, 242–243
    tapered loading absorbers, 243–244
  resonant absorbers, 240–242
    Dallenbach tuned layer absorbers, 240–241
    Jaumann layers, 242
    Salisbury screens, 241–242
MIL-G-83528, testing based on, 39
MIL-STD-285, testing based on, 38–39
Military electromagnetic compatibility standard, 10–11
Molding, gasket fabrication, conductive elastomer, 147
Mounting methods, gaskets, connectors, 122–123
Multicompartment cans, 190
Multigrounded shields, 271
Multilevel shielding, 190
Multiple reflection correction factor, 16–18
Multiple-slide stamping, 120

N

Nanocomposite shielding materials, electromagnetic interference shielding module, 309–310
Nanofiber reinforced polymer composites, 229
Natural rubber, conductive elastomer fabrication, 129
Neutron shielding design, materials selection, space power system radiation shielding, 282
Nickel nonnoble finish, 55
Nickel silver, 111
Noble finish selection, 53–54
  gold, 53
  noble metal alloys, 53–54
  palladium, 53
Noble metal alloys, 53–54
Nonmetallic materials, 28–29
Nonnoble finishes, 54–55
  nickel, 55
  silver, 55
  tin, 54–55
Nuclear shielding materials, 286–289
  electromagnetic interference shielding, 286–289

Index

O

Oriented wires, reinforced shielding gaskets with, 149
Oxidation protection, 53–55

P

Painted or plated plastic enclosures, 86
Palladium noble finish, 53
PCB layout, 178–179
PCB-level shielding. *See* Printed circuit board-level shielding
Permeability, 47–48
 characterization of, 48–49
Permittivity, 47–49
 characterization of, 48–49
Phosphor bronze, 110–111
Photoetching, 120
Poisson's ratio, 50
Pore corrosion, 63–64
Power decoupling, 179
Printed circuit board-level shielding, 305–307
 lightweight shields, 305–306
 mold-in-place combination gaskets, multicompartment shields, 306–307
 rotary form-in-place gaskets, 307
Problems with electromagnetic interference, 2–3
Progressive die forming, 118–119
Pyramidal absorbers, 242–243

R

Radiated electromagnetic interference emission, 5–8
Raw material selection, conductive elastomer fabrication
 base binder materials, 128–129
  butadiene-acrylonitrile, 129
  ethylene propylene diene monomer, 129
  fluorocarbon, 129
  fluorosilicone, 129
  natural rubber, 129
  silicone, 128–129
 conductive fillers, 129–130
Reflection loss, 15–16
Regulations, electromagnetic compatibility, 8–11
 European standard, 10
 Federal Communications Commission standard, 9–10
 international standard, 9
 military standard, 10–11

Reinforcements of composite shielding materials, electroless metal deposition for, 231–234
Resonant absorbers, 240–242
 Dallenbach tuned layer absorbers, 240–241
 Jaumann layers, 242
 Salisbury screens, 241–242
RoHS directive, 33–34
Roll forming, 119
Rubber, conductive elastomer fabrication, 129

S

Salisbury screens, 241–242
Screen printing, conductive elastomer gasket fabrication, 148–149
Shield with apertures, 20–23
Shielding effectiveness, 13–18, 37–45
 absorption loss, 14–15
 ASTM ES7-83, dual chamber test fixture based on, 42
 circular coaxial holder, 42
 dual mode stirred chamber, 39–40
 MIL-G-83528, testing based on, 39
 MIL-STD-285, testing based on, 38–39
 multiple reflection correction factor, 16–18
 reflection loss, 15–16
 transfer impedance methods, 42–45
  coaxial cables, 43
  conductive gaskets, 43–45
 transverse electromagnetic cell, 40–42
Shielding effects, 11–13
Shielding housing, enclosures, 27–29
 metal sheet, 28
 nonmetallic materials, 28–29
 ventilating panel, windows, 29
Shielding modeling, 18–24
 cavity resonance, 23–24
 electromagnetic interference gasket, 19
 metallic enclosure, 18–19
 shield with apertures, 20–23
Shielding tapes, 85
Shields grounded at one point, 270
Shock test, 65
Silicone, conductive elastomer fabrication, 128–129
Silver nonnoble finish, 55
Single-piece shielding cans, 186–189
SnapSHOT shield, 191–192
Software immunity, 311
Solderability, 55–57
Space power systems, radiation shielding for, 281–286
 earth neighborhood infrastructure, shielding materials for, 282–283

neutron shielding design, materials selection, 282
optimization of space radiation shielding, 285–286
space radiation, active electromagnetic shielding for, 283–285
Space radiation
active electromagnetic shielding for, 283–285
shielding optimization, 285–286
Spinodal decomposition, copper-nickel-tin spinodal alloys, 101–103
Springback during metal strip forming, 62
Stainless steel, 108–110
Standards, electromagnetic compatibility, 8–11
European standard, 10
Federal Communications Commission standard, 9–10
international standard, 9
military standard, 10–11
Stress relaxation, 50–51
Surface finish, 52–58
contact finishes, effects of mating cycles, operating environments, 57–58
corrosion, 53–55
noble finish selection, 53–54
nonnoble finishes, 54–55
galvanic corrosion, contact interface compatibility, 58
noble finish selection
gold, 53
noble metal alloys, 53–54
palladium, 53
nonnoble finishes
nickel, 55
silver, 55
tin, 54–55
oxidation protection, 53–55
solderability, 55–57
Surface mating assurance, 122–123

T

Tapered loading absorbers, 243–244
TEM cell. *See* Transverse electromagnetic cell
Temperature, 64
Testing standards, electromagnetic compatibility, 8–11
Thermal aging, 63
Thermal conductivity, 45–47

Thermal cycling, 63
Thermal interface materials, for thermal-enhanced board-level shielding, 208–211
Thermoformable alloys, 85
TIMs. *See* Thermal interface materials
Tin nonnoble finish, 54–55
Tin plating, 199–202
Tin whisker growth, 199–202
Trace separation, 179–180
Transfer impedance methods, 42–45
coaxial cables, 43
conductive gaskets, 43–45
Transverse electromagnetic cell, 40–42
Two-piece cans, 189

U

Ultrasoft sculpted fabric-over-foam, electromagnetic interference shielding module, 310
Unevenness, flange joint geometry, 138–139
Uniaxial tensile testing, 49–50

V

Vapor deposition, 205–206
Ventilation panel, conductive foam, ventilation structure, 159–165
Ventilation structure, conductive foam, 159–175
conductive plastic foam, vent panels, 165–168
fabric-over-foam, 169–175
integral electromagnetic interference window structure, 168–169
integral window structure, 165–169
metallized fabrics, 169–175
ventilation panel, 159–165
waveguide aperture, 159–165

W

Waveguide aperture, conductive foam, ventilation structure, 159–165
WEEE directive, 34–35

Z

Z-axis conductivity, conductive foam, electromagnetic interference shielding module, 308–309